# THE RISE O1 INFRASTRUCⲧⲢᴜⲢᴇ STATE

How US–China Rivalry Shapes Politics
and Place Worldwide

Edited by
Seth Schindler and Jessica DiCarlo

BRISTOL
UNIVERSITY
PRESS

First published in Great Britain in 2024 by

Bristol University Press
University of Bristol
1-9 Old Park Hill
Bristol
BS2 8BB
UK
t: +44 (0)117 374 6645
e: bup-info@bristol.ac.uk

Details of international sales and distribution partners are available at bristoluniversitypress.co.uk

© Bristol University Press 2024

British Library Cataloguing in Publication Data
A catalogue record for this book is available from the British Library

ISBN 978-1-5292-2077-3 hardcover
ISBN 978-1-5292-2078-0 paperback
ISBN 978-1-5292-2080-3 ePub
ISBN 978-1-5292-2079-7 ePdf

Cover design: Nicky Borowiec
Front cover image: Shutterstock_152814608

Bristol University Press uses environmentally responsible print partners.

Printed in Great Britain by CPI Group (UK) Ltd, Croydon, CR0 4YY

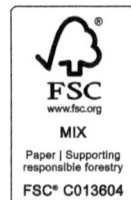

FSC
www.fsc.org
MIX
Paper | Supporting
responsible forestry
FSC® C013604

# Contents

# CONTENTS

# List of Figures and Tables

## Figures

## Tables

# List of Abbreviations
# and Acronyms

| | |
|---|---|
| 17+1 | A diplomatic forum connecting Central and Eastern European states with China (previously 16+1) |
| ADB | Asian Development Bank |
| ADR | Addis Ababa-Djibouti Railway |
| AFP | Armed Forces of the Philippines |
| AKP | Turkey's Justice and Development Party |
| ASEAN | Association for Southeast Asian Nations |
| B3W | Build Back Better World |
| BBB | Philippines Build Build Build programme |
| BRI | Belt and Road Initiative |
| CCP | Chinese Communist Party |
| CDB | China Development Bank |
| CGN | China General Nuclear Power Corporation |
| CNMC | China Nonferrous Metal Mining Company |
| CNPC | Chinese National Petroleum Company |
| CPP | Communist Party of the Philippines |
| CSG | China Southern Power Grid Company |
| DND | Department of National Defence of the Philippines |
| ECRL | East Coast Rail Link |
| EdL | Électricité du Laos |
| EdL-T | Électricité du Laos Transmission Company Limited |
| EIZ | Eastern Industrial Zone (Ethiopia) |
| EPRDF | Ethiopia People's Revolutionary Democratic Front |
| ERC | Ethiopian Railway Corporation |
| EU | European Union |
| EXIM | The Export-Import Bank of China |
| FDI | foreign direct investment |
| FONOPs | US-led Freedom of Navigation Operations |
| G7 | The Group of Seven is an intergovernmental political forum comprised of Japan, Canada, Italy, France, Germany, the UK, and the US |

| | |
|---|---|
| GMS | Greater Mekong Subregion (an ADB initiative) |
| HSR | high-speed railway |
| IDFC | International Development Finance Corporation |
| IFIs | International Financial Institutions |
| IMF | International Monetary Fund |
| IP | Industrial Park |
| IPDC | Industrial Parks Development Corporation |
| JICA | Japan International Cooperation Agency |
| LCEC | Laos–China Economic Corridor |
| LCR | Laos–China Railway |
| LPRP | Lao People's Revolutionary Party |
| MCC | Millennium Challenge Corporation (US) |
| MFEZ | multi-facility economic zone |
| MNC | multinational corporation |
| MoU | Memorandum of Understanding |
| MPI | Ministry of Planning and Investment of Laos |
| NATO | North Atlantic Treaty Organization |
| NGO | non-governmental organization |
| NPA | New People's Army (the CPP's military arm) |
| ODA | overseas development assistance/official development assistance |
| OECD | Organisation for Economic Co-operation and Development |
| OPIC | Overseas Private Investment Corporation |
| PCA | Permanent Court of Arbitration in the Hague |
| PP | Prosperity Party (Ethiopia) |
| PPP | public–private partnership |
| PQI | Partnership of Quality Infrastructure |
| PRC | People's Republic of China |
| SCS | South China Sea |
| SENPLADES | National Secretariat of Planning and Development (Ecuador) |
| SEZ | special economic zone |
| SOE | state-owned enterprise |
| SUPARCO | Space and Upper Atmosphere Research Commission (Pakistan) |
| SWAPO | South West Africa People's Organization |
| TCC | transnational capitalist class |
| TRACECA | Transport Corridor Europe-Caucasus-Asia |
| TWF | Türkiye Wealth Fund |
| UN | United Nations |
| UNCLOS | United Nations Convention on the Law of the Sea |
| US TCC | US-centred transnational capitalist class |

| | |
|---|---|
| USAID | United State Agency for International Development |
| V4 | Visegrád Group comprised of Czechia, Hungary, Poland and Slovakia |
| VFA | Visiting Forces Agreement |
| WSC | Wall Street Consensus |
| YPF | Yacimientos Petrolíferos Fiscales (Argentina) |
| ZCCZ | Zambia–China Economic and Trade Cooperation Zone |
| ZDA | Zambian Development Agency |

# Notes on Contributors

**Farzana Abdilashimova** is a feminist researcher and national gender specialist at Mountain Societies Development Support Programme public foundation in Bishkek, Kyrgyzstan. Her research interests encompass human rights and minority issues, feminism and gender, sexual and reproductive health, and rights in Kyrgyzstan and Central Asia.

**Mustafa Kemal Bayırbağ** is Associate Professor in the Department of Political Science and Public Administration of Middle East Technical University. His research interests are situated at the intersection of urban/ regional studies and public policy. State spatiality and political economy of social exclusion are two underlying themes of his research orientation, cutting across his past and future works on neoliberal urbanization, urban crisis and governance, the Belt and Road Initiative, local development, education policy and social policy.

**Brock Bersaglio** is Lecturer in the International Development Department at the University of Birmingham. Focusing on eastern and southern Africa, Brock's research critically engages with how natural resource management shapes and is shaped by human–nonhuman relationships in the context of biodiversity conservation, natural resource extraction, and sustainable development.

**Alvin Camba** is Assistant Professor at the Korbel School of International Studies at the University of Denver and a research fellow at the Climate Policy Lab at Tufts University. He studies Chinese investment in Southeast Asia. More information about his work can be found at alvincamba.com.

**Jerik Cruz** is a PhD student in political economy and computational methodology at the Massachusetts Institute of Technology, and a lecturer at the Department of Economics at the Ateneo de Manila University. His current research examines the dynamics of state building, urban infrastructure development, and the legacies of economic liberalization in developing democracies. He is a recipient of the MIT Homer A. Burnell Presidential

Fellowship, and has led the development of policy studies commissioned by the Asian Development Bank, the International Labour Organization, and the United Nations Development Program, among others.

**Meredith J. DeBoom** is Assistant Professor of Geography at the University of South Carolina. Her research analyses the geopolitics of resource extraction, development, and distribution, with an emphasis on Africa–China relations. She is particularly interested in how Africans engage with geopolitical and energy transitions to pursue domestic political goals. Meredith's research has been supported by the American Association of University Women, the National Science Foundation, and the University of Michigan Society of Fellows. She holds a PhD in Geography from the University of Colorado at Boulder.

**Jessica DiCarlo** is the Chevalier Postdoctoral Research Fellow in Transportation and Development in China at the University of British Columbia's School of Public Policy and Global Affairs, and holds a PhD in Geography from the University of Colorado Boulder. She writes on global China in Asia and her interests lie at the intersection of critical development studies, political ecology, and political economy. She has conducted ethnographic, qualitative, and quantitative research in Asia since 2008. Her expertise is centred in China, where she has worked in Yunnan, Liaoning, Tibetan regions, as well as Beijing and Shanghai, and her interest in borderlands led her to research in Nepal, India, and Laos. More details of her work are available at: jessicadiCarlo.org.

**Charis Enns** is a presidential fellow in the Global Development Institute at the University of Manchester. Charis' work examines the impacts of large-scale investments in land on rural lives and rural ecologies. Her current projects focus on the relationship between colonial settlement and ecological change in East Africa.

**Tom Goodfellow** is a senior lecturer in Urban Studies and Planning at the University of Sheffield. His research focuses on the political economy of urban development and change in Africa, particularly the politics of urban land and infrastructure. He is widely published in a range of leading journals, is co-author of *Cities and Development* (2016) and is the Treasurer of the IJURR Foundation.

**Ferenc Gyuris** is Associate Professor of Geography at the ELTE Eötvös Loránd University, Institute of Geography and Earth Sciences, Department of Social and Economic Geography, Budapest, Hungary. His research interests include the uneven geographies of communism and post-communism,

China's economic and political transformation along with its global impact, spatial inequality, and geographies of knowledge.

**Zhengli Huang** is a postdoctoral researcher in Urban and Rural Planning at Tongji University, Shanghai. She has worked as a research associate at the University of Sheffield in the UK and the Chinese University of Hong Kong. She was also Luce Visiting Scholar in Trinity College in the US. Her work focuses on African urbanization and China's impact on urban development, especially on housing, urban governance, and development finance.

**Micah Ingalls** holds a PhD from Cornell University and is the team leader of the Mekong Region Land Governance (MRLG) Project, which aims to improve the land tenure security of smallholder farmers in Cambodia, Lao People's Democratic Republic, Myanmar, and Vietnam. With nearly two decades of experience in Central, South, and Southeast Asia, Micah brings a depth of experience in the fields of resource governance, conflict management, and climate action. Prior to joining MRLG, Micah worked for the Centre for Development and Environment of the University of Bern, the Food and Agriculture Organization, United National Development Programme, and World Wide Fund for Nature, among others.

**Nicholas Jepson** is Leverhulme Research Fellow at the University of Manchester's Global Development Institute. His research examines the consequences of the rise of China for global capitalism and uneven development, particularly in Latin America, Sub-Saharan Africa and Central and Eastern Europe. His current project explores these issues in relation to sovereign debt and global financial governance.

**Wangui Kimari** is a junior research fellow at the Institute for Humanities in Africa at the University of Cape Town. Her work focuses on colonial planning regimes in Kenya, infrastructure, and movements for urban justice in Nairobi.

**Julie Michelle Klinger** is Assistant Professor in the Department of Geography and Spatial Sciences at the University of Delaware and Associate Director of the Minerals, Materials, and Society Programme, focusing on the dynamics of global resource frontiers and space-based technologies with particular emphases in China, Brazil, and the US. Dr Klinger has conducted extensive ethnographic, qualitative, and quantitative fieldwork over the past 15 years. She has published numerous articles on rare earth elements, natural resource use, environmental politics, and outer space. For current work, see https://klingerlab.com

**Cheng-Chwee Kuik** is Associate Professor and Head of the Centre for Asian Studies, Institute of Malaysian and International Studies, National University of Malaysia. He is concurrently a non-resident fellow at the Foreign Policy Institute, Johns Hopkins University. Dr Kuik's research concentrates on smaller state foreign policy, Asian security, and international relations. He is co-author (with David M. Lampton and Selina Ho) of *Rivers of Iron: Railroads and Chinese Power in Southeast Asia* (2020).

**Gediminas Lesutis** is a research associate at University of Cambridge and a research fellow at Darwin College, UK, and an incoming Marie Curie Research Fellow at University of Amsterdam, the Netherlands. His research is in the areas of political geography and global political economy, particularly in regard to everyday life, subjectivity, extractivism, mega-infrastructures, bio- and necro-political power, and the politics of development across Sub-Saharan Africa.

**Jessica C. Liao** is Assistant Professor of Political Science at North Carolina State University and Wilson China Fellow 2020–21. Her research interest includes China's foreign policy, the impact of China's rise on global governance, the international political economy of East Asia, China and Japan's investments in Southeast Asia and their environmental impact. For her current research, please see https://chass.ncsu.edu/people/cliao5/.

**Guanie Lim** is Assistant Professor at the National Graduate Institute for Policy Studies, Japan. His main research interests are comparative political economy, value chain analysis, and the Belt and Road Initiative in Southeast Asia. Lim is also interested in broader development issues within Asia, especially those of China, Vietnam, and Malaysia. His latest monograph — *The Political Economy of Growth in Vietnam: Between States and Markets* (2020) – details the catching-up experience of Vietnam since its 1986 Doi Moi (renovation) reforms.

**Masalu Luhula** is Advocate of the High Court of Tanzania and a land tenure specialist at Landesa, Tanzania office. He has experience in community land rights and land-based investment engagement.

**Jessica Neafie** is Assistant Professor of International Relations at Nazarbayev University. She writes on China's soft power, investment, and aid impact. She has conducted qualitative and quantitative research in Asia and Africa, and her expertise is centred at the intersection of investment, society, and the environment. Her work on investment and the environment is published in Politikon. She is currently working on several projects in Kazakhstan examining the impact of Chinese soft power and corporate environmentalism.

**Dinesh Paudel** is Associate Professor in the Department of Sustainable Development at Appalachian State University. His current research focuses on disaster capitalism, development and infrastructure geopolitics in Nepal and the Himalaya.

**Katharine Rankin** is Professor and Associate Chair in the Department of Geography and Planning at the University of Toronto. She has contributed broadly to scholarship on market and state formation through a decolonial, area-studies orientation engaging ethnographic approaches – and featuring studies of infrastructure development, post-conflict and post-disaster governance, commercial gentrification, microfinance, and the cultural politics of markets. Empirically these pursuits have transpired primarily in Nepal where Rankin has longstanding ties, but also in northern Vietnam and disinvested neighbourhoods in Toronto, Canada. Rankin is the author of *Cultural Politics of Markets: Economic Liberalization and Social Change in Nepal* (2004) and is currently Principal Investigator of a Canadian SSHRC-funded research project on Infrastructures of Democracy: State Building as Everyday Practice in Nepal's Agrarian Districts.

**Steve Rolf** is ESRC Research Fellow at the Digital Futures at Work Research Centre (Digit), University of Sussex Business School. His research examines the political economy of growth dynamics across various scales and contexts, with a particular focus on China and its relationship with the global economy. He is currently working on several projects examining the impact of digital platforms on markets, institutions, and society.

**Marcelo I. Saguier** is a researcher at the National Scientific and Technical Research Council, Argentina. He is Professor at the School of Politics and Government of the National University of San Martín where he also directs the MA/PhD programme in International Relations. His research interests and publications focus on the international political economy of the environment, visual global politics, and sustainable development. He is currently working on a project on 'Visual narratives of the global ecological crisis: towards an international political ecology of images'.

**Seth Schindler** is Senior Lecturer of Urban Development and Transformation at the Global Development Institute, University of Manchester. He is the Co-Director of Research of the African Cities Research Consortium. He previously coordinated the Global Studies Programme at Humboldt University of Berlin. His research is focused on deindustrialization and urbanization in the majority world and appears in leading urban studies journals.

**Maximiliano F. Vila Seoane** is a postdoctoral researcher at the National Scientific and Technical Research Council, Argentina. He is Professor at the School of Politics and Government of the National University of San Martín, where he teaches courses on international relations, cyberpolitics and international communication. His research interests and publications focus on digitalization, international politics, and development studies.

**Ulan Shamshiev** is a doctoral student in intercultural communications and comparative linguistics at the Kyrgyz National University in Bishkek. Ulan has conducted extensive anthropological and linguistic fieldwork in Kyrgyzstan and is an experienced translator and interpreter. He has been a visiting scholar in Germany and Slovakia.

**Mary Silaban** is an independent researcher in the energy and natural resources fields. Her research interest is in the geopolitics of energy and nature in the global climate change discussion. She is the co-founder of PERSPECTIVA, a data analytical, communications and business advocacy firm that covers various industries.

**Rune Steenberg** is a postdoctoral researcher for the Sinophone Borderlands project at Palacky University Olomouc. He studies Central Asia, Xinjiang and the Uyghurs, publishing on kinship, economic transformation, Uyghur intellectuals, and mass incarceration.

**Dorothy Tang** is Assistant Professor of Landscape Architecture at the National University of Singapore. Her work is concerned with the intersections of infrastructure and everyday life, especially in communities confronting large-scale environmental change. Her current research explores the landscape and urban impacts of Chinese overseas investments in the Mekong Region and Southern Africa. She was formerly Assistant Professor of Landscape Architecture at the University of Hong Kong, where she also directed the undergraduate programme.

**Angela Tritto** is Adjunct Assistant Professor at the Division of Public Policy and a Hong Kong Postdoctoral Fellowship Scheme recipient jointly appointed by the Institute of Emerging Market Studies and by the Division of Social Science at the Hong Kong University of Science and Technology. She is also a Fellow of the Global Future Council of Sustainable Tourism at the World Economic Forum. She is currently working on three interrelated research projects on the Belt and Road Initiative in Southeast Asia. Her research interests include management of innovation, environmental policies and technologies, heritage tourism management, and sustainable development.

# Preface and Acknowledgements

This project began to take shape in 2019 in Beijing during a workshop on the Belt and Road Initiative in South and Southeast Asia. We quickly noticed that we were conceptualizing a similar phenomenon that was concerned with the emergent multipolar global order – but at different scales. We realized that our research connects geopolitics and the shifting political economic landscapes of 'infrastructure states' and indeed should be put into conversation for both theoretical and practical reasons. This project began as a litmus test, as we shared our thoughts with friends and colleagues to see whether the framework resonated. Over the course of numerous conversations, we began to see that our conceptualization of the 'infrastructure state' had fascinating resonance across places and scales.

In the course of editing this volume we have refined our understanding of the US–China rivalry and infrastructural competition, and have learned much from the contributing authors. We are deeply grateful for the time and energy they put into their chapters and our discussions, which made editing this volume most enjoyable. We would also like to thank the Chinese Academy of Sciences and Regional Studies Association for convening the Beijing conference where this book began. We are also grateful for support from the University of Colorado Boulder, Boston University's Center for Global Development Policy, and the University of Manchester. Finally, thanks to everyone at Bristol University Press who helped us along the way.

# Introduction: Geopolitics, Infrastructure, and the Emergent Geographies of US–China Competition

*Jessica DiCarlo and Seth Schindler*

It is undeniable that great power rivalry has exploded and the unipolar world order that emerged after the collapse of the USSR has given way to a multipolar order with the US and China vying for hegemony (Jones 2020; Shambaugh 2020; Hung 2022). In 2021, US President Joe Biden announced an existential competition between authoritarianism and democracy. Chinese President Xi Jinping has quickly consolidated power and is cementing Beijing's status as a great power in the emergent multipolar world order. In this context, numerous commentators have asserted that tension between the US and China amounts to a 'new Cold War'.[1]

Despite the intensification of rivalry, the US and China remain deeply interconnected. Take the example of China's first passenger jet, the Comac C919, which signaled the country's first effort to break up the duopoly of Airbus and Boeing in the global passenger jet industry. Yet, looking more closely, the project is as much Western as it is Chinese. A substantial percentage of its supply chain is from the US (three-fifths) and Europe (one-third). Of the 127 linked suppliers identified for the C919, only 14 are from China, of which 7 are joint ventures.[2] The global nature of the production network behind the C919 highlights a paradox: although the US and China appear to have entered a prolonged period of geopolitical competition and confrontation, they remain coupled, with interests often aligned.

We argue that the key to unravelling the complexity of US–China rivalry is its territorial logic. In contrast to the Cold War, the US and China do not compete to establish blocs of loyal allies and client states. Instead, contemporary great power rivalry is geared towards the management of territorial integration, as both the US and China seek to establish positions of centrality in the networks of trade, production and consumption through which power will be projected. This explains why infrastructure finance and construction has become an arena of geopolitical competition – roads, ports, pipelines, energy grids, high-speed rail, and undersea cables orient places and people into Sino-centric networks or those of the US or regional actors.

Large-scale infrastructure has proliferated across continents in recent years, and trillions of dollars are earmarked for future projects. China has become the leader in infrastructure finance and construction since the 2013 announcement of the Belt and Road Initiative (BRI). The BRI now includes approximately 140 countries, and China's dominance in the field of infrastructure finance and construction has triggered alarm bells in Washington, Tokyo, and Brussels. Western commentary, policy analysis, and research on Chinese overseas development assistance tends to focus on geopolitics and international relations, and often interprets the BRI as a grand geopolitical strategy to reshape the world order and gain power through predatory lending. In June 2021, the G7 announced the *Build Back Better World* (B3W) initiative to compete with the BRI by providing countries with high-quality infrastructure and minimal negative social and environmental impacts. It is the latest in a series of similar responses to rival China in the field of infrastructure finance and construction, and in this context, we ask: how does US–China rivalry, with its competition to integrate territory through the construction of infrastructure, affect people and places around the world?

With this question we hope to shift the focus of academic debates in two ways: to geopolitical rivalry and to its effects at multiple scales. First, the past two decades have witnessed the proliferation of literature on China's global expansion and influence. Scholarship on the BRI and global China has moved beyond geopolitics to the complexities of how projects unfold on the ground (Lee, 2017; Klinger and Muldavin, 2019; Oliveira et al, 2020; Sidaway et al, 2020; DiCarlo, 2021). However, much of this work does not yet account for US–China or other geopolitical rivalry. Considering competition and geopolitics is essential to evaluate China's global expansion and its effects. Indeed, as historians would struggle to examine American or Soviet foreign policy in the 1950s in the absence of Cold War geopolitics, it is increasingly challenging to understand China's role in world affairs without accounting for geopolitical competition. Second, our analysis examines US–China competition from multiple scales – many of the chapters focus on specific infrastructure projects to show how contemporary geopolitical competition manifests in cities, towns, and rural areas. What emerges from these grounded cases is a complex picture in which infrastructure projects are more than manifestations of geopolitical competition – they are also constitutive of regional, national, and local visions and aspirations. Thus, Beijing and Washington may indeed compete to integrate territory, but they also must constantly adjust plans in accordance with the spatial aspirations of other governments, elite interests, political rivalries, and demands from civil society in host countries (Schindler and DiCarlo, 2022). While acknowledging the significance of great power rivalry, the chapters in this volume simultaneously attend to the agency of other countries, and show that the US and China are not free to compete as they like across the globe. This book draws attention to the risks and opportunities for states adapting to the emergent multipolar order. It also highlights the role played by middle and regional powers such as Japan, India and Vietnam. Ultimately this volume grounds the US–China rivalry and infrastructural competition in particular places and multiple scales, and highlights the multitude of actors and power dynamics at play, *simultaneously* detailing the contexts of great power rivalry and broadening perspectives beyond US–China bipolarity.

## State restructuring and the rise of the infrastructure state

Contemporary developmental objectives pursued by governments around the world are overwhelmingly spatial. Many low- and middle-income countries were subject to years of fiscal discipline at the hands of the Washington Consensus – from the 1980s to 2000s they were forced to embrace spatially blind policies that largely relegated infrastructure maintenance and construction to the private sector. After the 2008 financial crisis, new Keynesians staged a comeback in global development policy-making institutions, arguing that underinvestment in infrastructure created a significant infrastructure 'gap' or 'deficit' (see Goodfellow, 2020). Former World Bank President Justin Yifu Lin reasoned that addressing the infrastructure deficit could extricate the global economy from the doldrums of financial crisis. He advocated a 'global Marshall Plan' (Lin and Wang, 2013) and a 'global growth coalition' composed of international financial institutions (for example, the World Bank, African and Asian Development Banks), intergovernmental organizations (such as the OECD and G7), private and state-owned capital (for example, pension and sovereign wealth funds) coalesced to coordinate infrastructure investment (Schindler and Kanai, 2021). This regime of 'infrastructure-led development' is based on the premise that enhanced infrastructural connectivity will encourage foreign direct investment and ultimately lead to structural transformation of economies and societies from agricultural to advanced industrial (Schindler and Kanai, 2021).

Leaving aside the veracity of assumptions regarding the relationship between connectivity and structural transformation, infrastructure-led development appeals to governments because it allows them to pursue 'state spatial objectives' (Brenner, 2004). While governments assume significant risk (see Gabor, 2021), they are afforded scope to plan and implement grandiose spatial projects (Schindler et al, 2022). This is a welcome reversal to governments that were disciplined and disempowered by international financial institutions for nearly three decades. Many state spatial projects currently underpin industrial strategies – for example, the 'Make in India' initiative comprises five development corridors to build up manufacturing, while Kenya's *Vision 2030* is ostensibly a national spatial plan with industrial policy characteristics. The chapters in this volume demonstrate that governments worldwide have articulated spatial objectives that involve the construction of large-scale infrastructure.

The empowerment of governments to pursue spatial strategies has coincided with the intensification of the US–China rivalry. Competition to integrate territory is the flip-side to national state spatial strategies and, indeed, many countries have adroitly responded to the emergent territorial logic of great power rivalry by hedging their relationships with the US and China (Kuik, 2020). This has allowed some governments to undertake infrastructure projects that were previously unthinkable, yet often these projects are so complex and gargantuan that their achievement necessitates 'institutional calibrations, policy reorientations and regulatory experiments', or what Brenner (2009, 129) refers to as 'state restructuring'. This includes rebalancing power within national institutions or establishing new ones, introducing regulatory reform to fast-track infrastructure projects, streamlining land acquisition procedures and augmenting state capacity. Brenner introduced this term to describe how nation states in the North Atlantic adapted to the global shift from Fordist-Keynesianism to Post-Fordism by restructuring and rescaling their operations (Brenner, 2004). At present, state restructuring is a response to the dynamic political opportunity structure animated by the US–China rivalry, by governments 'attempting, at various spatial scales, to facilitate, manage, mediate, and redirect processes of geoeconomic restructuring' (Brenner, 2004, 61).

In some instances, state restructuring has resulted in the emergence of what we term the *infrastructure state* (Schindler et al, 2021). The infrastructure state exhibits significant variation from place to place but, in all cases, it seeks to address longstanding developmental challenges through the enhancement of connectivity. Therefore, for the infrastructure state, the expansion of large-scale infrastructure projects is the key to fostering social and economic transformation. The relationship between infrastructure and state building is well-documented (Mitchell, 2002; Hecht, 2011); however, throughout the neoliberal period, states largely ceded responsibility for infrastructure planning and construction to the private sector. This explains why the pursuit of spatial objectives has necessitated institutional reform or extensive state restructuring: decades of neoliberal reform left many states incapable of undertaking grand spatial projects. The reconstitution of state capacity in the context of great power politics has led to the emergence of the infrastructure state. Institutional reforms are often designed to enable or fast-track the completion of infrastructure projects. In some cases, such as Ethiopia and

Laos, new institutions and regulations have been created, while in others, such as Kazakhstan, there is minimal reform and a greater focus on hedging. Finally, the legitimacy of the infrastructure state is partly underpinned by the ability of the state apparatus to articulate and achieve spatial objectives.

This volume contributes to scholarship on state restructuring beyond the North Atlantic (Park, 2013; Bayırbağ, 2013; Klink, 2013; Kennedy, 2017; Lim, 2019; Williams et al, 2021), and shows that infrastructure states are shaped by negotiations among global and regional powers, and imperatives to pursue spatial objectives and align with networks of trade, production, and consumption. To these ends, contemporary infrastructure states seek to mobilize foreign capital for spatial projects that often promise to enhance transnational connectivity, with the hope that this will lead to foreign investment, industrial upgrading, and export-oriented economic growth. The contours of this narrative are evident around the world, from Eurasia to Africa and Latin America. However, ubiquity should not be mistaken for homogeneity. By localizing the US–China competition to integrate territory, the chapters in this volume explore the diverse modes of state restructuring, hedging, and agency.

Furthermore, they demonstrate that the US–China rivalry does not, by itself, determine spatial outcomes – in many cases infrastructure states pursue longstanding objectives that are shaped by domestic elites, bureaucrats and politicians, ordinary people, party politics and environments. Rather than an analysis on geopolitics or state power, several authors in this volume take an approach that, as Oakes (2021, 284) writes, engages the 'political as an effect of socio-technical configurations rather than policy pronouncements', revealing the often-unintended consequences and haphazard politics of infrastructure. The concept of the infrastructure state, thus, shines a light on the effects of US–China rivalry at multiple scales while it also underscores that decisions by actors in the Global South are influenced by local infrastructural histories and political economic dynamics. Quite simply, these local-level politics are shaped by and shape geopolitical competition.

## Situating infrastructural rivalry

Chapters are grouped according to their primary scale of analysis. The book begins in the offices of district-level bureaucrats in Tanzania and moves to national and international scales before ending in outer space. Part I analyses localized manifestations of infrastructural competition; Part II examines national politics with many chapters focused on instances of state restructuring; Part III shows how governments hedge between the US and China in pursuit of spatial objectives. Admittedly, grouping the chapters was an art rather than a science because each is multiscalar to some degree. The sections are separated by interludes that engage with themes that run through the chapters.

The first part, 'Grounding Infrastructural Rivalry', is situated most locally. Enns, Bersaglio, and Luhulu (Chapter 2) consider how local bureaucrats in Tanzania's Central Corridor mediate infrastructural visions and translate geopolitical rivalry across villages, state agencies, and with contractors. In the process their offices become 'everyday spaces in which global economic and geopolitical competition is negotiated'. Through their 'asphalt archaeology', Steenberg, Shamshiev, and Abdilashimova (Chapter 3) take us to roads in Kyrgyzstan to show now local politics drive infrastructure just as much as, if not more than, geopolitics. Like other chapters in this volume, they complicate simplistic narratives that track only US and Chinese influence in a place. They conclude by considering how amassing debt at the national level acts influences socio-political relations. Next, in Chapter 4 Tang's analysis of economic zones (EZs) in Lusaka, Zambia takes a historical view of the design and planning of Japanese and Chinese EZs. She compares two styles of economic diplomacy that reveal competing state interests and unsettled notions of Chinese hegemony. Chapter 5 examines multiple transportation infrastructure projects in Kenya, as Kimari and Lesutis develop the concept of 'symbolic geopolitical architectures'. Their analysis moves from Chinese and Kenyan development agendas to the public contestation that they engender.

Part II – 'Infrastructural Governance and State Restructuring' – moves to the arena of national politics. It is prefaced by a thematic interlude in which Rolf asks whether a Sino-centric transnational capitalist class is emerging. He concludes that in contrast to previous global regimes, there is scant evidence that self-interested national elites are augmenting Chinese hegemony, and this may, to an extent, check China's influence beyond its borders. This segues into Chapter 6 on elite politics in the Philippines, in which Camba, Cruz, and Lim show that elites leaned on Chinese capital to produce contradictory infrastructure, and that the Philippine Navy and Coast Guard exploit US–China competition to contrasting ends. Jepson (Chapter 7) then takes us to Ecuador to show how the US and China have been drawn into domestic rivalries. Even as Chinese finance enabled national policy sovereignty, many Chinese-funded infrastructure projects were portrayed as manifestations of corruption, while Former President Correa's ouster precipitated a renewal of relations with the US. Next, Huang and Goodfellow (Chapter 8) explore the sectors and spaces in which competition plays out to argue that Ethiopia received significant Chinese finance precisely because of its potential as an infrastructure state. Its centralized governance enabled swift state restructuring, yet recent events threaten to undermine an already fragmented polity. In Chapter 9, DeBoom traces the *longue durée* of uranium mining to evaluate the evolution of 'radioactive strategies' of successive Namibian governments. The analysis shows how the Namibian government has adapted its entanglements with foreign states and mining companies, and 'finally recast the uranium infrastructure of colonial, imperial, and neoliberal exploitation into infrastructure of its own design'.

Saguier and Vila Seoane (Chapter 10) also examine extractive industries – here oil, gas, and mining in Argentina – and how these sectors are positioned vis-à-vis the China–US rivalry. They portray a complex set of interactions among actors at various scales that defies simple classification. The chapter demonstrates that Argentina has resisted bipolarity and Chinese firms are most active in lithium mining, while US firms are most active in hydraulic fracking. Bayırbağ and Schindler (Chapter 11) show how Turkey's state spatial objectives have been aligned with the BRI. Rather than representing Ankara's subordination to Beijing, this allowed the Justice and Development Party to undertake infrastructure projects that underpin its patronage networks, particularly in the construction sector. The establishment of a sovereign wealth fund that mediates between Chinese sources of funding and Turkey's domestic banking system is an example of how state restructuring was undertaken to support spatial objectives. Finally, DiCarlo and Ingalls (Chapter 12) argue that, rather than a new Cold War framing, infrastructure development in Laos is more aptly characterized through a lens of 'mosaic geopolitics' that acknowledges the historical relationships and more-than-US actors involved. They also show how Laos has restructured with an eye toward Chinese-backed projects that align with longstanding domestic development objectives.

The third and final part entitled 'Geopolitics and State Spatial Strategies' zooms out again, this time to more geopolitically focused cases, particularly ones in which host states exhibit agency as they adeptly hedge between super and regional powers to pursue their own state spatial strategies. This section, too, begins with thematic interlude in which Kuik discusses host-state agency, receptivity of foreign-backed infrastructure, and hedging to introduce themes of the chapters that follow. In Chapter 13, Paudel and Rankin's contribution traces previous rounds of infrastructure competition in Nepal to argue that the current round creates opportunity for small, peripheral states to assert their own developmental agendas through infrastructure construction. They suggest that this reveals the limits of hegemony and may lead to new forms of solidarity among smaller countries. Tritto, Silaban, and Camba (Chapter 14) then show how Indonesia's hedging strategy is consistent with its longstanding ambivalence towards great power politics; through engagements with Japan and China, Jakarta attempts to remain aloof from US–China rivalry. This occurred terrestrially as the government actively pitted Chinese and Japanese bidders against one another for contracts to connect Jakarta and Bandung with a high-speed rail line; as well as in maritime spatial objectives, where it mirrored Japan's position on US-led Freedom of Navigation Operations. In nearby Vietnam, Liao (Chapter 15) tells a compelling and similar story of hedging between Chinese and Japanese economic statecraft. By examining the country's spatial strategies around its infrastructure policy, she illustrates Vietnam's evolving balancing act from the 1990s to the present. In Chapter 16, Gyuris traces Hungary's eastward turn after the 2008 financial crisis, in anticipation of accelerating growth through deepening ties with China. This is the only chapter focused on an EU member state, and it demonstrates the complexity of balancing Beijing and Brussels – as well as the US. The gains anticipated by a pro-Beijing policy have not materialized and a series of non-transparent infrastructure agreements with Chinese entities have stoked domestic opposition. The next election has become somewhat of a referendum on Budapest's warm relations with Beijing. Chapter 17 also presents a case of hedging, but here Neafie traces Kazakhstan's attempts to remain on cordial terms with Washington and China. Nur-Sultan has leveraged Kazakhstan's strategic location to secure concessions and support for infrastructure projects from both the US and China, but the case shows that this allows Kazakhstan to avoid state restructuring. In one case this resulted in the flagship Nur-Sultan light rail project becoming mired in corruption and collapsing. All of these chapters demonstrate the complex interplay between domestic politics and foreign policy, which shape the spatial projects that often implicate a bewildering array of state and non-state actors. Nowhere is this complexity more evident than outer space infrastructure. In Chapter 18, Klinger shows through several contexts in Central Eurasia how infrastructuring space is itself a spatial project undertaken by coalitions composed of public and private actors. While the US and China are involved in very different ways, she concludes that this 'final frontier' is not subject to bipolar rivalry and 'there is, quite simply, too much going on'.

Taken together these chapters show how the territorial logic of the US–China rivalry allows states to pursue spatial objectives that necessitate state restructuring and give rise to infrastructure states. Each chapter demonstrates how the infrastructure state takes different shapes and affects people and places uniquely based on the contexts and power dynamics within which spatial objectives are pursued. In some cases, states have considerable agency in shaping their territorial designs and transnational commitments, and in this context a host of middle and regional powers have actively expanded their participation in the global infrastructure sector. Some challenge the bipolar nature of new cold war rhetoric, while others point to historical precursors to nuance understandings of contemporary US–China competition. In the conclusion (Chapter 19) we situate contemporary US–China rivalry in a long history of geopolitical tension, noting that the last competition of this magnitude, the Cold War, led to the emergence of emancipatory anti-imperialist politics. We ask whether it is conceivable for the current round of geopolitical contestation to give rise to a version of 21st century Third Worldism.

### Notes

[1] *The Financial Times* devoted an entire section to the new Cold War, while the term has recently been used by Henry Kissinger and Robert Kaplan. Xi Jinping warned against a new Cold War, while Bernie Sanders struck a similar tone, albeit for different reasons.

[2] See www.airframer.com/aircraft_detail.html?model=C919 and www.csis.org/blogs/ trustee-china-hand/chinas-comac-aerospace-minor-leaguer.

## References

Bayırbağ, M.K. (2013) 'Continuity and change in public policy: redistribution, exclusion and state rescaling in Turkey', *International Journal of Urban and Regional Research*, 37(4): 1123–46.

Brenner, N. (2004) *New State Spaces: Urban Governance and the Rescaling of Statehood*, Oxford: Oxford University Press.

Brenner, N. (2009) 'Open questions on state rescaling', *Cambridge Journal of Regions, Economy and Society*, 2(1): 123–39.

DiCarlo, J. (2021) 'Grounding global China in Northern Laos: the making of the infrastructure frontier', Doctoral dissertation, University of Colorado Boulder.

DiCarlo, J. and Schindler, S. (2020) 'Will Biden pass the America LEADS Act and start a new Cold War with China?', *Global Policy Journal*. https://www.globalpolicyjournal.com/blog/10/12/2020/ will-biden-pass-america-leads-act-and-start-new-cold-war-china

Gabor, D. (2021) 'The Wall Street consensus', *Development and Change*, 52(3): 429–59.

Goodfellow, T. (2020) 'Finance, infrastructure and urban capital: the political economy of African "gap-filling"', *Review of African Political Economy*, 47(164): 256–74.

Hecht, G. (ed) (2011) *Entangled Geographies: Empire and Technopolitics in the Global Cold War*, Cambridge, MA: MIT Press.

Hung, H. (2022) *Clash of Empires: From 'Chimerica' to the 'New Cold War'*, Cambridge: Cambridge University Press.

Jones, B. (2020) *China and the Return of Great Power Strategic Competition*, New York: Brookings Institution.

Kennedy, L. (2017) 'State restructuring and emerging patterns of subnational policy-making and governance in China and India', *Environment and Planning C*, 35(1): 6–24.

Klinger, J.M. and Muldavin, J.S.S. (2019) 'New geographies of development: grounding China's global integration', *Territory, Politics, Governance*, 7(1): 1–21.

Klink, J. (2013) 'Development regimes, scales and state spatial restructuring: change and continuity in the production of urban space in metropolitan Rio de Janeiro, Brazil', *International Journal of Urban and Regional Research*, 37(4): 1168–87.

Kuik, C.-C. (2020) 'Hedging in post-pandemic Asia: what, how, and why?', *The Asian Forum*, 8(6). Available from: www.theasanforum.org/hedging- in-post-pandemic-asia-what-how-and-why/.

Lee, C.K. (2017) *The Specter of Global China: Politics, Labor, and Foreign Investment in Africa*, Chicago, IL: University of Chicago Press.

Lim, K.F. (2019) *On Shifting Foundations: State Rescaling, Policy Experimentation and Economic Restructuring in post-1949 China*, Chichester: Wiley for the Royal Geographical Society.

Lin, J.Y. and Wang, Y. (2013) 'Beyond the Marshall Plan: a global structural transformation fund'. Available from: www.post2020hlp.org/wp-content/uploads/docs/Lin-Wang_Beyond-the-Marshall-Plan-A-Global-Structural-Transformation-Fund.pdf.

Mitchell, T. (2002) *Rule of Experts: Egypt, Techno-Politics, Modernity*, Berkeley, CA: University of California Press.

Oakes, T. (2021) 'The Belt and Road as method: geopolitics, technopolitics and power through an infrastructure lens', *Asia Pacific Viewpoint*, 62(3): 281–85.

Oliveira, G., Murton, G., Rippa, A., Harlan, T., and Yang, Y.. (2020) 'China's Belt and Road Initiative: views from the ground', *Political Geography*, 82: 102225.

Park, B.-G. (2013) 'State rescaling in non-western contexts', *International Journal of Urban and Regional Research*, 37(4): 1115–22.

Schindler, S. and DiCarlo, J (2022) 'Towards a critical geopolitics of China-US rivalry: pericentricity, regional conflicts and transnational connections', *Area*. Available from: https://rgs-ibg.onlinelibrary.wiley.com/doi/pdf/10.1111/area.12812

Schindler, S. and Kanai, J.M. (2021) 'Getting the territory right: infrastructure-led development and the re-emergence of spatial planning strategies', *Regional Studies*, 55(1): 40–51.

Schindler, S., DiCarlo, J., and Paudel, D. (2021) 'The new cold war and the rise of the 21st century infrastructure state', *Transactions of the Institute of British Geographers*. https://rgs-ibg.onlinelibrary.wiley.com/doi/10.1111/tran.12480

Schindler, S., Alami, I., and Jepson, N. (2022) 'Goodbye *Washington Confusion*, hello Wall Street Consensus: contemporary state capitalism and the spatialisation of industrial strategy', *New Political Economy*. https://doi.org/10.1080/13563467.2022.2091534

Shambaugh, D. (2020) *Where Great Powers Meet: America and China in Southeast Asia*, Oxford: Oxford University Press.

Sidaway, J.D., Rowedder, S.C., Woon, C.Y., Lin, W., and Pholsena, V. (2020) 'Introduction: research agendas raised by the Belt and Road Initiative', *Environment and Planning C: Politics and Space*, 38(5): 795–802.

Williams, G., Mahadevia, D., Schindler, S., and Chattaraj, S. (2021) 'Megaprojects, mirages and miracles: territorializing the Delhi–Mumbai Industrial Corridor (DMIC) and state restructuring in contemporary India', *Territory, Politics, Governance*, www.tandfonline.com/doi/abs/10.1080/21622671.2020.1867630

PART I

# Grounding Infrastructural Rivalry

2

# Mediating the Infrastructure State: The Role of Local Bureaucrats in East Africa's Infrastructure Scramble

*Charis Enns, Brock Bersaglio, and Masalu Luhula*

When complete, the Central Corridor will link Tanzania, Burundi, Rwanda, Uganda, and the Democratic Republic of Congo to the port of Dar es Salaam through a multimodal transport infrastructure network consisting of inland waterways, port facilities, one-stop border crossings, roads, railways, and an oil and gas pipeline. This massive infrastructure corridor aims to integrate rural and remote areas of Tanzania, as well as neighbouring countries in the Great Lakes region, into global networks of production and trade. The corridor is meant to enable Tanzania and its neighbours to capitalize on the ongoing rare earth mining boom while also attracting investments in other sectors, such as agriculture, fishing, oil and gas, and tourism. As a spatial project, the Central Corridor embodies Tanzania's vision for national development, which involves the pursuit of regional and global integration through enhanced infrastructural connectivity.

Governments across East Africa are rushing to attract investment and grow their economies through infrastructure development (Carmody, 2017). This infrastructure scramble is fuelled, in part, by renewed emphasis on state-led national spatial planning as a means of producing territories that appeal to foreign investors (Kanai and Schindler, 2019). A key component of state-led territorial design is the demarcation of land suitable for uses that support export-oriented growth, such as special economic zones, free trade zones, export-processing zones, special agricultural zones, and industrial parks. National development plans also emphasize the construction and improvement of connective infrastructures, such as roads, railways, oil

13

and gas pipelines, and (air)ports to stitch territories together in ways that enhance flows of capital and commodities between spaces of extraction, production, and global markets (Schindler and Kanai, 2019). In this context, the Tanzanian government aims to leverage its geographical position as a 'traditional bridge' between countries in the Great Lakes region and the Indian Ocean to become a strategic centre for regional and global trade (URT, 2016, 34).

Evolving global economic and geopolitical dynamics increasingly shape the infrastructure scramble in East Africa. While the US withdrew from Tanzania during the Magufuli era, other states and development actors have continued or intensified their presence, including those from China, Kuwait, Turkey, and South Africa, as well as the World Bank. China's keen interest in infrastructure projects has provided many East African governments with access to capital necessary to implement national spatial plans and pursue infrastructure-led development (Vhumbunu, 2016; Onjala, 2018). In 2019, Tanzania tied with Kenya for the most infrastructure projects under construction on the continent (Edinger et al, 2019). This trend is likely to continue, with China poised to invest over US$1 trillion in foreign infrastructure projects through the BRI by 2027 (Chatzky and McBride, 2020; Han and Webber, 2020). By some estimates the initiative could increase East Africa's exports by US$192 million annually (Mukwaya and Mold, 2018). Tanzania has publicly embraced the BRI and has a long history of cooperation with China. In this regard, East Africa's infrastructure scramble is also fuelled by foreign powers, multilateral development banks, and investors searching for new territories of strategic investment.

Shifts in national spatial planning and global geopolitics may be fuelling the infrastructure scramble in East Africa, but local politics make projects im/possible. National spatial planning projects are not implemented on blank slates: they are enacted across diverse landscapes and land uses that span multiple jurisdictions. This chapter is concerned with the work of local bureaucrats who serve to 'fill the gap between project plans and on-the-ground realities' (Li, 2005, 391). Promises of regional integration and long-term national growth are not always enough to convince citizens to accept large infrastructure projects and the inevitable disruptions these projects cause to existing socialities and lifeworlds. As such, bureaucrats who interact with citizens on a routine basis often need to find creative ways to make projects 'implementable' on the ground (Funder and Mweemba, 2019) – for example, by tailoring grand visions of infrastructure to norms, priorities, and material realities unique to their constituents. In other words, local bureaucrats are intermediaries who do the work of fusing geopolitical aspirations, state-led national spatial planning strategies, and the socialities and lifeworlds of ordinary citizens to localized place-based regimes that serve wider economic and geopolitical interests.

Focusing on Tanzania's Central Corridor, we consider how local bureaucrats at the 'compromise level' of district government – 'which is said to allow the integration of top-down and bottom-up planning' (Gilson et al, 1994, 453) – translate geopolitical rivalries to contribute to the infrastructure scramble in East Africa. We argue that district-level bureaucrats participate in mediating different visions of infrastructure held by the many actors to which they have responsibility, ranging from local village assemblies to international contractors and state agencies. These acts of mediation are animated by the specific challenges and tensions that arise with the implementation of new infrastructure projects within their jurisdictions. As mediators, bureaucrats engage in various acts of negotiation that intervene in the original 'blueprints' of infrastructure projects and produce adapted infrastructures and spaces. Thus, we explore how district-level bureaucrats play an integral role in East Africa's infrastructure scramble and show that their offices are everyday spaces in which global economic and geopolitical competition is negotiated.

This chapter is informed by research started in 2017 in three districts along the Central Corridor in Tanzania: Manyoni, Itigi, and Uyui. When this research began, a small portion of the Central Corridor known as the Nyahua-Chaya Road that now connects Manyoni, Itigi, and Uyui was nearing completion. Key informant interviews and focus groups discussions were held with 115 participants along the Nyahua-Chaya Road; this includes six to eight representatives from each district government who participated in interviews or focus group discussions. The chapter concludes with reflections on how the work done by these local bureaucrats contributes to the infrastructure scramble in East Africa.

## Interface bureaucrats and spaces of citizen–state relations

The concept of the interface bureaucrat is rooted in anthropological literature on the state (Fineman, 1998; Ferguson and Gupta, 2002; Sharma and Gupta, 2009). The term describes bureaucrats who have regular, direct contact with citizens as they implement government initiatives in their jurisdictions on a day-to-day basis (Bierschenk and de Sardan, 2014). In the Tanzanian context, interface bureaucrats are usually located in the decentralized offices of government ministries established at the regional and district level (see Funder and Marani, 2015). These decentralized offices can be understood as arenas where citizens come face-to-face with the state, positioning bureaucrats at the 'interface' of citizen-state relations (Hagmann and Péclard, 2010; Willott, 2014).

Literature on interface bureaucrats offers insights into the contingent and discretionary manner in which grand visions of infrastructural connectivity are brought to life. For example, the implementation of many government initiatives is fraught with contestations that interface bureaucrats are responsible for navigating. These contestations may involve a range of actors – including citizens (of all sorts and subjectivities), community groups, companies, different government bodies, and international organizations – and usually require some form of negotiation for initiatives to proceed (Ribot, 2007; Radcliffe and Webb, 2015; Funder and Mweemba, 2019). Beyond managing relations with citizens, interface bureaucrats are often charged with making minor adjustments to initiatives to ensure they are locally appropriate (Funder and Mweemba, 2019). As our analysis in the following sections suggests, the agency expressed by interface bureaucrats during contestations results in adapted infrastructures, which take on slightly different appearances or meanings compared to the original plans of high-level actors, even if their main function remains fairly consistent.

Interface bureaucrats render government initiatives implementable on the ground in local contexts (Funder and Mweemba, 2019). They are, at times, responsible for upholding the legitimacy of both the state *and* its interventions. Within the institutional multiplicity that defines much of East Africa (see Mamdani, 2018), community groups, customary and informal institutions, local governments, non-governmental organizations, and different state agencies 'may oppose the interventions of local state agencies, and compete with them over service provision, authority, and even "stateness" itself' (Funder and Mweemba, 2019, 131–2). This makes the role of interface bureaucrats particularly challenging and important when it comes to the implementation of large infrastructure projects that span multiple jurisdictions.

In Tanzania, interface bureaucrats operate within a wider system of decentralized governance. Following independence from Britain in 1961, the trajectory of national development reoriented around the political and economic philosophy of *ujamaa*, which emphasized familyhood, socialism, and self-reliance (Malima, 1979). Decentralization was an integral component of *ujamaa* and was promoted, in part, as a measure to enhance public participation in development planning at a grassroots level. As a result, the district government has become a key space for development administration and planning. In the 1970s, locally elected district councils were abolished and replaced with representatives from the central government (Gilson et al, 1994), and further reforms in the 1990s tasked district governments with approving development plans prepared by village governments within their jurisdictions before submission to the prime minister's office (Greco, 2016). These later reforms created a decentralized bureaucratic environment in which the state could advance and pursue its visions of development with relative ease, despite an emphasis on grassroots planning (Greco, 2016). The district-level bureaucrats we discuss in this chapter continue to operate within this system and are seen as responsible for helping the central government achieve its evolving infrastructural and territorial ambitions.

Accordingly, district offices in Tanzania can be understood as arenas where the state comes face-to-face with citizens on development matters. At the same time, district-level interface bureaucrats also participate in governing spaces outside district offices that put them in contact with state and non-state actors. These spaces may include formal spaces of government, including village assemblies as well as higher level meetings and workshops, but when it comes to the implementation of infrastructure projects, farmers' fields, roadside shops, temporary worker camps, construction quarries, and forests may also serve as informal spaces where interface bureaucrats mediate relations between government organizations and citizens, contractors and customary institutions. Thus, the work interface bureaucrats do in mediating infrastructure is not just carried out in formal spaces of government; it is enacted through everyday, liminal spaces of citizen–state relations (Gupta, 1995).

## Grand visions for the Central Corridor

Tanzania's Central Corridor runs from the port of Dar es Salaam and links the country to Burundi, the Democratic Republic of Congo, Rwanda, and Uganda. Although the Central Corridor is not new, most existing transport infrastructure within the corridor has required replacement or maintenance for several decades. The original Central Corridor railway was completed in 1914 and became so dilapidated that it could no longer service Tanzania's trade or that of its neighbouring countries. Transportation of goods and people along the Central Corridor largely shifted to roads, even though sections of the road were also in poor condition or unpaved, making transportation slow and costly. Government plans to revive the Central Corridor involve improving and modernizing road and rail infrastructure, while also constructing new marine ports, airports, border crossings, and pipeline infrastructure to address challenges of inadequate connectivity between Tanzania's interior and the coast.

Tanzania's development approach is guided by *Tanzania Development Vision 2025*, which is being implemented through a series of Five-Year Development Plans (FYDPs). *Vision 2025* aspires to make Tanzania a middle-income country by 2025 and highlights the importance of transport infrastructures for developing a strong and competitive industrialized and export-oriented economy (URT, 2012). The current FYPD plan, called *Nurturing Industrialisation for Economic Transformation and Human Development* (FYDP II), highlights how improved transport infrastructure will enable Tanzania to leverage its strategic position both regionally and globally.

The Central Corridor is identified as a flagship project in FYDP II. Once complete, the government hopes it will position Tanzania at a 'place of physical intersection' for infrastructure networks that connect Eastern, Southern, and Central Africa to strategic ports along the Indian Ocean (URT, 2016, 34). The Central Corridor's railway will provide a 'bridge' between Central Africa and seaports along the continent's eastern seaboard (URT, 2016). However, such grand corridor visions are not held by the Government of Tanzania alone. Other regional and global actors are also attuned to the opportunities of this massive network of infrastructure. In 2017, the African Union dubbed the Central Corridor a 'smart corridor' and one of the continent's showcase projects (Were, 2019). Similarly, the Programme for Infrastructure Development in Africa identified the Central Corridor as a priority project with the potential to improve regional and continental trade.

Additionally, according to Were (2019) the Central Corridor carries the vision of a Pan-African approach to development, which 'deviates from the nation-specific projects of the 1950s and 1960s towards large inter-state regional connectivity infrastructural projects driven by hinterland economic demand and realization of the Pan-African vision of continental integration' (2019, 1). Given the extent of competition that regional infrastructure development has spurred – particularly between Tanzania and Kenyan as they vie for the position of East Africa's trade hub – it is difficult to say whether infrastructure is, in fact, actualizing Pan-African unity. It is safe to say though that projects like the Central Corridor are proving key to facilitating continental interconnectivity and bolstering continental trade in the process (Were, 2019). The importance and potential of growing continental trade networks has been substantiated by the global COVID-19 crisis, as regional value chains appear to have been more resilient than global value chains throughout the pandemic (Mold and Mveyange, 2020; see also Banga et al, 2020).

Chinese enterprises and investments have also played an important role in the realization of the Central Corridor. This is unsurprising given China's keen interest in expanding transport infrastructure globally through the BRI. As Sidaway et al (2020, 799) explain, the power of the BRI is 'in its capacity to align with its partners' bureaucratic state power, often stressing BRI's compatibility and complementarity with respective national visions and policies of development.' As a result of this complementarity, China has been highly supportive of Tanzania's 'corridorization' approach to infrastructure development overall, with dedicated finance coming from the Silk Road Fund, Export–Import Bank of China and China Development Bank. When it comes to the Central Corridor, Chinese firms and state-owned enterprises are at the forefront of winning construction contracts and investing in existing and emerging industries along the route.

China's interest in the Central Corridor is also linked to a geocultural agenda (Winter, 2020). As early as the 1970s, China supported the construction of rail infrastructure linking Tanzania and Zambia (Tazara Railway), with the stated ambition of providing Zambia with access to global markets for its exports that avoided passing through South Africa during apartheid. The Export-Import Bank of China has more recently lent Tanzania US$7.6 billion to finance the refurbishment of the Central Corridor railway while also providing support and funding for new locomotives and carriages. The Government of Tanzania also signed a deal with two Chinese groups, China Civil Engineering Construction and China Railway 15th Bureau, for the construction of a 3.2 km bridge over part of Lake Victoria. Plans for this bridge date back to the 1970s when President Nyerere hoped to promote integration, movement, and trade among countries around the lake.

Other bilateral and multilateral donors that recognize the importance of the Central Corridor include the World Bank and the Kuwait Fund for Arab Economic Development. With such diverse sources of finance accessible for different components of the corridor, there is no reason to believe that its design and implementation will be particularly beholden to Chinese interests. Yet, other global actors have recently withdrawn somewhat from Tanzania in light of the late President Magufuli's nationalist economic agenda. US policy towards Tanzania has become stagnant in recent years (Harris, 2021). This is at least partly because: 'The United Republic of Tanzania, according to Government officials, welcomes foreign direct investment (FDI) as it pursues its industrialization and development agenda. However, in practice, government policies and actions do not effectively keep and attract investment' (DOS, 2020, 4). While the US and other powerful actors pull away from Tanzania, China is deepening its relationship with the country, with recent commitments made between Foreign Minister Wang Yi and the late President Magufuli to 'jointly endorse multilateralism [and] oppose foreign interference' (Harris, 2021). Thus, complementary strategic interests in opening remote and rural areas to global markets, alongside shared cultural and historical narratives, are important in understanding China's role in shaping grand visions for the Central Corridor.

## Mediating grand and everyday corridor visions

Our insights into how local bureaucrats in the districts of Manyoni, Itigi, and Uyui contribute to mediating both grand and everyday visions for the Central Corridor are derived from conversations with bureaucrats whose duties and areas of responsibility are relevant to the implementation of the corridor. This includes District Executive Directors and various heads of department, such as engineers, environment officers, community development officers, land officers, legal officers, and planning officers. Furthermore, we also spoke with village assemblies, village leadership, and Village Executive Officers positioned along the Nyahua-Chaya Road, who offered insights into how district-level bureaucrats interact with citizens when their jurisdictions become enrolled in massive infrastructure projects.

## Adjusting visions of infrastructure

Recognizing that the promise of Pan-African and global connectivity alone is likely not enough to convince Tanzanians to accept massive infrastructure projects cutting through villages, the government of Tanzania has sought to highlight the Central Corridor's localized benefits. Small-scale farming is the most common livelihood activity along the corridor, with farmers producing crops such as cassava, cotton, groundnuts, millet, maize, pigeon peas, rice, sunflowers, and tobacco. The official narrative is that these livelihood activities will benefit from improved access to markets, linking producers to new value chains and fostering investment in industry and wage labour (URT, 2016).

For those with small plots of land who only produce crops for subsistence purposes or who practise non-farming livelihood activities, the promise of new market linkages is not necessarily convincing. With this in mind, district officers in Manyoni, Itigi, and Uyui have devised and conveyed slightly amended visions of the corridor to their constituents. For example, Uyui District recently allocated 8,000 acres of land for cashew production to local investors. In areas where timber and nontimber forest products contribute to people's livelihoods, district officers formulated plans to market, package, and sell local forest products in shops along the Nyahua-Chaya Road. In Uyui, subsequent investment in bee keeping led to the construction of a honey processing facility in Kigwa. Additionally, in Itigi, where the endemic Itigi-Sumbu Thicket forms a unique ecoregion, some district officers have considered initiating conservation and tourism activities as travel through their district becomes easier and volumes of traffic grow.

Another opportunity that district officers highlighted to convince their constituents of the benefits of the Central Corridor relates to local transport. As a transnational, multimodal infrastructure network, the Central Corridor has not been designed with local transport needs in mind. Instead, plans emphasize rail transport – which is bound to have high tariffs in order to recover costs as well as relatively long distances between stations – and large motorways built for lorries moving between sites of extraction and markets across the region. Local bureaucrats, however, developed their own plans for affixing local transport services to the corridor project. These include using district-level funds to construct 'last mile' infrastructure, bus stops, car parks, and rest stops for lorries. District officers in Itigi described their vision to eventually become the main stop-over point for people and freight between the west and east of the country. In addition to enabling people to travel more easily and affordably along the corridor route, positioning these districts as local transport hubs is perceived as a way to create new income generating activities and jobs.

Importantly, district officers reported that when higher level government officials visited their offices to inform them of plans for the Central Corridor and provide a timeline for construction, there was no consultation or discussion about corridor plans. Although district officers were not provided opportunities to directly adjust or offer feedback on the original blueprints for the Nyahua-Chaya Road or wider corridor, they have utilized other opportunities to creatively amend and augment this massive infrastructure project. As Funder and Mweemba explain, 'interface bureaucrats are typically charged with carrying out interventions that have been conceived and devised elsewhere. ... Ongoing adjustment is therefore required to make interventions "implementable" in local circumstances' (2019, 131). It is here where the role of interface bureaucrats becomes so important: by taking the original vision of a road and moulding it to fit local contexts, interface bureaucrats translate and articulate grand visions for infrastructure in ways that resonate with ordinary citizens within their jurisdictions.

## Managing contentious relationships

Even when interface bureaucrats reimagine and adapt infrastructure plans in ways that better speak to local constituents, actual benefits of such projects remain anticipatory. In other words, constituents must wait for infrastructure to be operational and, while waiting, they are left to watch as land is acquired for development and agricultural land and forests are cleared during construction. For this reason, corridor projects almost inevitably become spaces of contestation between high-level government actors, investors and contractors, and citizens, including wage labourers involved in construction (Sulle, 2020). District officers – who might also be implicated in contestations – are tasked with handling the political fallout.

A recurring point of contention along the Central Corridor relates to land acquisition and compensation. Farmers with land subsumed by corridor construction were often not provided with a clear timeline for when they had to move or where they would be resettled. In some cases, farmers were told that their land was being acquired for the road immediately and that they should cease all agricultural activity, but more than three years passed before construction started. As a result, these farmers lost multiple growing seasons for seemingly no reason. Furthermore, affected farming communities along the corridor route reported a series of problems with the compensation process. Although compensation was meant to be provided for titled land and for land with permanent structures, this did not always happen in practice.

Another sticking point for constituents impacted by construction of the Nyahua-Chaya Road related to jobs. The Chinese company contracted to build the portion of the road between Manyoni, Itigi, and Uyui did hire some local labourers. However, local labour was often hired on a short-term basis with no contract or job security. Labourers reported pay between 6,000 and 10,000 Tanzanian Shillings (TSh) (US$2.50–4.30) per day for casual labour and said that due to high food prices and low food availability during construction, they rarely returned home with more than TSh2,000 (US$0.86) each day if they fed themselves properly while working. They explained that their pay was too little to feed their families and one labourer exclaimed that, "it was better that you go home without eating, otherwise your wife will think you have a concubine, because you are bringing home such little money after each day's work" (Nyahua–Chaya Road, April 2018).

To manage these concerns and related disputes, district officers reported lobbying and politicking on behalf of their constituents for better compensation and working conditions. For example, in one village, letters of complaint were sent to the prime minister's office to demand payment for damage to crops due to construction. Shortly after letters were sent, the District Commissioner was sent to accompany an official from TANROADS (The Ministry of Works, Transport, and Communications) to meet with the village council. Two weeks after the District Commissioner's visit, compensation was paid out to village members who suffered damages. In this regard, some contestations may result in negotiations involving district officers acting as intermediaries between higher levels of government and their constituents, and thus in compromises to appease those affected. Such negotiations in decentralized offices of government ministries established at the district level reveal the importance of local politics in the process of corridor-making (Sulle, 2020).

Without district officers playing this integral role of interface bureaucrats, it might be difficult for projects like the Central Corridor to proceed. Indeed, in other regional corridor projects complaints are left unresolved and unmediated, which has spurred political mobilization, protest, and resistance. For example, the Mtwara Corridor in southern Tanzania saw significant delays as a result of contestations over land acquisition and compensation, as well as disputes over local-level benefits of this investment. These contestations escalated and led to accusations of state-directed violence, and subsequent threats of secession. Along the Mtwara Corridor, local bureaucrats decided to actively side with their constituents – rather than mediate grand and everyday infrastructure visions – due to the lack of public consultation and neglect for public interest in corridor design and implementation. Such examples point to the influential role of local bureaucrats in landing and seeing through major investments in infrastructure.

# Conclusion

This chapter examined the multiscalar politics of implementing the Central Corridor in Tanzania. Consistent with Tanzania's current vision for national development, the Central Corridor aims to drive development through regional infrastructural connectivity – leveraging Tanzania's strategic geographic position to become a globally competitive transport and logistics centre. Numerous influential global actors have provided support for the achievement of this development vision, encouraging and financing infrastructure projects, including the World Bank, the African Development Bank, and Programme for Infrastructure Development in Africa. However, China has been at the forefront of infrastructure development in Tanzania. This is because Tanzania's aim of regional and transnational integration aligns with China's objective of pursuing integration into the global economy through the BRI, and because China and Tanzania share a history of collaborative infrastructure development. With complementary economic, strategic, geopolitical, and geocultural objectives, it is unsurprising that these two countries remain motivated to work together.

Yet, the infrastructural visions of national and global actors are not simply implemented as imagined 'on the ground'. Instead, a range of efforts are required to 'fill the gap between project plans and on-the-ground realities' (Li, 2005, 391). In this chapter, we explored how massive infrastructure investments – which carry grand geoeconomic and geopolitical ambitions – are translated and articulated at the local level. The Central Corridor usefully illustrates how different actors participate in adjusting top-down visions of infrastructure development for implementation in unique local contexts, resulting in infrastructures that take on slightly different appearances or meanings in the process. Here, bureaucrats working at the interface of citizen–state relations play an important role in adjusting corridor blueprints to better fit within their jurisdictional contexts. Although these bureaucrats 'rarely operate with sovereign power' (Funder and Mweemba, 2019, 131), they are still capable of creatively expressing their agency through acts of negotiation that enable infrastructure projects to proceed.

By adapting visions of infrastructure and managing contentious relationships, interface bureaucrats also contribute to upholding grand visions for infrastructure that circulate at the national, regional, and global level. With each effort to adjust the Central Corridor to better align with the experiences and realities of ordinary citizens within their jurisdictions, district officers help to ensure the corridor is seen as legitimate by those they govern. Moreover, by managing and mediating complaints from constituents about corridor design and construction, bureaucrats diminish the chance of conflict escalation, which could threaten to stall or block infrastructure projects from proceeding. In this sense, local-level bureaucrats participate in mediating the infrastructure state by taking visions for infrastructural connectivity designed by high-level actors in faraway places and translating them into a vernacular that gives projects meaning – and therefore aids in their implementation – in local contexts. These acts of mediation simultaneously add legitimacy to grand visions for corridor projects.

## References

Banga, K., Keane, J., Mendez-Parra, M., Pettinotti, L., and Sommer, L. (2020) 'Africa trade and COVID-19: the supply chain dimension', Working Paper 586, London: Overseas Development Institute and African Trade Policy Centre.

Bierschenk, T. and de Sardan, J.O. (eds) (2014) *States at Work: Dynamics of African Bureaucracies*, Leiden: Brill.

Carmody, P. (2017) *The New Scramble for Africa*, Hoboken, NJ: John Wiley & Sons.

Chatzky, A. and McBride, J. (2020) 'China's massive Belt and Road Initiative', 28 January, Council on Foreign Relations. Available from: www.cfr.org/backgrounder/chinas-massive-belt-and-road-initiative.

Edinger, H., Ngulube, N., and Ntsoane, M. (2019) *Capital Projects in a Digital Age: Africa Construction Trends Report 2019*, Johannesburg: Deloitte.

Ferguson, J. and Gupta, A. (2002) 'Spatializing states: toward an ethnography of neoliberal governmentality', *American Ethnologist*, 29(4): 981–1002.

Fineman, S. (1998) 'Street-level bureaucrats and the social construction of environmental control', *Organization Studies*, 19(6): 953–74.

Funder, M. and Marani, M. (2015) 'Local bureaucrats as bricoleurs: the everyday implementation practices of county environment officers in rural Kenya', *International Journal of the Commons*, 9(1): 87–106.

Funder, M. and Mweemba, C.E. (2019) 'Interface bureaucrats and the everyday remaking of climate interventions: evidence from climate change adaptation in Zambia', *Global Environmental Change*, 55: 130–38.

Gilson, L., Kilima, P., and Tanner, M. (1994) 'Local government decentralization and the health sector in Tanzania', *Public Administration and Development*, 14(5): 451–77.

Greco, E. (2016) 'Village land politics and the legacy of ujamaa', *Review of African Political Economy*, 43(1): 22–40.

Gupta, A. (1995) 'Blurred boundaries: the discourse of corruption, the culture of politics, and the imagined state', *American Ethnologist*, 22(2): 375–402.

Hagmann, T. and Péclard, D. (2010) 'Negotiating statehood: dynamics of power and domination in Africa', *Development and Change*, 41(4): 539–62.

Han, X. and Webber, M. (2020) 'From Chinese dam building in Africa to the Belt and Road Initiative: assembling infrastructure projects and their linkages', *Political Geography*, 77: 102102.

Harris, M. (2021) 'Unfinished business: Magufuli's autocratic rule in Tanzania', Center for Strategic and International Studies, 5 February. Available from: www.csis.org/analysis/unfinished-business-magufulis-autocratic-rule-tanzania.

Kanai, J.M. and Schindler, S. (2019) 'Peri-urban promises of connectivity: linking project-led polycentrism to the infrastructure scramble', *Environment and Planning A: Economy and Space*, 51(2): 302–22.

Li, T.M. (2005) 'Beyond "the state" and failed schemes', *American Anthropologist*, 107(3): 383–94.

Malima, K. (1979) 'Planning for self-reliance: Tanzania's third five-year development plan', *Africa Development*, 4(1): 37–56.

Mamdani, M. (2018) *Citizen and subject: Contemporary Africa and the Legacy of Late Colonialism*, Princeton, NJ: Princeton University Press.

Mold, A. and Mveyange, A. (2020) 'Crisis? What crisis? COVID-19 and the unexpected recovery of regional trade in East Africa', Brookings, 28 September. Available from: www.brookings.edu/blog/africa-in-focus/2020/09/28/crisis-what-crisis-covid-19-and-the-unexpected-recovery-of-regional-trade-in-east-africa/.

Mukwaya, R. and Mold, A. (2018) 'Modelling the economic impact of the China Belt and Road Initiative on countries in Eastern Africa', Paper presented at the 21st Annual Conference on Global Economic Analysis, Cartagena, Colombia.

Onjala, J. (2018) 'China's development loans and the threat of debt crisis in Kenya', *Development Policy Review*, 36: O710–O728.

Radcliffe, S.A. and Webb, A.J. (2015) 'Subaltern bureaucrats and postcolonial rule: Indigenous professional registers of engagement with the Chilean state', *Comparative Studies in Society and History*, 57(1): 248–73.

Ribot, J.C. (2007) 'Representation, citizenship and the public domain in democratic decentralization', *Development*, 50(1): 43–9.

Schindler, S. and Kanai, J.M. (2019) 'Getting the territory right: infrastructure-led development and the re-emergence of spatial planning strategies', *Regional Studies*, 55(1): 40–51.

Sharma, A. and Gupta, A. (eds) (2009) *The Anthropology of the State: A Reader*, Hoboken, NJ: John Wiley & Sons.

Sidaway, J.D., Rowedder, S.C., Woon, C.Y., Lin, W., and Pholsena, V. (2020) 'Introduction: research agendas raised by the Belt and Road Initiative', *Environment and Planning C: Politics and Space*, 38(5): 795–802.

Sulle, E. (2020) 'Bureaucrats, investors and smallholders: contesting land rights and agro-commercialisation in the Southern agricultural growth corridor of Tanzania', *Journal of Eastern African Studies*, 14(2): 332–53.

United Republic of Tanzania (URT) (2012) 'The Tanzania long term perspective plan (LTPP), 2011/12-2025/26: the roadmap to a middle-income country'. Dodoma: President's Office, Planning Commission.

United Republic of Tanzania (URT) (2016) *National Five-Year Development Plan 2016/17–2020/21: Nurturing Industrialisation for Economic Transformation and Human Development*, Dodoma: Ministry of Finance and Planning.

United States Department of State (DOS) (2020) 'Investment climate statements: Tanzania'. Available from: www.state.gov/reports/2020-investment-climate-statements/tanzania/.

Vhumbunu, C.H. (2016) 'Enabling African regional infrastructure renaissance through the China–Africa partnership: a trans–continental appraisal', *International Journal of China Studies*, 7(3): 271–300.

Were, E. (2019) 'East African infrastructural development race: a sign of postmodern Pan-Africanism?', *Cambridge Review of International Affairs*. DOI: 10.1080/09557571.2019.1648382.

Willott, C. (2014) 'Factionalism and staff success in a Nigerian university: a departmental case study', in T. Bierschenk and J.O. de Sardan (eds) *States at Work: Dynamics of African Bureaucracies*, Leiden: Brill, pp 91–112.

Winter, T. (2020) 'Geocultural power: China's Belt and Road Initiative', *Geopolitics*, 26(5): 1376–99.

# 3

# Roads, Debt, and Kyrgyzstan's Quest for Geopolitical Kinship

*Rune Steenberg, Ulan Shamshiev, and Farzana Abdilashimova*

## Introduction

In October 2020, protests broke out in Bishkek over alleged vote buying and electoral fraud during the recent parliamentary elections. For the third time in 20 years a Kyrgyz president was forced to step down. Shortly after protests began on 11 October, former Kyrgyz ambassador to the US and Canada, Kadyr Toktogulov, published an opinion piece on CNN calling for 'Western support for Kyrgyzstan' (Toktogulov, 2020). Referring to Kyrgyzstan as an 'island of democracy and free speech in Central Asia', he stressed the country's importance for strengthening democracy in Central Asia. It had previously enjoyed support from the EU, Switzerland, and South Korea to improve its electoral system. However, Toktogulov (2020) continued, Kyrgyzstan's road to democracy also requires a different kind of support:

> [C]onsiderable economic assistance and cooperation are also needed to strengthen Kyrgyzstan's relations with the European Union and other Western countries. It is important for Kyrgyzstan to be able to balance its relations with China, which holds a significant part of the Central Asian nation's debt, and Russia, its most important strategic ally in economy and security.

His rationale for why the West should be interested in providing such assistance stays within the double-speak of the Washington Consensus, which lauded moves toward democracy while facilitating Western economic, geostrategic, and military benefits. It promoted the International Monetary Fund's (IMF) 'shock therapy', to which Kyrgyzstan was subjected in the early phase of its independence that included selling state assets, hosting US military operations, and protecting the entrance of exploitative Western companies into the fragile Kyrgyz economy. This double-speak masked as increased freedom resulted in severe poverty, inequality, and reduced living standards (Kuehnast and Dudwick, 2004).

Toktogulov's jargon may be calculation or naïveté. Either way, his call for assistance encapsulated the current geopolitical and economic situation in Kyrgyzstan. It betrays a longing of certain Kyrgyz elites for Western economic and political support. It also captures the reality that the 1990s promises and hopes of freedom and prosperity never materialized and today seem as utopian as ever. Western engagement in Central Asia has declined leaving Russia and China as the most influential political and economic actors in the region. Although Toktogulov recognizes this, he also sees new potential for Western re-engagement with Kyrgyzstan, particularly in the context of the intensifying rivalry with Russia and China, born out of the shifting power dynamics between these great powers:

> Western democracies should ensure that Kyrgyzstan is given the resources and opportunity to reduce its reliance on Russia and China, given how greatly different the two powers' democratic trajectories are compared to Kyrgyzstan. The struggle of the people of Kyrgyzstan for freedoms and rights and aspirations to live in a democratic republic should not be in vain. (Togtogulov, 2020)

Central Asia, the historical scene of the so-called Great Game between imperial Britain and Russia, is no stranger to big-power geopolitics and proxy conflicts. Two analytical models have been prominently employed to capture Kyrgyzstan's geopolitical in-between positioning. The country has been described as pursuing a 'multi-vector' foreign policy strategy (Sarı, 2012) that plays great powers against each other and engages with them in turn to fulfil different needs and balance its multiply peripheral position. A more sceptical position suggests the multi-vector metaphor as naive in terms of assessing the actual agency of poor, indebted states with heavy foreign, non-governmental organization (NGO), and donor involvement. Boris Pétric (2005) questioned Kyrgyzstan's functioning as a state, instead calling it a 'globalised protectorate' with limited autonomy. These two approaches both point to important aspects of the country's peripheral in-between position while placing different emphasis on its voluntariness and the agency of the country's elites.

A central and telling measure of engagement of big powers in Kyrgyzstan is the construction and funding of infrastructure. This chapter introduces a historical account of Kyrgyzstan's infrastructure development in the late 20th and early 21st century. The established narrative presents the shifting dominance of the Soviet Union, then US-led development banks, and finally China as the main infrastructural builders and investors. In the 21st century, US investments declined, and Chinese rose. Yet this linear meta-narrative becomes complicated when we look more closely. Multiple actors are involved, local perspectives are pivotal, and decision making is influenced by not only big-power politics but also domestic Kyrgyz political rivalries, established institutional patterns, and strong local sentiments. These complications suggest that sustainable infrastructure development in areas marginal to the world capitalist system, such as Kyrgyzstan, calls for more holistic, as opposed to purely monetary engagements. To capture this analytically, it therefore becomes necessary to add a third model to the ones proposed: Kyrgyzstan as pursuing a multi-vector foreign policy strategy or as a globalized protectorate. We suggest a praxeological approach (Bourdieu, 1977) that allows for a holistic view on society and captures the positioned perspectives of actors within the structure without losing sight of the structure. Such an approach is found in kinship studies and economic anthropology in form of a model of debt as an investment into relations of broader engagement facilitating 'mutuality of being' (Graeber, 2011; Sahlins, 2013).

## A brief history of Kyrgyzstan's infrastructure

Kyrgyzstan's modern infrastructure, its paved roads, electrical grid, water pipes, heating system, industry, transportation, as well as state education and healthcare were developed under Soviet rule. During this period of 'high modernism' (Scott, 1998) Kyrgyzstan was regarded as part of a larger region, and its infrastructural connections to neighbouring republics reflects this. The water management of the Syrdarya and Amudarya Rivers that flow from Kyrgyzstan to the heavily irrigated agricultural plains of Uzbekistan and Turkmenistan caused conflict when the republics became separated (Bichsel, 2009). Rail lines and roads spanning inner-Soviet borders now cross international ones; this has led to severe traffic disruptions and eventually the shut-down of the rail line from Bishkek in the north to Osh in the south of Kyrgyzstan because it passed through Uzbekistan. A diversion of a large stretch of road in the Ferghana Valley that parallels the border between the same two countries likewise had to be augmented with a second road further inland. Industrially too, the region was integrated into the Soviet Union and each republic struggled to disembed from the planned economy and re-embed within the capitalist world system. In the early 1990s, the World Bank, IMF, and later the Asian Development Bank (ADB) along with several national development agencies from Western countries (prominently the Swiss and United State Agency for International Development, USAID) began to support – and demand – institutional reforms meant to transform

Kyrgyzstan into a capitalist 'democratic' system. Western-dominated international donor organizations along with a myriad of NGOs following in their wake pushed a 'shock therapy' style transition. The reforms were almost all 'shock' with little 'therapy', and Kyrgyzstan's economic output, standard of living, and levels of education dropped significantly. Rather than constituting a viable state based on democratic decisions and laws the political scene became dominated by particular elite groups (Jones-Luong, 2002), so-called 'clans' (Gulette, 2010), and networks of the privileged, along with capital-heavy foreign corporations and organizations. The Canadian mining company Center Gold's takeover of Kyrgyzstan's largest gold mine Kumtor is one of the most prominent examples of this phenomenon.

Kyrgyzstan joined the World Trade Organization (WTO) in 2002, shortly after China, and received a high level of loans and investment from the World Bank and the ADB, not least for roads and energy infrastructure. After the US-led invasion of Afghanistan, the US Army established a military base near the capital, Bishkek, in 2001.[1] By this time, China had become a main player in Kyrgyzstan's economy, primarily through an inflow of cheap consumer goods, some of which were re-exported throughout Central Asia. In the mid-2000s, China, embarking on its Go West strategy, began to finance roads in this mountainous country along with hydropower and heating plants. As China's role grew, Western involvement declined, especially following the Wall Street initiated World Financial Crisis in 2007. In 2014, the US Army was forced to close its base in Kyrgyzstan due to numerous scandals and popular protests.

By the end of the second decade of the 20th century, China's influence in Kyrgyzstan eclipsed that of the US. China is by far Kyrgyzstan's largest infrastructure investor. Between 2016 and 2020 China pledged more than US$1 billion in loans to Kyrgyzstan for roads and another billion for energy infrastructure, making China Kyrgyzstan's number one creditor, overtaking the World Bank, the ADB, and all Western countries. As the main foreign direct investor (Mogilevskii, 2019[2]) China has the largest number of enterprises in the country at around 400 (Ryskulova, 2019), and has grown out of a singular role as the source of cheaply produced goods to a diversified economic actor in Kyrgyzstan, powered by its spectacular internal growth and focus on infrastructure. As of 2019, the Kyrgyz government owed 42 per cent of its external debts (US$1.7 billion) or 24 per cent of the country's GDP to the Export-Import Bank of China (EXIM) (Mogilevskii, 2019). Comparatively low interest rates of 2 per cent make these loans cheaper than the 10 per cent that national Kyrgyz banks offer, but this still means an annual interest payment of about US$340 million. Much like the case of loans from the large international banks and Western development aid, capital provided by Chinese loans and grants has often found its way back to companies based in the donor country, who 'send their companies and their workers to do the work' (Mogilevskii, 2019). The rationale for accepting Chinese capital is multifaceted. In one sense, it is due to the drying up of other sources, especially from the US, other Western donors, as well as Russia. In addition, though often less transparent than Western loans, Chinese loans do not come with conditionalities, have low interest rates, less bureaucratic barriers to overcome, and are more negotiable (Ormushev, 2020).

## Complicating the narrative: a contemporary asphalt archaeology

The meta-narrative of the US taking over from the Soviet Union in the 1990s and being replaced by China in the 2010s – while true to a degree – oversimplifies Kyrgyzstan's infrastructural developments. Many other actors have been significantly involved in infrastructure besides the US and China. Most significantly, the meta-narrative does not do justice to the continued centrality of Russia. Toktogulov's call for Western assistance was not heeded. Without much support or indeed notice from Western countries, on 11 November 2020, President Sooronbay Jeenbekov resigned and opposition politician Sadyr Japarov was freed from prison and instated as interim prime minister and president. In January 2021 he was elected Kyrgyzstan's sixth president, and changed the mode of Kyrgyz government from a parliamentary to a presidential system (AlJazeera, 2021). His first foreign visit was to Russia, as had been the case for his two most recent predecessors – a gesture that confirmed Russia as Kyrgyzstan's world-political centre of gravity. Unlike the US, Russia retained a military base in northern Kyrgyzstan. In 2014, the same year the US base was closed, the Russian base received a 15-year extension, and when violence broke out in Osh in 2010 between ethnic Kyrgyz and minority Uzbek, the Kyrgyz government called on Russia rather than the North Atlantic Treaty Organization (NATO), the UN, the US, or China for military assistance. While today Kyrgyzstan is a WTO member, a partner to the China-initiated Shanghai Cooperation Organisation and in 2013 joined China's BRI, in 2015 it also became a member of the Eurasian Economic Union (EEU) led by Russia, which eased trade with former Soviet republics while complicating customs procedures with China. Perhaps the most important connection to Russia provides the backbone of much of the Kyrgyz economy: in the 2010s it is estimated that between 500,000 to 1 million Kyrgyz were working abroad, the majority by far in Russia, and their remittances accounted for up to 30 per cent of Kyrgyzstan's GDP (Reeves, 2012; Sagynbekova, 2017). The Soviet legacy of education and Russian language use also engender consumption of Russian media and a feeling of familiarity and sympathy with Russian culture and society despite the colonial past and undeniably violent, discriminatory, and exploitative treatment of many Central Asian labour migrants (Eraliev and Urinboyev,

2020). Russia thus maintains a central role politically, culturally, and within the livelihood strategies for much of the Kyrgyz population.

China too displays a degree of continuity in its engagement with Kyrgyzstan. While early involvement in Central Asian infrastructure was devoted to pipeline building, it played a key role in the 1990s' hydropower and road building. Trade between the two countries flourished in the early 2000s, when the US was still establishing itself as the military hegemon in the region and, contrary to conventional perception, Chinese FDI to Kyrgyzstan peaked in 2015 and has since decreased substantially – trade between the two countries even peaked in 2008 (Mogilevskii, 2019). After 2015 several economic indicators declined as the effects of the Global Financial Crisis hit Chinese export business, Kyrgyzstan joined the EEU, and securitization in Xinjiang strangled cross-border trade and entrepreneurship. Events in Xinjiang also negatively influenced local attitudes towards China leading to resistance against Chinese projects and a rising suspicion of China.

## Polyester roads

The past 30 years of major road building in Kyrgyzstan offers examples of both the influence of Chinese firms as well as other infrastructural actors. Prior to 1951, there was no road connecting the country's two largest cities of Bishkek in the north with Osh in the south. A gravel road was built that year and in 1957 Soviet engineers began to pave it. It was mainly used by workers on the new hydroelectric plants built at the Toktokul Dam in the 1950s and 1960s and the main north–south connection continued to be the railway line through Kazakhstan and Uzbekistan. When the Soviet Union collapsed, the road began to crumble due to lack of maintenance. Between 1999 and 2005, 483 km of the 670 km road was reconstructed by Turkish, Chinese, Russian, Iranian, and construction companies from several other countries, financed mainly by the ADB with assistance from the Japan International Cooperation Agency and the Islamic Development Bank. After 2012, with financing by the ADB and the European Development Bank, the last stretches were completed by Chinese and Azerbaijani construction firms.

After Osh, the road forks leading south-west to Batken and the Ferghana Valley and east to Sary-Tash, from where separate roads connect to Tajikistan and China respectively. The China side is locally known as the Polyester Road, a pun on the ancient Silk Road, as this is one of the main arteries for the import of cheap textile and other consumer goods rolling in on large trucks from China. Many products continue to Uzbekistan and beyond. The section to Sary-Tash was constructed in 2007–09, with the remaining 71 km to the Chinese border built in 2010–12. About 43 km of the 550 km was built by a Turkish contractor, and the rest by China Road and Bridge

Corporation (a state-owned enterprise, SOE), with funding from the ADB and China EXIM Bank.

The second point of entrance for Chinese goods into Kyrgyzstan is Torugart by Naryn. From here the road passes Issyk Kul Lake and continues to Bishkek. All but one section of this 497 km road was built between 2010–14 by Chinese construction companies and more than half of it financed by China EXIM Bank with other parts financed by the ADB. A 93 km section is financed by Arab Coordination Group and constructed by Kuwaiti and Chinese companies.

A third important road, known as the Alternative North–South Route leading from Balikchy across the Kyrgyz mountains to Jalalabad, is under construction. Around 250 km of the 450 km Alternative North–South Route is currently being built and funded with support from China EXIM Bank and China Road and Bridge Corporation. The rest is funded by the ADB and the Islamic Development Bank and yet to be constructed. A similar picture can be observed in relation to the road across the southern-most province of Batken, where the Chinese funded and built sections were completed around 2015–16 while the parts financed by others remain largely unfinished or even not started.

Over the past 30 years of road building in Kyrgyzstan, there has been a gradual increase in Chinese engagement. While it began with construction, soon Chinese firms became involved in financing as well. Still, even though Chinese-led projects tend to be completed faster, international organizations (especially the ADB) tend to finance major traffic arteries while smaller and more peripheral roads are funded through Chinese banks. This is changing due to China's central role in the Alternative North–South Route. Banks and development agencies from Japan and Arabic countries are also heavily involved while Russia gives aid and loans directly to the government rather than earmarking money for specific infrastructure projects. Chinese involvement and funding peaked around 2014–17 but has since declined.

## More complications: roadside gravel

The involvement of additional actors and a differentiated take on development trajectories complicate the simple narrative of the decline of the US and the rise of China in Kyrgyzstan. Yet, this still does not do justice to the situation on the ground, as we discovered embarking on a tour along the two main roads from Bishkek to Osh and to Naryn respectively in the summer of 2020. Here, elite priorities, local concerns, established institutional patterns and popular sentiments, and stories influence infrastructural decision making to a degree that has national and international significance, yet is often overlooked.

The Bishkek–Osh Road skirts the banks of the Toktogul reservoir created by a large hydropower dam constructed during Soviet times. While the western bank seems more straightforward, the road follows the eastern bank much longer than seems necessary without bridging across even very narrow parts. According to locals, the road was placed here because politicians lobbied to have their villages included. From the reservoir down into the Ferghana Valley the road travels to the lowest geographical point in Kyrgyzstan, Shamalde Say at 500 m above sea level, a centre for fruit and vegetable production. Here the roadsides are extended with extra gravel to enable the many cars stopping at stalls to buy produce that is cheaper and fresher than in the urban centres.

The very lack of such roadside gravel along the northern main traffic artery connecting Naryn and Issykköl has become a topic of contention in the Kyrgyz parliament. On this stretch, the Chinese constructors did not add a proper roadside resulting in a vertical drop of 30 cm on both sides which makes stopping almost impossible for regular cars. This seemingly had not been specified in the contract. On the same road, along the part between Bishkek and Issykköl, the surface is well paved allowing for speedy driving and its generous roadside permits easy stopping for picture taking. This was made a priority due to its importance for the local and national tourism industry. Local economic concerns also motivated political support for the Alternative North–South Route construction mentioned earlier. It started in 2014 when traffic had become increasingly congested and damage from landslides on the Bishkek–Osh Road halted north–south traffic for days. Further, organized demonstrations against the Canadian-run Kumtor mine, demanding proper reparations for deadly cyanide spills and its nationalization, targeted this road, blocking it for days and causing serious economic damage.

The road was in particularly bad condition between the towns Madaniyat and Turpakbel. This section had not undergone renovation since Soviet times. A local resident aired the belief that the money for this part of the road had been channelled to the Akse district after violent riots there in 2002 in order to calm the population there. He blamed this on then President Askar Akaev (1991–2005). Yet, people also expressed gratitude to him for building the Bishkek–Osh Road while the later road from Bishkek to Torugart is credited to President Kurmanbek Bakiev (2005–10). Large public infrastructure, for better and for worse, is attributed to national or local politicians and pointed to as achievements or scandal. They are referred to in the inner-Kyrgyz political struggles, particularly between the north and the south. Foreign organizations and banks involved are secondary to most people. Many blamed Almazbek Atambayev (2011–17) for the corruption scandal around the renovation of Bishkek's heating plant, rather than the Chinese contractors and investors. Yet, in areas of private business that contribute less directly to public infrastructure, this is less the case.

The Canadian mine company running Kumtor has often been the target of public protests and in 2020 one Chinese gold mine was forced to close and the construction of a large commercial centre near the Chinese border discontinued, as the result of local resistance.

## Distrusting China

At an academic BRI workshop in London in the autumn of 2017, Kyrgyz anthropologist Cholpon Chotaeva presented a survey conducted among business leaders, politicians, professionals, and academics in Kyrgyzstan. She asked interviewees about their views on Chinese involvement in Kyrgyzstan. The responses were overwhelmingly positive. A doctor reported how Chinese medical aid had helped their hospital improve its service and research, and the great possibilities offered by Chinese investment were lauded by all respondents. This positivity reflects two things: first, elites in Kyrgyzstan have a comparatively benevolent attitude towards China and other donors as they disproportionately benefit from investments and opportunities and face fewer negative consequences such as land loss, pollution, and underemployment than the poorer segments of society. Second, 2017 was within the peak of Chinese investment, the US and other Western actors were gradually pulling out, and, importantly, the mass incarceration and massive human rights abuses of local minority populations in Xinjiang were not yet widely known. Three years later the situation had markedly changed.

In spring 2020, local protests led the Kyrgyz government to cancel the Kyrgyz–Chinese Ata-Bashi Free Economic Zone Joint Venture that was planned to be established at the Naryn border crossing between the two countries. Around the same time, local resistance forced the Chinese mining company Zhong Ji to end its operations in the Soltun Sary gold mine near Naryn. This incident served as inspiration for the Kyrgyz-produced feature film *Meken* (Fatherland), released later that year which narrated the story of a courageous Kyrgyz village rising up against a Chinese mining company. The film was first censored and later banned from movie theatres for 'inciting ethnic hatred', but aired on YouTube[3] and other online platforms. The same argument has been made against Uyghur cultural organizations and Xinjiang activists in Kyrgyzstan and Kazakhstan in order to silence criticism of China (Pannier, 2020b). Today, although the Kazakh and Kyrgyz governments still refrain from confronting China, the countries' elites seem to be gradually retracting support. Kazakhstan has refused to extradite Xinjiang refugees to China and Kyrgyzstan has delayed participation in the planned China–Kyrgyzstan–Uzbekistan rail line because several Kyrgyz towns were not included (Pannier, 2020a).

Beijing seems aware of the need to improve its image in Central Asia. In 2019, an unpopular Chinese mining project was reprimanded for having wrongfully portrayed itself as part of the BRI and was forced to delete this reference. Billboards advertise the BRI and the free, bi-weekly, Chinese-owned Russian language journal Silk Road (*Solko Repuch*) promotes 'friendship between the peoples' of Kyrgyzstan and China. Still, on social media and in quotidian discourse, China is viewed critically and sometimes with outright hostility (Mardell, 2020), increasingly connected to knowledge of the atrocities in Xinjiang (Djanibekova, 2018) but also feeding on Kyrgyz fears of being colonized by the much larger neighbour often referring to the Kyrgyz government's acquiescence to Chinese claims on disputed mountainous pastureland at Üjüngükuch near Naryn in 2002. These fears are regularly preyed on by politicians like Adakhan Maduramov, who reportedly claimed that 70,000 Chinese had married Kyrgyz women in the past years. Official estimates listed a total of 70,000 Chinese residing temporarily or permanently in Kyrgyzstan and according to a local researcher the actual number of Chinese citizens (probably including ethnic Uyghur and Kyrgyz) legally married to Kyrgyz is about 160 individuals.

Internationally, reports of transferred ownership of ports in Greece and Sri Lanka to Chinese companies as compensation for unpaid loans has caused anxiety in Kyrgyzstan. The country holds US$1.7 billion debt with China, which is likely to grow in the coming years, not least due to planned infrastructure. Among the growing China critics in Kyrgyzstan this has spurred the trope that 'every newborn Kyrgyz already has US$800,000 of debt to China'. In late October 2020, China relieved some debt for a number of countries because of the COVID-19 pandemic, but Kyrgyzstan was not among them. This provoked anger and worry in Kyrgyzstan, which is accustomed to having its debts to Russia routinely forgiven and those to international development banks restructured. Many Kyrgyz concluded that even though Chinese loans are relatively cheap, easy, fast, and with minimal red tape or conditionality, being in debt to China may carry more severe consequences for Kyrgyzstan than the debt to Russia, the US, the World Bank or the ADB (Bakirova, 2020).

## Conclusion

This chapter traced the established narrative of infrastructure development in Kyrgyzstan as first dominated by the Soviet Union, followed by the US through international development banks in the 1990s, and most recently by China. We complicated this narrative by diversifying our scales of analysis and examining who is actually engaged in infrastructure planning and construction, thereby recognizing a multitude of actors and non-linear trajectories development trajectories. Contrasting two oft-cited frameworks used to describe Kyrgyzstan's role between several powerful forces – as a 'globalised protectorate' or consciously as pursuing a multi-vector foreign policy strategy – we examined the country's degree of ownership over its infrastructural futures. Kyrgyzstan possesses some agency in manoeuvring between China, Russia, and the US as well as other important actors like the IMF, the World Bank, the ADB, Uzbekistan, Kazakhstan, Iran, Switzerland, Japan, Saudi Arabia, and a number of large international corporations and NGOs. Yet, in light of the country's weak political and economic position and as loans amass, Kyrgyz agency is limited. This is exemplified by the Canadian mining company at Kumtor extracting millions of dollars in profit. Current President Japarov was willing to go to prison to protest this exploitation when part of the opposition, but now – in power – seems to have accepted it.

We provided on-the-ground road ethnography and asphalt archaeology to complicate these models and narratives, and capture aspects of the complex reality in Kyrgyzstan. Our contemporary asphalt archaeology shows that infrastructures have their own stories on the ground, which guide Kyrgyz politicians' decisions to pursue them and the people's judgement over them. Local perception is central, for elections but also particularly for protests and popular uprisings. They have affected huge projects like the Kumtor mine, main roads, US military bases, Chinese mining and dry ports, bridges, high-speed railways, and are often related to political factions and elite competition.[4] Popular sentiments as well as established patterns of administration and the political north–south divide influence who builds and finances Kyrgyz infrastructure. Many main roads are credited to ousted President Bakiev and thus to the southern factions, the great heating plant corruption scandal is attributed to Atambaev and thus the northern factions, unfinished roads are blamed on rebellious villages, and pollution and land grabs on certain Chinese and Western companies. All Kyrgyz politicians navigate these narratives at least to the same degree that they navigate the great powers and their expectations.[5]

Turning to debt, the Kyrgyz government has repeatedly warned its population against celebrating with large and elaborate weddings and feasts because expenses drive households into crippling debt, severely hurting the economy. Indebtedness leads to obligations and clientelist structures that may compromise adherence to the law, enhancing economically irresponsible behaviour, nepotism, and tribalism. Yet, for many people in Kyrgyzstan and other contexts on the margins of the capitalist world system, large weddings provide spaces for investment into social capital, which are crucial for household livelihoods and social security – precisely because they create dependencies. To quote Sahlin's (2013) famous definition: kinship debt contributes to tying people and households together in a 'mutuality of being'.

In parallel to the logic of large weddings, amassing debt can also be part of a state strategy for the international integration and relation building. This has, to a degree, functioned for Kyrgyzstan in relation to Russia, a number of international donor banks, and in the 1990s the US. The meaning of a debt hinges not only on the amount but also on its conditions, to whom it is owed, and the ties it creates with the creditor. This is true for a household as well as for a country. Amassing national debt can be the acceptable outcome of an investment deemed necessary, such as in infrastructure. Yet, with a modest GDP dominated by a Canadian-owned gold mine and remittances from migrant workers to Russia, with little potential for securing a more central role in the capitalist world system anytime soon, the billions of dollars invested into Kyrgyzstan's roads are unlikely to pay for themselves. Indeed, they have not done so since the 1990s regardless of financier and in all cases loans have been accompanied by other forms of engagement. Interest payments, tenders for foreign companies, and access to Kyrgyzstan's markets have contributed to keeping the country in a dependent position. Yet, this very dependency has cultivated an interest by creditors in maintaining stability in the economy and society through grants, humanitarian aid, and technical support. Russia has granted debt relief several times. The World Bank, IMF, and ADB approved multi-million-dollar relief packages related to COVID-19, while Russia provided PCR tests, and China has sent ventilators, vaccines, and deferred US$35 million in debt payments, although it seems reluctant to give proper debt relief (Bartlett, 2020). Theorizing debt, Graeber (2011, 124–6) cites 16th-century French monk and satirical writer François Rabelais to make the point that there is nothing like owing someone a lot of money to make them interested and invested in guaranteeing your longevity and prosperity. Such relations may also be established with China, but over the past five years the process has become increasingly complicated and local trust in China's intentions has been shaken among both elites and non-elites as they question the political implications of growing credit and interest payments. Minimal mutuality is currently felt with China or with the West, and despite its colonial history and relative economic weakness, Russia remains the closest to such a kinship tie.

The US has made some efforts to engage with Central Asia in Barack Obama's Pivot to Asia Strategy, the Trump administration's first US strategy for Central Asia 'in two decades' (Starr and Cornell, 2020), and the annual C5+1 summit talks with the foreign ministers of the five post-Soviet republics which was in 2020 again attended by a US secretary of state after two years of less prominent participation. Yet, US advisors are blunter about their focus on global geo-power in these engagements than Toktogulov in his plea: 'In an era where great power competition is seen as the most serious challenge to national security, the United States should care about countries sandwiched between Russia, China, India, Iran, and Pakistan' (Starr and Cornell, 2020). Whether this can lead to the establishment of a mutuality of being depends on the actual long-term importance attributed to the region by large geopolitical actors. As has been the case since the 1990s,

Kyrgyzstan may just not be important or central enough for the US or China to prioritize the development of a multi-dimensional close relations and mutuality of being.

## Notes

1   The Manas Air Base was in 2009 renamed Manas Transition Center to downplay its military aspect.
2   See also Nacstatkom statistics: www.stat.kg/ru/publications/sbornik-kyrgyzstan-v-cifrah/.
3   www.youtube.com/watch?v=63_sxMT2ezM.
4   Similarly, IPAC (2019) has convincingly shown how Indonesia's stand on China's abuse of the Uyghurs does not merely depend on global geo-political consideration, but just as much reflects national, inner-political rivalries between the governing party and the opposition.
5   Such navigations are presented by Jennifer Sehring (forthcoming) who, looking at Central Asian water management across borders, argues that the countries' strategies and the success of coordination are not primarily decided by large scale national political considerations as much as by local established practices among the diplomats and their hinterlands.

## References

AlJazeera (2021) 'Kyrgyzstan voters back presidential rule in referendum', 11 April. Available from: www.aljazeera.com/news/2021/4/11/kyrgyzstan-voters-back-presidential-rule-in-referendum.

Bakirova, A. (2020) 'На китайском сайте опубликовали статью: «Кыргызстан был землями Китая»' ['An article was published on the Chinese site: "Kyrgyzstan was the lands of China"'], Kloop, 17 April. Available from: https://kloop.kg/blog/2020/04/17/kitajskoe-izdanie-opublikovalo-statyu-kyrgyzstan-byl-zemlyami-kitaya-pochemu-eto-vazhno/

Bartlett, P. (2020) 'China offers cash-strapped Kyrgyzstan a glimmer of hope on debt', Nikkei, 3 December. Available from: https://asia.nikkei.com/Politics/International-relations/China-offers-cash-strapped-Kyrgyzstan-a-glimmer-of-hope-on-debt.

Bichsel, C. (2009) Conflict Transformation in Central Asia: Irrigation Disputes in the Ferghana Valley, London: Routledge.

Bourdieu, P. (1977) Outline of a Theory of Practice, Cambridge: Cambridge University Press.

Djanibekova, N. (2018) 'Freeze turns to hot fury over Bishkek's power plant failure', Eurasianet, 30 January. Available from: https://eurasianet.org/kyrgyzstan-freeze-turns-to-hot-fury-over-bishkeks-power-plant-failure.

Eraliev, S. and Urinboyev, R. (2020) 'Precarious times for Central Asian migrants in Russia', Current History, 119(819): 258–63.

Graeber, D. (2011) Debt: The First 5000 Years, New York: Melville House.

Gullette, D. (2010) The Genealogical Construction of the Kyrgyz Republic: Kinship, State and 'Tribalism', Kent: Global Oriental.

IPAC (2019) 'Explaining Indonesia's silence on the Uyghur issue', IPAC Report No. 57, 20 June, Jakarta, IPAC.

Jones-Luong, P. (2002) *Institutional Change and Continuity in Post-Soviet Central Asia*, Cambridge: Cambridge University Press.

Kuehnast, K.R. and Dudwick, N. (2004) 'Better a hundred friends than a hundred rubles? Social networks in transition – the Kyrgyz Republic', World Bank Working Paper No. 39, Washington, DC: World Bank.

Mardell, J. (2020) 'Fear of the Middle Kingdom in Central Asia', *Berlin Policy Journal*, 16 January. Available from: https://berlinpolicyjournal.com/sinophobia-in-central-asia/.

Mogilevskii, R. (2019) 'Kyrgyzstan and the Belt and Road Initiative', IPPA Working Paper No. 50, Naryn: Institute of Public Policy and Administration.

Ormushev, K. (2020) 'Китай использует проблемы Кыргызстана в своих интересах. Статья Калнура Ормушева' ['China uses the problems of Kyrgyzstan to its advantage'], Kaktus Media, 1 May. Available from: https://kaktus.media/doc/412111_kitay_ispolzyet_problemy_kyrgyz stana_v_svoih_interesah._statia_kalnyra_ormysheva.html.

Pannier, B. (2020a) 'Kyrgyzstan the missing link in China's railway to Uzbekistan … and beyond', RFE/RL, 29 June. Available from: www.rferl.org/a/kyrgyzstan-the-missing-link-in-china-s-railway-to-uzbekistan-and-beyond/30697016.html

Pannier, B. (2020b) 'Why are Central Asian countries silent about China's Uyghurs?', RFE/RL, 22 September. Available from: www.rferl.org/a/why-are-central-asian-countries-silent-about-china-s-uyghurs-/30852452.html.

Pétric, B.-M. (2005) 'Post-Soviet Kyrgyzstan or the birth of a globalized protectorate', *Central Asian Survey*, 24(3): 319–32.

Reeves, M. (2012) 'Black work, green money: remittances, ritual, and domestic economies in Southern Kyrgyzstan', *Slavic Review*, 71(1): 108–34.

Ryskulova, N. (2019) 'Вдолгу у большого соседа. Почему в Кыргыз стане боятся "китайской экспансии."' ['I owe a big neighbour. Why Kyrgyzstan is afraid of "Chinese expansion"'], BBC, 7 December. Available from: www.bbc.com/russian/features-50678871.

Sagynbekova, L. (2017) 'International labour migration in the context of the Eurasian Economic Union: issues and challenges of Kyrgyz migrants in Russia', IPPA Working Paper 39, Naryn: Institute of Public Policy and Administration.

Sahlins, M. (2013) *What Kinship Is – and Is Not*, Chicago, IL: University of Chicago Press.

Sarı, Y. (2012) 'Foreign policy of Kyrgyzstan under Askar Akayev and Kurmanbek Bakiyev', *Perceptions: Journal of International Affairs*, 17(3): 131–50.

Scott, J. (1998) *Seeing Like a State: How Certain Schemes to Improve the Human Condition Have Failed*, New Haven, CT: Yale University Press.

Sehring, J. (forthcoming) 'Masculinity and water diplomacy in Central Asia', in M. Schmidt, R. Steenberg, M. Spies, and H. Alff (eds) *Beyond Post-Soviet: Layered Legacies and Transformations in Central Asia*, Augsburg: Geographica Augustana.

Starr, F. and Cornell, S. (2020) 'A new strategy for Central Asia', *The Hill*, 18 February. Available from: https://thehill.com/opinion/international/483511-a-new-strategy-for-central-asia.

Toktogulov, K. (2020) 'US and West need to stand solid behind Kyrgyzstan, Central Asia's only democracy', CNN, 11 October. Available from: https://edition.cnn.com/2020/10/10/opinions/kyrgyzstan-democracy-intl-hnk/index.html.

# 4

# Chinese Investment Meets Zambian Policy: The Planning and Design of Multi-Facility Economic Zones in Lusaka

*Dorothy Tang*

Towards the end of 2020, Zambia missed an interest payment on Eurobonds and became the first African nation to default during the global pandemic. Media outlets estimated that Zambia's foreign debt was US$12 billion, of which US$3 billion is owed to Chinese policy banks and other Chinese entities (Strohecker, 2020). In the midst of increased tensions between the US and China, American media revisited narratives of Chinese debt-trap diplomacy and the implications for Zambia and the developing world (see Steinhauser and Wallace, 2020). While many of these reports border on sensational, there is legitimate cause for concern. By 2021, China had committed US$9.9 billion in loans to Zambia, with 70 per cent dedicated to infrastructure development such as transportation, power, water, and information and communication technology (CARI and GDPC, 2021). In contrast to other forms of bilateral or multilateral financing, China has embraced a 'non-interference' policy and therefore does not require additional fiscal discipline or institutional adjustments to access loans. This is one reason Chinese lending has become more popular in developing nations (Bräutigam, 2011). Scholars have argued that Chinese lending practices and overseas investment have alternative rationales: to prioritize long-term returns such as increasing market share or secure critical resources instead of short-term financial gains (Lee, 2018; Kaplan, 2019). For example, some Chinese-financed infrastructure projects are contingent on the use of Chinese experts, contractors, or materials in order to export excess construction capacity (Huang and Chen, 2016). However, borrowing nations increasingly exert agency in these scenarios and actively negotiate the degree of Chinese involvement (Kaplan, 2019).

Zambia is one such case. In the midst of debt restructuring, the government borrowed an additional US$1.5 billion to purchase the struggling Mopani Copper Mine from the Anglo-Swiss mining conglomerate Glencore, defying speculation that this mine would fall into Chinese hands. This turn to resource nationalism purportedly takes advantage of rising copper prices driven by China's post-pandemic economic growth, rather than selling off a valuable asset in its entirety (Bariyo and Wexler, 2021). Zambia is the second-largest producer of copper in Africa, and its history of development straddles an uneasy relationship between dependency on copper extraction and a desire to diversify its economy. Since the last major debt crisis in the 1980s caused by declining copper prices and the subsequent enforcement of Structural Adjustment Programs by the World Bank and IMF, Zambia has actively pursued alternative avenues for economic development. The Japanese International Cooperation Agency (JICA) advised Zambia on new industrial policies that incentivize foreign and domestic investment in other sectors through programmes such as multi-facility economic zones (MFEZ). In the mid-2000s, the Zambian government leveraged new Chinese investments in its copper resources and interest in economic diplomacy in Africa to finance two MFEZs (Figure 4.1), one in the copper mining region and the other in the capital city of Lusaka. However, within a few years, the Zambian government undermined the Chinese zones by establishing a larger MFEZ in Lusaka that featured more reliable and extensive road, water, and electricity infrastructure. The Zambian zone was developed with technical assistance from JICA and a masterplan by a Malaysian consultant under the Triangle of Hope programme, one of JICA's first experiments in South-South cooperation models.

Within this context, Chinese-funded MFEZs in Zambia are paradigmatic cases of the complex processes of negotiations between state interests, industrialization, and urban development. As instruments of Chinese foreign policy, state-led investments, and globally accepted best practice for economic development, economic zones require land, infrastructure, and a development strategy that is loosely aligned with priorities of their host countries. Rather than conceiving of these MFEZs as isolated Chinese enclaves, I situate them within a longer spatial and temporal trajectory of geopolitics to unpack how urban design and planning practices are also subject to such negotiations. The remainder of the chapter has two main parts: first, I contextualize the history of Zambian economic planning that led to the establishment of MFEZs in relationship to Chinese and Japanese foreign policies. Second, I compare the planning and design of the Zambia–China Economic and Trade Cooperation Zone (ZCCZ) and the Lusaka South MFEZ – funded by the Zambian government with assistance from Japan and Malaysia – to analyse the underlying spatial and urban logics of

**Figure 4.1:** Multi-facility economic zones in Lusaka, Zambia

ZCCZ LUSAKA EAST MFEZ & LUSAKA SOUTH MFEZ
LUSAKA, ZAMBIA

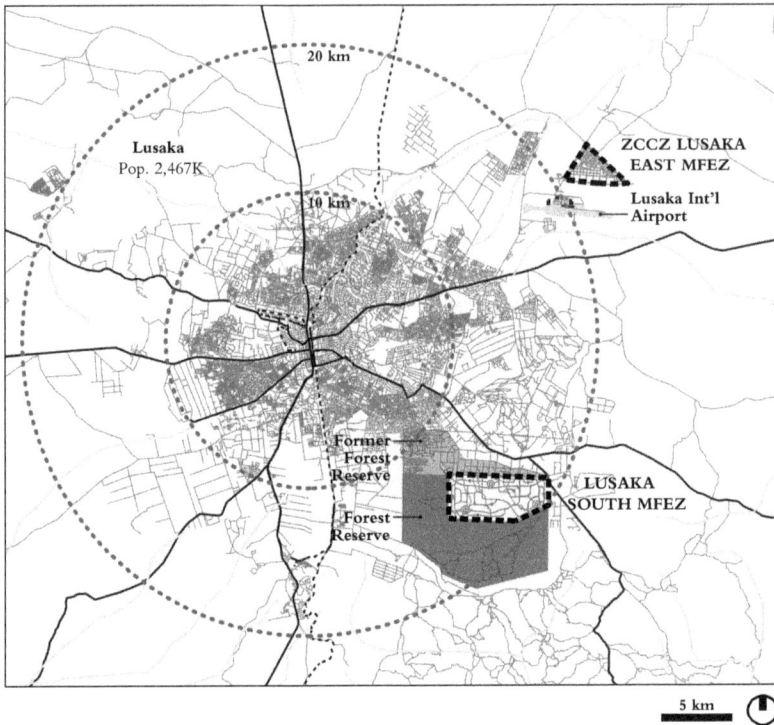

Note: All maps and illustrations in this chapter are based on publicly available spatial data. Sources include: Open Street Maps, Google Earth, GADM, and Natural Earth. Masterplans are redrawn from 中国境外经贸合作区, www.cocz.org/news/content-243508.aspx, and 'Lusaka Multi-Facility Economic Zone (LS-MFEZ)', presented at the Japan-Africa Business Forum 2014, http://ab-network.jp/wp-content/uploads/2014/07/Economic-Zone-Projects.pdf.

two state-led economic development projects. Empirical evidence is drawn from archival research, unstructured interviews with MFEZ employees, JICA personnel, and MFEZ planners in Tokyo, Shanghai, and Lusaka in 2018. Additionally, I rely on publicly available planning documents, media reports, documentary films as well as open-source spatial data to chart the spatial relationships between the MFEZs and their urban context.

## Zambian economic planning and the entry of Chinese capital

There is a long history of state-level interaction between Zambia and China, especially with the construction of Tazara Railway in the mid-1970s. Over this period, the two nations have experienced vastly different trajectories,

**Figure 4.2:** Zambian copper production, global prices, and significant political milestones

Source: Adapted from Fraser and Larmer (2010)

but meet at crucial junctures of economic and political change (Figure 4.2). Zambia is located along the Central African Copperbelt, which has the largest high-grade copper deposits in the world, and its economic and political fortunes have been intricately tied to the boom and bust cycles of global copper demands (Fraser, 2010). In 2016, the mining sector in Zambia accounted for 12 per cent of its GDP, 70 per cent of total export value, and 62 per cent of foreign direct investment (FDI; World Bank, 2016). Zambia was a middle-income country when it became independent in 1964, but recognized the precarious dependency on volatile and outward-flowing copper markets established during the colonial era. Zambia's first National Development Plan of 1964 established a framework where mining revenues would fund other forms of rural investment to diversify the economy over time. By the 1970s copper mines were nationalized and became the core institutions that subsidized social welfare for the nation (Larmer, 2010). The Tazara Railway, completed in 1975, was a crucial piece of infrastructure that connected landlocked Zambia to the port of Dar es Salaam, and facilitated export of Zambian copper. The Tazara Railway was also China's first large-scale foreign-aid project and symbolized China's commitment to socialist solidarity with Zambia and Tanzania (Song, 2015). The railway was completed despite China's domestic political turmoil and economic hardships during the Cultural Revolution.[1]

However, by the 1980s, copper prices and plunged and the debt crisis forced Zambia into Structural Adjustment Programs[2] and the gradual privatization of state-owned enterprises (SOE) and utilities. Between 1997 and 2000, the copper mines and their capital assets were unbundled and sold to a 'motley' set of multinational companies from India, China, Canada, Switzerland, Israel, Australia, and South Africa (Haglund, 2010). It is within this shifting economic and political context that China Nonferrous Metal Mining Company (CNMC) entered Zambia through the purchase of the Chambishi Copper Mines in 1998. CNMC is a high-level SOE with significant political and economic influence in China. In the 1990s, China marketized SOEs and encouraged them to invest overseas as part of the Go Out policy.[3] The Chambishi Copper Mine was one of the first overseas mineral assets that the Chinese SOE acquired. In 2009, CNMC was hand-picked by the Zambian government to purchase a majority stake in the troubled Luanshya Copper Mine nearby (Shacinda, 2009). Sociologist Ching Kwan Lee describes in great detail how CNMC's status as a Chinese SOE affects their behaviour in Zambia, and the impacts of a different rationale for capital accumulation due to their dual mandate as a profit-making enterprise and diplomatic representation of the Chinese state (Lee, 2018). As a consequence, CNMC has prioritized long-term capital investments in mining operations over short-term profits and has taken some pains to appease local demands for better labour practices.

Collaboration between the Zambian state and CNMC is also evident in the 2006 establishment of Chinese-funded MFEZs. Following the triennial Forum on China–Africa Cooperation in Beijing, the Chinese government committed to funding a series of special economic zones in the African continent. Chinese subsidies to build overseas special economic zones would be granted based on the quality of a joint proposal between a Chinese enterprise and an African government. This timing coincided with enactment of the Zambian Development Agency Act in 2006. The new industrial policies, developed with technical assistance from JICA, aimed to diversify the economy and encourage FDI. MFEZs were seen as an important instrument to achieve these goals. The Zambian government quickly approached CNMC to propose two MFEZs – one in Chambishi to elevate copper-based industries by expanding to associated sectors, and the other in Lusaka near the airport to incubate logistics industries and agro-processing. The Chinese government approved both zones but bundled them into 'one project' for the appearance of fairness and provided subsidies for CNMC to proceed (Bräutigam and Tang, 2012; Lee, 2018). The ZCCZ in Chambishi opened in 2007, and ZCCZ-Lusaka East followed in 2009.

## Special economic zones as economic statecraft

The Zambian MFEZs follow a long trajectory of industrial policy that utilizes industrial enclaves to incubate industrialization. Special economic zones (SEZs) were originally conceived in the mid-20th century as an instrument to jump start domestic economic development through tax incentives, enhanced infrastructure, and exceptions to labour regulations (Ong, 2006; Easterling, 2014). Japan developed a series of state-led industrial clusters that evolved from import substitution to export-oriented industrialization. JICA later promoted this model of economic planning as 'industrial estates' via technical assistance in Southeast Asia, notably in Thailand (Söderberg, 1996). Other experiments of using SEZs for export-oriented industrialization in Taiwan and Korea were very successful, and China's economic reforms in the 1980s are largely credited to the establishment of SEZs to attract FDI and bolster domestic industrialization. The implementation of early SEZs relied on strong state intervention and industrial policies, as seen in the developmental state approach in East Asia,[4] and have seen less success in weaker states such as India. Nonetheless, SEZs are now considered an international best practice by many multilateral agencies such as UNIDO and the World Bank. In particular, China's extraordinary success with SEZs and economic growth since the 1980s is an aspirational model for many developing economies. Analysts estimate there are currently 182 Chinese-financed zones in 52 countries (Li et al, 2019) with approximately 20 officially sanctioned by the Ministry of Commerce that receive subsidies from the Chinese government (COCZ, 2018).

The translation of the Chinese experience of SEZs from domestic economic policy to economic statecraft is not straightforward. Bräutigam and Tang (2011, 2012, 2014) have examined Chinese state interests through extensive analyses of Chinese-funded SEZs in Africa. They argue that the SEZ is a central platform for the Chinese principle of 'mutual benefit' by financing infrastructure projects while encouraging Chinese private enterprises to invest overseas (2011). All the initial zones in Africa were chosen through a competitive process as a partnership between a private Chinese developer and a host country. The Chinese state heavily subsidizes capital investment of SEZs abroad with grants and long-term loans, and also partially reimburses expenses incurred by Chinese enterprises moving into the zones. Yet, they have taken a hands-off approach and give the private Chinese developers significant leeway for the planning and implementation of the zones. Currently, there are five Chinese-funded SEZs in Sub-Saharan Africa, each managed by different types of enterprises ranging from central-level SOEs to privately held companies.

However, Bräutigam and Tang (2012) found that, even with generous subsidies from the Chinese government, Chinese SEZs in Africa have not performed well. Many are not profitable, but more importantly the desired 'spillover' effects of employment and urban development for local communities have not materialized. And yet the commitment for Chinese-funded SEZs has not abated. This implies that beyond commercial interests, SEZs have a greater political function for the Chinese government. The diplomatic importance of SEZs can be seen in instances when the Chinese government officially intervened in a business dispute between a private zone developer and the local government in Nigeria, or in the importance of ceremonial and diplomatic functions held at various SEZs. Bräutigam and Tang (2012) conclude that African SEZs are much more valuable to the Chinese government as instruments of foreign policy than effective tools for economic gain.

These findings are certainly relevant in understanding the performance of Chinese MFEZs in Zambia. For example, ZCCZ-Lusaka East is planned primarily as a light industry, logistics, and agro-processing cluster. However, between 2009 and 2018, only 28 per cent of the master planned area was developed, with eight operational firms and the creation of 207 jobs (Figure 4.3). Many circumstances contribute to the lagging development of the zone. Zone managers lament turning away polluting investors due to local environmental regulations and their responsibility to portray Chinese investment in the best light.[5] Inadequate electricity provisioning severely limits the industrial capacity of the zone (Zeng, 2016). In addition, ZCCZ executives specifically point to the Zambian state-developed Lusaka South MFEZ – with ample infrastructure capacity and more favourable investment incentives – as a major threat to the success of Chinese efforts.[6] This is further constrained by the Zambian government's intention that the ZCCZ zones target Chinese investors (Lee, 2018), while Lusaka South MFEZ is open to both domestic and international investors. Before a detailed comparison of the urban design and planning of the two competing zones in Lusaka, it is important to first trace the history of JICA's technical assistance programmes in Zambia to contextualize the significance of Lusaka South MFEZ.

**Figure 4.3:** Multi-facility economic zones development progress in Lusaka, Zambia, 2018

**ZCCZ LUSAKA EAST MFEZ**
**Developer:** CNMC
**Masterplan:** 5.7 km²
**Implemented:** 1.6 km²
**Opened:** 2009

US$20 Million Pledged
14 Firms Approved
8 Firms Operational
207 Jobs

**LUSAKA SOUTH MFEZ**
**Developer:** Zambian Development Agency
**Masterplan:** 21 km²
**Implemented:** 4.14 km²
**Opened:** 2013

US$443.8 Million
US$1.5 Billion Pledged
22 Firms Approved
10 Firms operational
1000 Jobs Projected

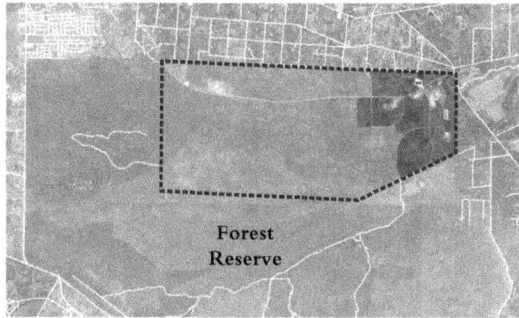

Lusaka Int'l Airport

Forest Reserve

1 km

# Japanese technical assistance and Zambian industrial policy

Japanese overseas development assistance began with post-war reparations in Southeast Asia and has since become the fourth-largest donor globally. Early Japanese aid was characterized by the principle of self-help[7] and an interest in developing international markets for Japanese industry (Potter, 2012). This approach was built on Japan's own experience as a developmental state and the close ties between government and industry. After the end of the Cold War, Japan sought to be a leader of global development assistance and prioritized cooperation in the African continent. Unlike the institutional reform requirements of other multilateral development agencies during the same period, the Japanese were interested in how the successful experiences of other developing economies in Asia, such as Malaysia, could inform a programme for South–South cooperation. They collaborated with the retired deputy director of the Malaysian Industrial Development Authority, Dato' J. Jegathesan, to learn from Malaysia's experience of advancing from a 'Least Developed Country' to a dynamic middle-income country. When Japan was invited to advise the Zambian government on industrial policy in 2004, Jegathesan and JICA set in motion the Triangle of Hope programme, a set of development principles that emphasized three equilateral forces: strong political will, efficient administration, and a dynamic private sector (Jegathesan and Ono, 2013). The product of this joint effort was the 2006 Zambian Development Agency (ZDA) Act, which established a new quasi-governmental agency to facilitate interaction between government institutions and the private sector. ZDA's main impetus was to attract FDI, foster micro, small, and medium enterprises, and promote MFEZs (Zambia Development Agency, no date).

The timing of the ZDA Act coincided with the 2006 Forum on China–Africa Cooperation summit and the first set of commitments by the Chinese government to fund SEZs overseas. Although the two Chinese-funded ZCCZ zones were the first MFEZs to be formally established to attract industrial capital in Zambia, the ZDA continued to work with JICA to develop a Zambian zone, which resulted in a JICA-funded a feasibility study for a 21 km$^2$ MFEZ located in the southern periphery of Lusaka. Subsequently, JICA hired a Malaysian consultancy – the in-house urban design team of Kulim Hi-Tech Industrial Park – to develop a comprehensive masterplan that included industry, housing, retail, research and development institutions, schools, and other services (JICA, 2009). The masterplan is loosely based on the experience of Kulim Hi-Tech Industrial Park, which JICA also advised in the early 1990s and has since became a model for industrial development in Malaysia (Phelps and Dawood, 2014).

## Planning and designing MFEZs in Lusaka

Land acquisition and siting decisions point to the political disparity between the Chinese-funded ZCCZ-Lusaka East MFEZ and the Zambian-backed Lusaka South MFEZ (Figure 4.4). ZCCZ-Lusaka East is located in former agriculture land near Lusaka International Airport. Without the authority of the state to appropriate land, ZCCZ had to purchase a parcel of land large enough for an MFEZ. In contrast, Lusaka South is located on a recently decommissioned forest reserve owned by the Zambian state. Despite resistance from local environmental groups concerned about groundwater recharge for Lusaka (*Lusaka Times*, 2014), the JICA feasibility study (2009) identified the site as a degraded forest area with no intrinsic ecological value, and thus suitable for a large-scale industrial and urban development. Public lands were easily dedicated to the development of the project with little dispute.

The 21 km$^2$ masterplan for Lusaka South MFEZ resembles the planning for Kulim Hi-Tech Park. The Kulim model is a satellite town development sited in an underdeveloped region, and are designed to be self-contained communities to foster a knowledge economy. Located on southern periphery of Lusaka, Lusaka South is planned as a mixed-use development that includes

**Figure 4.4:** Planned land uses in multi-facility economic zones, Lusaka

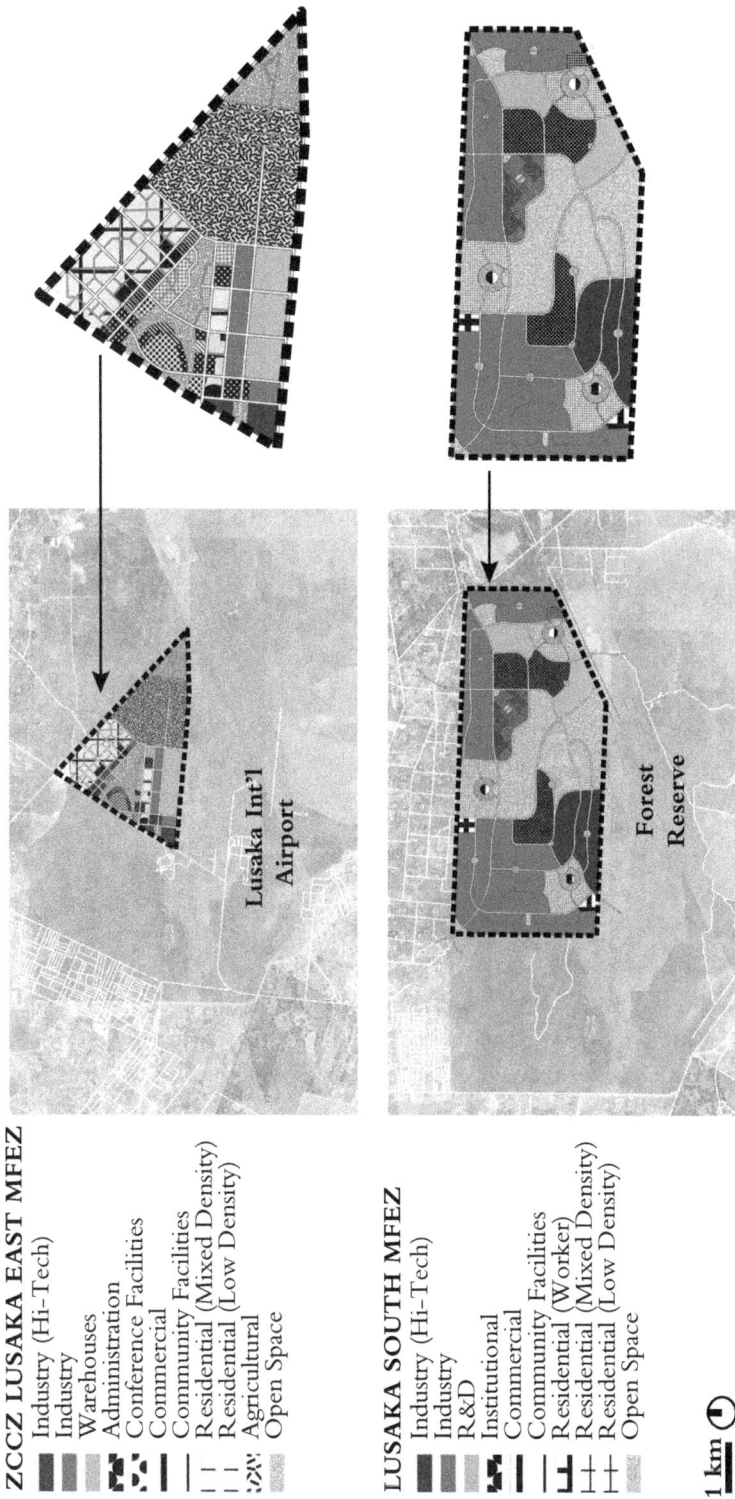

**ZCCZ LUSAKA EAST MFEZ**

- Industry (Hi-Tech)
- Industry
- Warehouses
- Administration
- Conference Facilities
- Commercial
- Community Facilities
- Residential (Mixed Density)
- Residential (Low Density)
- Agricultural
- Open Space

**LUSAKA SOUTH MFEZ**

- Industry (Hi-Tech)
- Industry
- R&D
- Institutional
- Commercial
- Community Facilities
- Residential (Worker)
- Residential (Mixed Density)
- Residential (Low Density)
- Open Space

Lusaka Int'l Airport

Forest Reserve

1 km

industry, commercial, and residential development organized around three 'poles' and planned to be implemented over five phases (LSMFEZ, 2021). The residential development caters to the managerial class of the intended inhabitants with very little area designated as 'workers housing' located adjacent to commercial areas and research and development facilities. The zone utilizes mega-block structures that suggest it would recruit developers to implement the majority of the zones, with the Zambia Development Agency providing major infrastructural connections, including a dedicated electric sub-station, new water supply mains, and roads. The zone targets a large range of sectors, including agribusiness, packaging, agro-processing, pharmaceuticals, electronics, research and development, professional and medical services. By 2021, the zone had secured investments from major Zambian corporations and multinationals and had created over 900 jobs.

In contrast, the 5.7 km$^2$ plan for ZCCZ-Lusaka East is much more modest. The MFEZ was planned to leverage its proximity to the international airport and focused on agro-processing, pharmaceuticals, logistics and warehousing, and international commerce (Zeng, 2016). In addition to a small residential area, mostly employee dormitories, a large portion of the site is designated as an 'Ecological Agricultural Demonstration District', that presumably is related to China's commitment to technical assistance. In examining the evolution of the plans, the current land-use plan has a significantly lower proportion of land designated for warehousing and logistics, which suggests the lag of infrastructural provision to the site has resulted in the shifting of development priorities.[8] These lofty plans, however, have not been attractive to Chinese investors. When I visited the MFEZ in 2018, the prefabricated factories housed light manufacturing of household goods, shoes, and construction materials. The most prominent structure was a mushroom growing facility and a small pharmaceutical facility was under construction. However, the majority of the structures were vacant, with small vegetable gardens and a flock of ducks occupying the surrounding landscape. Empty roads were perfect shortcuts for neighbouring farmers to herd their livestock from one grazing ground to another.

The difference in infrastructure provisions further reveals the unequal political agency of the two MFEZs. With strong backing from the Zambian state and JICA, Lusaka South is planned as a new city with dedicated infrastructure. On the other hand, while ZCCZ-Lusaka East is also a bilateral state-sanctioned project, it is funded and managed by a private Chinese corporation. While Chinese planners[9] were sought out to translate their experiences of planning for Chinese SEZs in an overseas setting, there are some fundamental differences. Chinese domestic SEZs are state-led with major public investments in infrastructure and political impetus in various levels of government. This translates into infrastructure planning that anticipates growth over a longer period of time, and systematic development in Chinese SEZs. Overseas

SEZs are state-sanctioned by the host countries, but the zone developers – in most cases private corporations – are responsible for securing utilities and building necessary infrastructure. Chinese overseas SEZs are essentially private industrial real estate investments in a foreign country and, without strong state intervention, must behave like a private investment. Without a strong assurance of returns, ZCCZ's internal road and drainage infrastructure is modest in comparison with its state-led counterparts, whether in China or Zambia. Furthermore, ZCCZ relies on the goodwill of the Zambian government to provide the necessary road connections and stable electricity to fulfil their plan to become a major logistics hub, which have fallen short even according to Chinese accounts (Zeng, 2016; Zhang and Song, 2017).

## Conclusion

Chinese and Japanese economic diplomacy have both resulted in the establishment of MFEZs in Zambia. However, the planning and urban design of these MFEZs, whether location, infrastructural connections, land-use planning, and even the scale of development parcels, provide evidence to how competing state interests are played out and reasserts the agency of host countries. As an instrument of diplomacy, Chinese actors and JICA have utilized the development model of special economic zones very differently. While the aspiration for a Chinese economic miracle is desirable for the Zambian state, the financing arrangement through subsidizing private industrial development limits the ability for Chinese actors to access familiar policy and development mechanisms that were successful within China. Japanese actors, on the other hand, provided development assistance through policy innovation that eventually enabled the Zambian state to undermine Chinese MFEZs. One might conclude that China is considered a partner of convenience by the Zambian state: the coincidental timing of Zambian industrial policies and Chinese outward investment incentives have resulted in sustained collaborations in copper mining, the establishment of the first MFEZs, and increased Chinese presence in the construction industry. Yet, the Zambian state does not invite Chinese participation out of deference, but in fact seeks to learn from these experiences to bolster their own development projects – with assistance from Japan. The two distinct styles of economy diplomacy unsettle the commonly held notion of Chinese hegemony in Zambia and point our attention to how various types of state power are embedded in the material and spatial forms of infrastructure.

The question at stake here is whether the presence of Chinese infrastructure financing in Zambia is a grand geopolitical strategy, or are the Chinese merely a partner of convenience. How does Japanese technical assistance factor into these dynamics? Is there competition between Chinese and Japanese actors in Zambia? This line of questioning, however, conflates commercial interests with geopolitical interests. More importantly, it ignores the ambitions of the Zambian state in pursuing its own economic development. The establishment of the ZCCZ MFEZs in Lusaka and the Copperbelt region confirms the assertive role of the Zambian state in leveraging Chinese geopolitical interests for economic development. While ZCCZ-Lusaka East has not met original expectations, their parent company CNMC has gained additional access to copper resources after cooperating with the Zambian government. However, despite the commercial interests of CNMC in Zambia, the rhetoric and actions of ZCCZ managers decry their poor performance as a function of diplomatic representation. Conversely, JICA officials from the Zambia team recalled that initially their engagement with Zambia was to pave the way for Japanese commercial activity, but there was no interest from the private sector.[10] This led to their pivot to South–South cooperation models and attempts to facilitate Malaysian investments in Zambia – a diplomatic effort. China and Japan both have long histories of blending commercial and political interests through economic diplomacy. Because of this unique trait – in contrast to the ideologically driven geopolitical strategy of the West – it would be erroneous to conclude that there is clear competition between Chinese and Japanese interests in developing MFEZs. Should the development of Lusaka South MFEZ flourish, despite the current economic setback, the real winner is Zambia.

## Acknowledgements

Funding for this project was provided by the MIT Center for International Studies Summer Studies Grant and the Rodwin Fund for International Travel administered by the MIT Department of Urban Studies and Planning. I thank MIT colleagues in the 11.800 writing workshop led by David Hsu for initial feedback. Richard Samuels was instrumental in making connections with interlocuters at JICA, and Hou Li connected me with Chinese planners involved in African SEZ development. Last but not least, I am grateful for guidance from Eran Ben-Joseph and Gabriella Carolini throughout the research process.

## Notes

1   The Cultural Revolution (1966–76) was a violent political movement that aimed to purge imperialism and capitalism in China through mass mobilization. It was Mao Zedong's attempt to consolidate political power after a disastrous modernization campaign that led to one of the deadliest famines in Chinese history. It ended with Mao's death in 1976.

2   Structural Adjustment Programs were loans provided by the IMF and the World Bank for countries in economic crisis. They were contingent on a number of domestic policy adjustments which encouraged the privatization of State-owned assets or utilities, liberalizing trade, and managing government deficit.

3   China's Go Out policy was launched in 1999 to encourage outward FDI. SOEs were heavily incentivized to participate in the program to diversify their market reach and to secure critical resources for China's domestic economic development.

[4]    The developmental state refers to state-led economic planning in the late 20th century, which is largely credited for the rapid development of East and Southeast Asian economies. The state intervenes in the market economy through labour regulation, state investment and subsidies, and collaboration with major industries. Japan's post-war industrialization is considered one of the 'original' developmental states (Johnson, 1982), and this model was replicated in Asia through emulation and Japanese development assistance programs. See Haggard (2018) for further reading on the relationship between industrial policies and the developmental state.

[5]    Chinese employee of the Planning and Investment Promotion Department, ZCCZ-Lusaka East, 2 August 2018. Personal interview.

[6]    Wu Wenjun, Deputy General Manager of ZCCZ, 5 November 2017. Television interview (Zhang and Song, 2017).

[7]    The principle of 'self-help' prioritizes self-initiated requests from recipient countries and demonstrated commitment to fund in-country activities. Similar to Chinese overseas development aid, early Japanese aid did not interfere with domestic political or institutional processes.

[8]    I drew this conclusion by comparing the masterplan published on the official ZCCZ website (ZCCZ, no date) and a more recent masterplan posted on the COCZ website (COCZ, no date). The COCZ is the official agency that coordinates overseas SEZs.

[9]    Xia Nankai, chief planner and urban designer, Shanghai Tongji Urban Planning Design Institute, 18 June 2019. Personal interview.

[10]   Group interview in Tokyo, 17 July 2018.

# References

Bariyo, N. and Wexler, A. (2021) 'After default, Zambia's outsized bet on copper could play into China's hands', *Wall Street Journal*, 27 April. Available from: www.wsj.com/articles/after-default-zambias-outsized-bet-on-copper-could-play-into-chinas-hands-11619514520.

Bräutigam, D. (2011) *The Dragon's Gift: The Real Story of China in Africa*, Oxford: Oxford University Press.

Bräutigam, D. and Tang, X. (2011) 'African Shenzhen: China's special economic zones in Africa', *The Journal of Modern African Studies*, 1: 27.

Bräutigam, D. and Tang, X. (2012) 'Economic statecraft in China's new overseas special economic zones: soft power, business or resource security?', *International Affairs*, 88(4): 799–816.

Bräutigam, D. and Tang, X. (2014) '"Going global in groups": structural transformation and China's special economic zones overseas', *Economic Transformation in Africa*, 63(C): 78–91.

China Africa Research Initiative (CARI) and Boston University Global Development Policy Center (GDPC) (2021) China Africa Research Initiative Loans Database, Version 2.0. Available from: https://chinaafricaloandata.bu.edu/.

COCZ (2018) 中国境外经贸合作区 [Chinese Overseas Cooperation Zones]. Available from: www.cocz.org/news/content-243519.aspx.

COCZ (no date) 中国境外经贸合作区 [Chinese Overseas Cooperation Zones]. Available from: www.cocz.org/news/content-243508.aspx.

Easterling, K. (2014) *Extrastatecraft: The Power of Infrastructure Space*, London: Verso.

Fraser, A. (2010) 'Introduction: boom and bust on the Zambian copperbelt', in A. Fraser and M. Larmer (eds) *Zambia, Mining, and Neoliberalism: Boom and Bust on the Globalized Copperbelt*, New York: Palgrave Macmillan, pp 1–30.

Fraser, A. and Larmer, M. (eds) (2010) *Zambia, Mining, and Neoliberalism: Boom and Bust on the Globalized Copperbelt*, New York: Palgrave Macmillan.

Haggard, S. (2018) *Developmental States*, Cambridge: Cambridge University Press.

Haglund, D. (2010) 'From boom to bust: diversity and regulation in Zambia's privatized copper sector', in A. Fraser and M. Larmer (eds) *Zambia, Mining, and Neoliberalism: Boom and Bust on the Globalized Copperbelt*, New York: Palgrave Macmillan, pp 91–126.

Huang, Z. and Chen, X. (2016) 'Is China building Africa?', *The European Financial Review*. Available from: www.europeanfinancialreview.com/?p=6110.

Jegathesan, D.J. and Ono, M. (2013) 'Strategic action initiatives for economic development', in K. Ohno and I. Ohno (eds) *Eastern and Western Ideas for African Growth: Diversity and Complementarity in Development Aid*, London: Routledge, pp 187–203.

JICA (2009) 'The study on master plan of Lusaka South multi-facility economic zone in the Republic of Zambia', Japan International Cooperation Agency. Available from: http://libopac.jica.go.jp/detail?bbid=0000246516.

Johnson, C.A. (1982) *MITI and the Japanese Miracle: The Growth of Industrial Policy, 1925–1975*, Stanford, CA: Stanford University Press.

Kaplan, S.B. (2019) 'The rise of China's patient capital: a tectonic shift in global finance in developing countries?' SSRN. Available from: https://ssrn.com/abstract=3108215.

Larmer, M. (2010) 'Historical perspectives on Zambia's mining booms and busts', in A. Fraser and M. Larmer (eds) *Zambia, Mining, and Neoliberalism: Boom and Bust on the Globalized Copperbelt*, New York: Palgrave Macmillan, pp 31–58.

Lee, C.K. (2018) *The Specter of Global China: Politics, Labor, and Foreign Investment in Africa*, Chicago, IL: University of Chicago Press.

Li, H., Wu, M., Niu, Z., and Li, Q. (2019) 'Information data set of China's overseas industrial parks from 1992 to 2018 (1992–2018 年中国境外产业园区信息数据集)', *China Scientific Data*. doi: 10.11922/797.

LSMFEZ (2021) '*Lusaka South MFEZ*'. Available from: https://lsmfez.co.zm/.

Lusaka Times (2014) 'Zambia: Lusaka South multi-facility economic zone (MFEZ) is illegal-environmental campaigners', *Lusaka Times*, 24 February. Available from: www.lusakatimes.com/2014/02/24/lusaka-south-multi-facility-economic-zone-mfez-illegal-environmental-campaigners/.

Ong, A. (2006) 'Zoning technologies in East Asia', in *Neoliberalism as Exception: Mutations in Citizenship and Sovereignty*, Durham, NC: Duke University Press, pp 97–120.

Phelps, N.A. and Dawood, S.R.S. (2014) 'Untangling the spaces of high technology in Malaysia', *Environment and Planning C: Government and Policy*, 32(5): 896–915.

Potter, D.M. (2012) 'Japan's official development assistance', in H.S. Kim and D.M. Potter (eds) *Foreign Aid Competition in Northeast Asia*, Sterling, VA: Kumarian Press, pp 13–35.

Shacinda, S. (2009) 'Zambia picks China firm to run Luanshya copper mine', Reuters, 8 May. Available from: www.reuters.com/article/ozabs-minerals-zambia-luanshya-20090508-idAFJOE54709620090508.

Söderberg, M. (1996) 'Road to development in Thailand', in M. Söderberg (ed) *The Business of Japanese Foreign Aid: Five Cases Studies from Asia*, London: Routledge, pp 89–125.

Song, W. (2015) 'Seeking new allies in Africa: China's policy towards Africa during the Cold War as reflected in the construction of the Tanzania–Zambia railway', *Journal of Modern Chinese History*, 9(1): 46–65.

Steinhauser, G. and Wallace, J. (2020) 'Africa's first pandemic default tests new effort to ease debt from China', *Wall Street Journal*, 18 November. Available from: www.wsj.com/articles/africas-first-pandemic-default-tests-new-effort-to-ease-debt-from-china-1160561967.

Strohecker, K. (2020) 'Zambia on track for protracted debt overhaul as creditors slam lack of engagement', Reuters, 16 November. Available from: www.reuters.com/article/zambia-debt-creditors-idUSL8N2I22D4.

World Bank (2016) 'How can Zambia benefit more from mining?' Available from: www.worldbank.org/en/news/feature/2016/07/18/how-can-zambia-benefit-more-from-mining.

Zambia Development Agency (no date) 'Multi-facility economic zone (MFEZ)'. Available from: www.zda.org.zm/?q=content/multi-facility-economic-zone-mfez.

ZCCZ (no date) 赞比亚中国经济贸易合作区发展有限公司 [Zambia China Economic and Trade Cooperation Zone Development Co., Ltd.]. Available from: http://zccz.cnmc.com.cn/detailtem.jsp?column_no=070205&article_millseconds=1349751417515.

Zeng, D.Z. (2016) *Multi-facility Economic Zones in Zambia: Progress, Challenges and Possible Interventions*, Washington, DC: World Bank. Available from: http://documents.worldbank.org/curated/en/720981495115586647/pdf/115143-WP-PUBLIC-Feb-2016-GTCCS-ZambiaMFEZ.pdf.

Zhang, Z. and Song, B. (2017) 中国第一个境外经贸区？ [China's first overseas economic zone?], 龙行'天下 [*Odyssey of the Dragon*]. Hong Kong: Phoenix TV.

5

# Infrastructure as Symbolic Geopolitical Architecture: Kenya's Megaprojects and Contested Meanings of Development

*Wangui Kimari and Gediminas Lesutis*

China has expanded economic ties across Africa, and its infrastructural influence in Kenya is now unrivalled. Articulated primarily through road and rail networks, bilateral relations with Kenya are allowing for the realization of state agendas – Kenya's *Vision 2030* as well as China's Belt and Road Initiative (BRI). Certainly, Kenya–China relations have a long history. China was one of the first countries to open an embassy in Kenya after it gained independence in 1963 (Alden and Alves, 2008; Farooq et al, 2018; Kimari, 2021), yet these ties have strengthened, principally, over the last decade. Though these events highlight the expanding influence of China within the region, it is not uncontested; it has been eyed with circumspection by local and international publics and criticized sharply by the US government.

China's involvement in the region has augmented the Kenyan state's ability to pursue national planning blueprints. A key government developmental strategy *Vision 2030*, for example, is explicitly based on transport, energy, and water infrastructures that purportedly function as 'critical interventions that are required to jump start the economy and vault the country to middle-income industrialized status'. This national strategy is complemented by regional infrastructural visions such as the Programme for Infrastructure Development in Africa, which builds on the African Union's (2010) promotion of 'development' through continental infrastructural integration. Furthermore, against the backdrop of 'infrastructure-led development' (Kanai and Schindler, 2019) pursued by the Kenyan state (Lesutis, 2021a, 2021b), the national and regional infrastructure agendas are framed as noteworthy complements to the BRI. Highlighting this, Li Jiuping, the general manager of Afristar, the operator of the recently built Kenyan Standard Gauge Railway, stated that: '[with] Africa having a huge infrastructure deficit hindering its economic development, trade and industrialization, the BRI provides an opportunity that the continent could strategically take advantage of' (Xinhua, 2020).

Despite the enthusiasm with which infrastructure projects are undertaken by the Kenyan state, they continue to prompt socio-political contestation. Highlighting this, our chapter details how, even with complementary Kenyan, African, and Chinese state-led development visions for infrastructure projects, what they promise and deliver is problematized at both national and geopolitical levels. Undoubtedly, local challenges to megaprojects trouble the state visions and international alliances that make infrastructure-led development possible, while anchoring personal and collective anxieties about national development in the concrete articulations of megaprojects. While local voices most prominently contest both the material outputs and the multilateral agreements that underpin large-scale infrastructures across Kenya, they are also entangled with geopolitical competition between the US and China that manifests in politicized media discourses about which infrastructural arrangements are more 'beneficial' for the Kenyan people. For these reasons, we argue that megaprojects function as *symbolic geopolitical architectures* through which the public evaluates and engages multiscalar 'development', national policies, and the geopolitical alliances that enable them.

To these ends, looking at the Thika Superhighway, the JKIA–Westlands expressway, and the Standard Gauge Railway, we discuss how the Kenyan public reframes state/regional/BRI-sanctioned infrastructural visions and practices, while also debating the meanings of 'development' symbolized by these architectures. Concomitantly, we also map out the geopolitical anxieties between the US and China that are and will continue to be territorialized in Kenya's reach for future infrastructure.

## Concrete ambassadors and 'our national pride'

The Thika Superhighway was inaugurated on 9 November 2012 and referred to by the then President Mwai Kibaki as 'our national pride' (African Development Bank, 2012). The International Consortium for Infrastructure in Africa (ICAF) dubbed the highway 'the most ambitious infrastructure project in Kenya's history' (ICAF, 2012). In both public and media discourses, this eight-lane throughfare achieved iconic status, and its intended impact on both the national landscape and psyche was aptly registered in the media eight years after its launch: 'The first benefit of the construction was not monetary but psychological. A boost in the nation's morale 50 years after independence to imagine and a major infrastructure project. Kenyans have frequently compared the superhighway with others around the continent in their social media battles with neighbouring countries' (Mohammed, 2020).

Despite the infrastructural challenges that continue to plague this road, as well as the lack of transparency and adherence to environmental regulations that marked its construction processes (Klopp, 2012 Barczewski, 2013), Thika Superhighway has, certainly, fuelled the Kenyan public's imagination. Mbataru (2018), however, discusses its more practical imperatives: the reality that Kenya was losing billions of shillings every year because of road wear and tear as well as traffic jams. Furthermore, as 'connectivity' and 'development' were complementary justifications for this thoroughfare, then improvement and expansion would be the main requirements for its successive iteration. Ultimately, these desires for infrastructure aligned with China's BRI, in this way interlinking national, regional, and international objectives (Plummer, 2019).

Between 2009 and 2012, the Thika Superhighway was constructed by several Chinese firms with funding from the African Development Bank, China Export–Import (EXIM) Bank of China, and the Kenyan government.[1] Many Kenyans upheld the project as a source of national esteem, and some even used pictures of the highway to 'battle' with citizens from neighbouring countries on social media platforms over whose development was more impressive (see Mohammed, 2020). In addition, such road construction ventures, according to the Chinese embassy in Kenya, 'boosted China's image among Kenyans', linking it to 'development' and 'roads', while inevitably functioning as concrete 'ambassadors' for China (Embassy of China in Kenya, 2013).

Notwithstanding the national pride or ambassadorial role for China the project engendered, it was not immune to critical 'counter-channels of discourse', which included rumours that Chinese prisoners were involved in building the highway, and frustration on the part of local construction companies that were not given a substantive role in the fabrication of this major transport artery (Plummer, 2019). The Kenya Alliance of Residents Associations (KARA) and The Center for Sustainable Urban Development (CSUD) further document mixed reactions to the project, and especially the views of those who would count as non-motorized transport. They highlight citizen concerns about the lack of public participation at all stages of its development, as well as the absence of adequate footbridges and signage along the highway: ultimately 'misgivings about the project planning, design and implementation' prompted by the road safety, health, and socioeconomic impacts of this infrastructure (KARA and CSUD, 2012, 8–12). Similarly, challenging the initial national spatial logics put forward for its construction, Mbataru (2018) has also questioned whether the Thika Superhighway brought about an improvement in the livelihoods of the small-scale traders who were positioned throughout this major road (see Plummer, 2019). However, it is worth noting that land and property owners along the 42 km of the superhighway benefitted significantly from the increase in land prices, a direct result of the proximity of their assets to this thoroughfare (KARA and CSUD, 2012).

In 2014, two years after the inauguration of the new road, the blueprints for a national Standard Gauge Railway (SGR) were articulated by the central government, and the SGR was incorporated as a 'flagship' project into Kenya's *Vision 2030* over five years after the launch of this proposal. Despite these state actions, the project was questioned from the outset by both individual citizens and civil society. Barczewski (2013, 10) details that 'the Ministry of Roads and KeNHA [Kenya National Highways Authority] were reluctant to engage NGOs and community organizations during the planning phase of the NTHIP [Nairobi Thika Highway Improvement Project] because they did not want to attract any scrutiny that would slow down the planning process'. Similarly, the SGR was characterized by a lack of transparency throughout the lifecycle of the project. To date, the contractual agreement with the implementing partner – the China Road and Bridge Corporation – has not been made public (Ndii, 2018; Kimari and Ernstson, 2020; Lesutis, 2021b).

As the largest national infrastructure project since independence, the SGR accounts for at least a third of the country's national debt. Consequently, President Uhuru Kenyatta's assertion that it signalled a 'new dawn' (BBC, 2017) has been undermined by its annual financial losses and related racialized labour contestations. Summarizing the public conversations surrounding these megaprojects, Plummer (2019, 681) notes that: 'In the cases of the Thika highway project and the SGR, two Kenyan infrastructural projects, there was a robust discourse that often took the form of hearsay and rumour that exposed a different interpretation of China's role in rebuilding Kenya's transport networks.'

Despite the government and the CRBC's concerted public relations exercise (such as rebranding the SGR as the 'Madaraka [Freedom] Express'), they have been unable to dispel anxieties about what is seen as the imposition of infrastructure-led development. These are, at once, preoccupations with both the governance that enables these costly and contractually opaque infrastructure projects, and their impact on ordinary lives that lead to multiple and often varying perceptions of Chinese presence in the reconstruction of Kenyan transport networks (Plummer, 2019; Kimari, 2021). Klopp (2012, 3) summarizes large-scale infrastructure governance in Kenya as influenced by: 1) the 'large and distorting role of external actors', 2) 'fragmentation in institutions, policymaking and projects', 3) 'closed and top-down planning processes', and 4) 'the absence of mobilization for policies and projects that serve many city residents, especially the poorer segments'. Both the Thika

Superhighway and the SGR demonstrate these four characteristics, remain embedded within corresponding national, regional, and international agendas and spatial plans, such as Kenya's *Vision 2030*, the African Union's *Agenda 2063* and the BRI. The purchase of this uneven infrastructure governance, constituted by the aforementioned four qualities, is widespread: it even animates minor architectures such as smaller inter-county roads, of which the Kisumu–Kakamega highway is an emblematic example (Kenya News, 2019). However, since the launch of the SGR in 2017, it has surfaced in a more prominent fashion following the commencement of construction for another mega road in Nairobi: The Jomo Kenyatta International Airport (JKIA)–Westlands expressway.

On 16 October 2019, many Kenyans were surprised to hear that President Kenyatta would inaugurate yet another mega-infrastructure project. Though the traffic on Nairobi-Mombasa Road was notorious and had long been a key complaint for many city residents, the revelation that a toll expressway was the government's solution was unexpected. As the details about this project came to light, public discontent, both online and offline, prompted the County of Nairobi and the Kenya National Highways Authority (KeNHA) to convene participatory consultations (Kimari, 2021). A number of critical apprehensions seemed to orient the immediate negative responses to the new expressway: what one news reporter called 'too much borrowing and a lack of accountability',[2] as well as ongoing fears about the environmental impact of the project. In this regard, initial proposals showed that the highway would impact green urban spaces, and, in particular, at least an acre of the popular Uhuru Park in central Nairobi would be hived off to create room for it.

Building on contentions that emerged during the construction of both the Thika Superhighway and the SGR, this proposed road exacerbated popular concerns about debt, sovereignty, national development, and the government's lack of consideration for most city residents, the vast number of whom do not own cars. While nominally aligned with *Vision 2030*, urban planners took issue with the fact that this proposal was decided seemingly out of nowhere, since it was never included in the 2014 Master Plan for Nairobi. At least one case has been lodged in the courts to try and halt the JKIA–Westlands expressway, but its construction continues unabated (Walter, 2020). Specifically, critics of the expressway challenge the government's assertion that it will foster 'development', and highlight the mixed results of both the SGR and the Thika Superhighway to validate their concerns.

Even as the Thika highway, SGR, and airport expressway are the most iconic examples of the medley of visions – national, regional, and international – that inform Kenya's state spatial strategies (Brenner, 2004), they continue to be subject to withering criticism from ordinary citizens, who question the viability of infrastructure-led development; its very necessity, and the governance processes that animate it. Certainly, Kenyans continue to discursively reframe these architectures, doubting their capacity to deliver, and the following section elaborates in more detail how such projects are contested, and the very meaning of 'development' is called into question.

## The Standard Gauge Railway and socio-political contestations

In late May 2017, the Kenyan government celebrated the completion of Phase I of the SGR that now connects Mombasa with Nairobi; Phase IIA travels 120 kilometres further north-west of the capital, reaching the town of Naivasha. The SGR – celebrated by the Kenyan government as a promise of 'development' (Lesutis, 2021a) – was designed to connect Asian commodity markets with Kisumu, a Kenyan port city on Lake Victoria, and make inroads further into East Africa: Rwanda, Uganda, and South Sudan. Connectivity is meant to enhance the strategic role of Kenya in facilitating global trade in the region. Despite state aspirations for national and regional prosperity, the SGR has become an object through which the Kenyan state, its industrialization visions, and alliances with global architectures of development articulated through China's BRI,[3] is scrutinized by Kenya's public.

The SGR starts in the port city of Mombasa – historically the national gateway through which commodities and people travelling across the Indian Ocean entered and exited the country. Mombasa is situated at the centre of the regional economy, providing employment in cargo handling, port maintenance, and administrative and security work. With regional unemployment rates as high as 44 per cent (Rift Valley Institute, 2017), the city has attracted people in search of work in or around the port. The port's operations have expanded in the past decade, and private small- and medium-sized business enterprises have increasingly taken over freight and storage services. This extended logistics and port-related activities beyond the official boundaries of the port (see Lamarque, 2019). This outgrowth has further facilitated the integration of the port into the urban fabric; besides the port itself, in Mombasa there are other economic sectors, including food services, guesthouses, drinking establishments, and an informal sex work industry, all of which accommodate populations coming and going with the movement of ship cargo.

In April 2019, two years after the completion of the SGR line, the Kenyan government passed a national decree requiring all container cargo entering Mombasa Port to be transported using the new railway. Considering Mombasa's dependence on cargo-carrying trucks, it was only a matter of time before the economies that serve its port would feel the drastic effects of this change. The SGR decree first hit freight-handling services in Mombasa;

since all container cargo must be shipped by the SGR railway line from the Mombasa Port, incoming and outgoing containers could only be stored within the port area, and thus the facilities located outside of the port that previously offered storage services for cargo-carrying trucks saw a drastic reduction in their business. As a direct result, approximately 80 per cent of storage units went bankrupt, and around 12,000 trucks were put out of business. Thus, although Mombasa remains a key gateway, the bypassing of its local economy has resulted in widespread discontent. Many residents fear that it is a 'dying city that will soon be a ghost town' (see Lesutis, 2021b).

Local representatives of freight and handling services in Mombasa decided to take matters into their own hands by organizing weekly protests of the SGR directive. Called 'Black Mondays', these protests were held at key locations around the city – in front of the governor's office, the main road intersection that connects Mombasa with the airport, or the SGR terminal. The protesters highlighted specific problems with the SGR and how the legal context that regulates its use endangered livelihoods. Carrying wooden coffins and placards with phrases such as 'Mombasa is bleeding' and 'RIP Mombasa economy', the protesters indicated how the crumbling economy of Mombasa also means social death for people implicated in the port's economic activities. In this way, across the politically uneven landscape of Kenya, megaprojects like the SGR activate discursive and material forms of contestation that challenge Kenya's national development vision.

The SGR project has also resulted in public anxiety about Kenya's sovereignty – that the country is increasingly under the influence of Chinese 'colonialism', particularly through China's BRI, within which this railway is implicated. As the BRI attempts to orient global trade and international finance governance towards China, the initiative has accelerated a wide range of discursive and material transformations in the economic, social, and political spheres of life in different parts of the world (see Murton and Lord, 2020; Szadziewski, 2020). Against this background, Chinese Vice Foreign Minister Zhang Ming visited Kenya and remarked that the SGR train 'bears the Kenyan people's dream for this century of striving for national development and prosperity. This also shows China's firm support for Kenya achieving independent and sustainable development' (quoted in Freytas-Tamura, 2017).

Despite this rhetoric embraced by Chinese officials, as well as the Kenyan government's emphasis on the SGR as a symbol of development, the project reactivated public anxiety about the country's financial and territorial sovereignty. This is principally expressed in relation to the costs of the railway, amounting to US$4 billion, which, issued in the form of a loan from China EXIM Bank, constitutes a third of Kenya's national debt (see Ndii, 2018). The Kenyan public has condemned this high level of debt, fearing that it will undermine national sovereignty and the possibilities of a good life for both current and future populations. As one small business owner observed, "even the babies that we are yet to have in Kenya will be paying for the cost of the SGR – the government has really sold us to the Chinese, they are our colonial masters now".[4] Similarly, in settlements around SGR infrastructure, including service and passenger stations, workers compounds, or other railway facilities, residents point out how the SGR belongs to China and not to Kenya. In rural areas, it is not the SGR train but 'China train [sic]', it is not 'national development' but 'China men [sic] controlling land' (for details, see Lesutis, 2021b).

Local anxiety about increasing foreign influence and Kenya's diminishing independence became particularly prominent in April 2019 when several national media outlets announced that the Kenya Port Authority's assets, including the Mombasa Port facilities, will be taken over by China EXIM Bank of Kenya defaults on its loans. While the validity of these claims was deliberated in various public forums – for the actual financial agreement between the Kenyan government and China EXIM Bank is confidential and not available for public scrutiny – President Kenyatta denounced it as propaganda. However, the public discourse, and particularly the collectively shared anxiety about China's 'colonialism' that it has sustained, highlights how the SGR has become an object through which the Kenyan state and its place in the world is continuously (re)evaluated by local citizens, particularly in relation to the imagined and real power and structural influence of China (see Lesutis, 2021b).

Although these anxieties about Chinese colonialism operate at a symbolic level and need to be understood in the context of changing global power dynamics often articulated through the language of orientalism (see Liu, 2017; Kimari, 2021), fears about the detrimental influence of external actors to Kenya's visions of development are not unfounded. Public preoccupations about neo-colonialism were reenergized when the regional railway-based transport development strategy was called into question in May 2019, after China EXIM Bank decided to freeze the funding for Phase IIB of the SGR project (the line intended to reach Uganda). It made this decision considering numerous challenges experienced during the construction of the SGR Phase I and Phase IIA, as well as regional rivalry prompted by Tanzania's announcement that it would build its own SGR. The withholding of this loan has jeopardized the financial viability of the whole SGR project; without links to regional transportation networks, the national railway system is unlikely to generate enough revenue to pay off the public debt (see Ndii, 2019). Currently, after leaving Nairobi, Phase IIA stops abruptly in Naivasha, only 120 km from the capital city, leading some Kenyans to describe this part of the project as the 'railway to nowhere'. In this context, China's ability to 'unexpectedly' withhold loans has been interpreted as a form of colonialism that haunts Kenya's development priorities and aspirations.

While the Uhuru Kenyatta-led government aims to deliver 'development' through megaprojects, the SGR has become a material and symbolic site through which Kenya's current infrastructure-based dash for prosperity is evaluated and critiqued. Interpreted in local public vernaculars, the project is no longer a signifier of progress. Instead, it has become an object of critique through which state-led infrastructural visions and their connections to global architectures of development, and even the state itself, are discursively contested and undermined. Against this background, US representatives portray their proposed involvement in Kenya's large-scale infrastructure sector as a benevolent means of catalysing development.

## 'Total RUBBISH!': US–China competition

In late November 2019, an article appeared in *The Star* newspaper titled: 'Chinese eye Mombasa-Nairobi expressway as US ditch project' (see Mwita, 2019). This account detailed two key events: first, that the US had, ostensibly, pulled out of rebuilding one of the longest and most utilized inter-county roads in the country, a 485-km road that connects Nairobi and the port city of Mombasa, because of 'inflated costs' presumably linked to corruption within the government tendering processes. Second, that Chinese firms and SOEs were increasingly interested in bidding on the project, as they could rely on the strong Chinese state support that had, bolstered by strong China–Kenya relations, secured several other large-scale projects in the country (Mwita, 2019).

In a response conveyed via Twitter, the Trump appointed American ambassador to Kenya, Kyle McCarter, declared that 'there is so much wrong with this article I don't know where to start to respond. Total RUBBISH! Journalistic malpractice.' This reaction – dubbed 'heated' (see Olander, 2019) – alerted Kenyans to the reality that national infrastructure projects were an additional field in which the US–China rivalry was playing out and being territorialized. What is more, McCarter's assertions earlier in 2019, about his government's intention to build a highway that provided 'value for money for Kenyans', and that 'Kenya has a challenge of debt and we are wary of burdening Kenyans' (Olander, 2019), demonstrated how local preoccupations with China–Kenya infrastructural arrangements were refracted through American tropes intended to challenge China's dominance in the infrastructure sector. Ultimately, Olander (2019) asserts that what was being conveyed by McCarter's formal and informal messaging was that:

- The new [Mombasa – Nairobi] expressway will be built in a timely manner.
- It will not saddle the Kenyan government with excessive debt.
- It will be of higher quality than the work of Chinese contractors.

McCarter's concerns build on local narratives about the debt, loans, corruption, and questions about infrastructure quality registered in various ways and through multiple signifiers by Kenyan problematizations of Chinese construction in the country (see also Kimari, 2021). Meanwhile, work on this highway is yet to begin four years after the government announced that a US firm would undertake its construction, and Chinese firms continue to secure multiple mega-infrastructure projects in the country. Thus, while the US is positioning itself as a benevolent alternative that remains 'fully committed' (Herbling, 2019) to the project, it appears that, for now, despite local contestations and 'heated' responses from American diplomatic representatives, joint China–Kenya spatial projects will continue to be the symbols of both development and its actual delivery.

## Conclusion

Through investments in large-scale connective infrastructures in Kenya since 2009, this chapter revealed how megaprojects emerge from both national and regional development visions, while simultaneously functioning as material and symbolic articulations of geopolitical influence – as well as their contestation – within transcontinental inter-state arrangements. Their importance on a local and global scale enables them to represent what we call *symbolic geopolitical architectures* of development, which are advanced through China's BRI but, simultaneously, also pursued as a direct result of US anxiety to effectively counter China's influence. Here, we specifically demonstrated how, while complementing infrastructural visions are prioritized to show the alignment of Kenyan, African, and Chinese state-led development agendas, such infrastructural developments encounter contestation at multiple levels. This includes politicized public attempts to confront China's global influence, most prominently by US actors, even if the material effects of such strategies are yet to materialize in Kenya.

Highlighting how citizens reframe infrastructures, the state-articulated imaginaries of 'progress' that anchor them, as well as the *actual* ability of these architectures to foster prosperity, we showed that megaprojects are *symbolic geopolitical architectures* through which citizens, non-citizens, and even representatives of foreign states narrate their anxieties. This demonstrates how Kenya's infrastructure development functions as a symbolic geopolitical field in which China and the US are set to compete in the future. The actual infrastructures, therefore, unfold through complex multiscalar politics, registered by the national critiques of megaprojects and the emerging discursive responses from the US that map onto and are interrelated with these public debates. Ultimately infrastructure in Kenya is not only the articulation of competition between global powers, but it is also shaped by a domestic politics directed by ordinary citizens.

## Notes

[1] Some of these companies include the China Road and Bridge Corporation (CRBC), China Wu Yi Company, Sinohydro Corporation and Sheng Li Engineering Construction.

[2] From 'Expressway causes uproar as government maintains Uhuru Park will not be touched', KTN News, 31 October 2019. Available from: www.youtube.com/watch?v=TRS2ZMHNv5M&feature=emb_logo.

[3] The SGR project is funded by financial loans from the Export–Import Bank of China and built by Chinese contractors.

[4] Interview in Mombasa, June 2019.

## References

AfDB (2012) 'AfDB-funded Thika Superhighway: a masterpiece for East Africa 'a national pride' – President Mwai Kibaki', 6 November. Available from: www.afdb.org/en/news-and-events/afdb-funded-thika-superhighway-a-masterpiece-for-east-africa-a-national-pride-president-mwai-kibaki-9986.

African Union (2010) *Programme for Infrastructure Development in Africa: Interconnecting, Integrating and Transforming a Continent*, Addis Ababa: African Union.

Alden, C. and Alves, C. (2008) 'History and identity in the construction of China's Africa policy', *Review of African Political Economy*, 35(115): 43–58.

Barczewski, B (2013) 'How well do environmental regulations work in Kenya? A case study of the Thika highway improvement project', Center for Sustainable Urban Development, Columbia University. Available from: www.landportal.org/library/resources/how-well-do-environmental-regulations-work-kenya-case-study-thika-highway.

BBC (2017) 'Kenya opens Nairobi-Mombasa Madaraka Express railway', 31 May. Available from: www.bbc.com/news/world-africa-40092600.

Brenner, N. (2004) *New State Spaces: Urban Governance and the Rescaling of Statehood*, Oxford: Oxford University Press.

Embassy of China in Kenya (2013) 'Road construction projects boost China's image in Kenya', 5 July. Available from: http://ke.china-embassy.org/eng/zkgx/t1060248.htm.

Farooq, M.S., Tongkai, Y., Jiangang, Z., and Feroze, N. (2018) 'Kenya and the 21st century maritime silk road: implications for China–Africa relations', *China Quarterly of International Strategic Studies*, 4(3): 401–18.

Freytas-Tamura, K. (2017) 'Kenyans fear Chinese-backed railway is another "lunatic express"', *New York Times*, 21 May. Available from: www.nytimes.com/2017/06/08/world/africa/kenyans-fear-chinese-backed-railway-is-another-lunatic-express.html.

Government of the Republic of Kenya (2007) *Kenya Vision 2030*, Nairobi: Government of the Republic of Kenya. Available from: https://vision2030.go.ke/.

Herbling, D. (2019) 'US denies \$3.5 billion road risks debt stress for Kenya', Bloomberg, 5 November. Available from: www.bloomberg.com/news/articles/2019-11-01/u-s-denies-3-5-billion-road-risks-debt-stress-for-kenya.

Infrastructure Consortium for Africa (ICAF) (2012) 'Launch of flagship road project in Kenya: Thika superhighway', 10 November. Available from: www.icafrica.org/en/news-events/infrastructure-news/article/launch-of-flagship-road-project-in-kenya-nairobi-thika-superhighway-3481/.

Kanai, J.M. and Schindler, S. (2019) 'Peri-urban promises of connectivity: linking project-led polycentrism to the infrastructure scramble', *Environment and Planning A*, 51(2): 302–22.

Kenya Alliance of Resident Associations (KARA) and Centre for Sustainable Urban Development (CSUD) (2012) 'Thika Highway Improvement Project: the social/community component of the analysis of the Thika Highway Improvement Project'.

Kenya News (2019) 'Kisumu-Kakamega highway contractor gives notices over delayed dues', 24 September. Available from: www.kenyanews.go.ke/kisumu-kakamega-highway-contractor-gives-notice-over-delayed-dues/.

Kimari, W. (2021) '"Under construction": everyday anxieties and the proliferating social meanings of China in Kenya', *Africa*, 91(1): 135–52.

Kimari, W. and Ernstson, H. (2021) 'Imperial remains and imperial invitations: Centering race within the contemporary large-scale infrastructures of East Africa', *Antipode*, 52(3): 825–46.

Klopp, J. (2012) 'Towards a political economy of transportation policy and practice in Nairobi', *Urban Forum*, 23(1): 1–21.

Lamarque, H. (2019) 'Profitable inefficiency: the politics of port infrastructure in Mombasa, Kenya', *The Journal of Modern African Studies*, 57(1): 85–119.

Lesutis, G. (2021a) 'Infrastructure as techno-politics of differentiation: socio-political effects of mega-infrastructures in Kenya', *Transactions of the Institute of British Geographers*: 1–13.

Lesutis, G. (2021b) 'Infrastructural territorialisations: mega-infrastructures and the (re)making of Kenya', *Political Geography*, 90: 1–11.

Liu, X. (2017) 'Look beyond and beneath the soft power: an alternative analytical framework for China's cultural diplomacy', *Cambridge Journal of China Studies*, 12(4): 77–94.

Mbataru, P. (2018) 'An analysis of the influence of road infrastructure implementation on local development: the case of Thika Superhighway in Kenya', *Africa Journal of Public Sector Development and Governance*, 1(1): 82–9.

Mohammed, A. (2020) 'How a 50km road built property millionaires', *The Standard*, 17 July. Available from: www.standardmedia.co.ke/home-away/article/2001378974/how-a-50km-road-project-built-property-millionaires.

Murton, G. and Lord, A. (2020) 'Trans-Himalayan power corridors: infrastructural politics and China's Belt and Road Initiative in Nepal', *Political Geography*, 77.

Mwita, M. (2019) 'Chinese eye Mombasa-Nairobi expressway as US ditch project', *The Star*, 1 November. Available from: www.the-star.co.ke/business/2019-11-01-chinese-eye-mombasa-nairobi-expressway-as-us-ditch-project/.

Ndii, D. (2018) 'SGR by the numbers – some unpleasant arithmetic', *The East African Review*, 21 July. Available from: www.theelephant.info/op-eds/2018/07/21/sgr-by-the-numbers-some-unpleasant-arithmetic/

Ndii, D. (2019) 'From game changer to railway to nowhere: the rise and fall of Lunatic Line 2.0', *The Elephant*, 2 November. Available from: www.theelephant.info/op-eds/2019/11/02/from-game-changer-to-railway-to-nowhere-the-rise-and-fall-of-lunatic-line-2-0/.

Olander, E.C. (2019) 'Will China or the US build the new Mombasa–Nairobi expressway? Confusion swirls amid conflicting headlines', The China Africa Project, 4 November. Available from: https://chinaafricaproject.com/analysis/will-china-or-the-u-s-build-the-new-mombasa-nairobi-expressway-confusion-swirls-amid-conflicting-headlines/.

Plummer, A. (2019) 'Kenya and China's labour relations: infrastructural development for whom, by whom?', *Africa*, 89(4): 680–95.

Rift Valley Institute (2017) 'Big barrier: youth unemployment at the coast', meeting report. Available from: http://riftvalley.net/publication/big-barrier-youth-unemployment-coast.

Szadziewski, H. (2020) 'Converging anticipatory geographies in Oceania: the Belt and Road Initiative and Look North in Fiji', *Political Geography*, 77.

Walter, D. (2020) 'Petition filled seeking to stop JKIA-Westlands expressway corruption', *Citizen Digital*, 28 January. Available from: https://citizentv.co.ke/news/petition-filed-seeking-to-stop-jkia-westlands-expressway-construction-315995/.

Xinhua (2020) 'Kenya's modern railway cements trade, infrastructure relations with China', *Xinhua*, 30 August. Available from: www.xinhuanet.com/english/2020-08/30/c_139329255.htm.

# *Interlude*: The Emergence of a Sino-Centric Transnational Capitalist Class?

*Steve Rolf*

In her path-breaking study of Chinese state-owned copper mining firms in Zambia, Ching Kwan Lee (2017) observes that a spectre of 'global China' is haunting the world. As the rise of China's state capitalist economy continues apace and Belt and Road Initiative (BRI) projects begin to shape and take shape on every continent, Western policy makers fret not only about China's military power but about its capabilities in building roads and bridges, power-generating infrastructure, and rolling out undersea internet cabling and 5G equipment (Sum, 2019). Underlying this anxiety is a fear that China is becoming a global hegemon. This chapter explores the crucial role capitalist classes play in shaping geopolitics and international relations, and considers whether global China is giving rise to a Sino-centric transnational capitalist class.

The history of capitalism is one of cycles of accumulation anchored by a hegemon – the Dutch, followed by the British, and now, the United States (Arrighi, 1994). These hegemonic powers were underpinned by transnationally-oriented commercial classes which spanned territorial borders, but were connected to the hegemon's markets, acting to secure the conditions for the expanded reproduction of the system. It follows that China's ability to reorient the global economy and establish itself as a hegemon likely depends upon the emergence of a Sino-centric transnational capitalist class whose pursuit of self-interest underpins Chinese hegemony. In other words, it is not enough to simply construct infrastructure or extend credit; but conditions must emerge within which 'domestic' capitalist classes prosper while being rendered dependent upon or enmeshed within China's domestic market, commercial networks, and/ or financial institutions.

My purpose in this interlude is to consider whether such conditions are emerging. The following section introduces theories of transnational capitalist class (TCC) formation, which I illustrate with a focus on the US-centred transnational capitalist class (US TCC). I then propose a schema for analysing Sino-centric class formation – along economic, political, and ideological lines – and introduce cases in Venezuela, Serbia, and New Zealand. I find that despite China's growing global political economic power, there is scant evidence of a Sino-centric TCC operating as a class-for-itself. For now, this substantially limits China's ability to become a global hegemon anchoring a distinctively Chinese-centred system of capital accumulation.

## Transnational capitalist classes

The 'Amsterdam School' and the 'Global Capitalism School' represent the most prominent attempts to theoretically and empirically demonstrate the significance of a transnational section of the capitalist class in the contemporary world economy (see Jessop and Overbeek, 2018). For both schools of thought, the TCC represents a section of the capitalist class dependent on the globalized and liberalized world economy for its capacity to accumulate and self-reproduce. The current TCC emerged in the post-war period under the United States' stewardship of the world economy, and hence, the US TCC comprises interlocking sets of global elites whose self-interest is dependent on the preservation of the world order. The US TCC manages a significant share of globalized circuits of production, distribution, and consumption through the coordination of:

> foreign direct investment, cross-border financial flows, cross-border mergers and acquisitions, foreign investment activity by sovereign wealth funds, the transnational character of stock ownership, foreign affiliates, tax havens, global assembly lines, intra-firm trade, the vast network of national sub-contractors tied to transnational corporations (TNCs), the ratio of foreign-owned assets, employment and sales to similar national figures, and the percent of foreign revenues and profits. (Harris, 2014: 314)

The extent to which the emergence of the US TCC rendered states and national capitalist classes superfluous is a subject of longstanding debate. Panitch and Gindin (2012) argue that far from undermining the state system, transnationally-oriented capitalist classes have historically 'depend[ed] on equal national treatment by many states; [while] these states were also internationalized in the sense of coming to take on more and more responsibility for creating and strengthening the conditions for non-discriminatory accumulation within their borders'. Indeed, notwithstanding important exceptions (Robinson, 2004; Harris, 2014), TCC scholarship has recognized the growing significance of state institutions despite an increasing transnationalization of production and finance, implying strong nation states are compatible with a globalizing economy (for example, Van Apeldoorn et al, 2012; Carroll, 2013; Babic et al, 2020).

In this view, the US hegemonic order established in 1945 did not arise simply because of the state's 'power resources' (GDP, military spending, innovative capacity, and so on). Equally if not more significant was the US ability to threaten recalcitrant states with geopolitical discipline and isolation, while offering compliant states access to its sizeable domestic market, financial liquidity, and productive investment and technology transfer from its multinational corporations (MNCs; Arrighi, 1994). National elites outside of the Soviet sphere of influence were incentivized to open their economies to US firms and embrace ownership structures and (later) control structures dominated by US MNCs. The rise of the post-war miracle economies in West Germany and Japan, followed by the East Asian Tigers and newly industrialized countries as competitors to the United States was not 'due to the "failure" of the United States to remain "ahead of the pack", but to the success of the US-led Keynesian system of international accumulation' (Saad-Filho and Ayers, 2008).

This system amounted to an 'informal' American empire, characterized by interlocking corporate directorships, an expansive network of US-owned branch plants, and ownership shares by US business elites. As early as the 1970s, observers like Hymer (1972) and Poulantzas (1975) noted the rise of a US-centred 'international bourgeoisie' that actively reshaped domestic 'internal bourgeoisies' across Europe and beyond, through social, economic, and corporate networks (see Fennema and Heemskerk, 2018). This is evocative of the 'comprador' bourgeoisies that augmented British capitalist interests in the age of imperialism. Gallagher and Robinson (1953, 7–10) assert that in Argentina and Brazil: 'There was no need for brusque or peremptory interventions on behalf of British interests ... [since] the classes whose prosperity was drawn from that [foreign] trade normally worked themselves in local politics to preserve the local political conditions needed for it.' Conversely, they note that 'British expansion sometimes failed, if it gained political supremacy without effecting a successful commercial penetration'. For example, '[t]here were spectacular exertions of British policy in China, [which] did little to produce new customers'. Thus, TCC formation requires more than simple economic integration: it requires a complex articulation of a particular economic incentive structure, political incorporation, and ideological dominion to cement the integration of dominant bourgeoise fractions into hegemonic geopolitical, geoeconomic and geocultural networks.

## China's capitalist class and its global expansion

Two world-historic transformations that took place in the 1970s are of significance here: the rise of global neoliberalism, and China's integration into the capitalist world economy. First, the transition to neoliberalism encompassed a shift in the predominant US capitalist fraction to those controlling finance. This led to an attendant shift from productivist forms of investment (based upon infrastructure and manufacturing) to extractivist economic activities centred upon hot money flows and predatory lending (Gowan, 1999; Rude, 2009). Despite sluggish growth in the world economy since the 1980s, US corporations and financial institutions cemented ownership and control over substantial parts of foreign companies, and came to dominate financial networks and intellectual property (Starrs, 2018). Since the 2008 global financial crisis, this 'network centrality' has been weaponized to maintain US predominance and compel states to accept US power – a process manifest in the widespread use of sanctions, and various formal and informal mechanisms for upholding them such as the Magnitsky Act, control over the SWIFT interbank payment network, and the 'Clinton list', among many others (Van Der Pijl and Yurchenko, 2015). Although these disciplinary mechanisms have bolstered US power, they have also incentivized states that fall foul of them to seek alternatives.

Meanwhile, many countries sought to insulate themselves from the worst vicissitudes of the turbulent global economy and extractive phase of US hegemony (Stuenkel, 2017; Bond and Garcia, 2020). Chinese state officials in particular remained wary of permitting full integration with the US TCC. The liberalization of China's economy since the 1980s was staggered and uneven, and cannot be considered 'complete', at least insofar as China represents a radically distinct variety of capitalism. The hybrid economy combines elements of deep global economic integration through embeddedness in manufacturing global production networks and holdings of US Treasury bonds, on the one hand, with powerful capital controls, state-owned enterprises (SOEs), and expansive industrial policies on the other (Rolf, 2020). This accounts for three significant idiosyncrasies of Chinese capitalism. First, the Chinese Communist Party has increasingly merged with state institutions. Second, while powerful structural interdependencies between state and capital exist elsewhere, the near total political subordination of China's capitalist class to the imperatives of the party state is remarkable. Third, the national ownership of SOEs, the financial sector, and powerful systems of capital controls amount to 'financial repression', and grant the state extraordinary leverage over the direction of surpluses for reinvestment (Heilmann, 2018).

This regime has contributed to unprecedented growth. In 2020 China had more firms than the US in the Forbes Global 2000 list (124 vs 121) (Murphy et al, 2021), virtually all of which are owned and controlled by the central government's asset management arm, the State-owned Assets Supervision and Administration Commission of the State Council (SASAC). Those that are private (like tech giant Alibaba) are subject to intensive oversight by political and regulatory institutions. Meanwhile, unlike most other large economies, foreign MNCs and banks are strictly limited in their ownership and control of Chinese corporations (Starrs, 2017). Although China is by no means a closed or unitary state-society complex, the key mechanisms of control and incorporation exerted by the US TCC elsewhere are only very partially operable in China. The result has been the emergence of a *state* capitalist class firmly centred within China's domestic market, and overwhelmingly owned and financed by Chinese elites and (state-owned) banks. This state capitalist class is extending its reach overseas and is at the core of the BRI.

Chinese firms have internationalized over the past two decades since the advent of a 'going out' strategy (Li and Cheong, 2019). Barely registering as a global source of FDI in 2000 (US$916 million), China provided US$143 billion and US$117 billion in 2018 and 2019, respectively, ranking it among the world's largest sources of investment (UNCTAD, 2020). Nearly 90 per cent of such investment activity is pursued by subnational SOEs, which are governed at some distance from the central institutions of the party state and focus overwhelmingly on maximizing returns (Jones, 2020; Breslin, 2021). Even giant nationally-owned SOEs typically operate as corporatized, rather than political, organizations, notwithstanding greater efforts to accumulate political capital and ingratiate themselves with host states (Lee, 2017). As a result, many scholars have questioned the extent to which the Chinese capitalist class poses a substantive threat or alternative to the US-led international order (Jones and Zeng, 2019; Babones et al, 2020; Bräutigam, 2020). However, the weak penetration of the US TCC within China's economy combined with the internationalization of Chinese firms *does* undoubtedly serve to functionally integrate other economies into distinctively Sino-centric business networks. In turn, Chinese economic penetration overseas *potentially* offers emerging economies an alternative developmental paradigm to US TCC incorporation (Jepson, 2020). This is true in the case of development aid and infrastructure finance because Chinese investments typically come with few conditionalities (unlike loans from multilateral development banks). But beyond aid and finance, there are three ways in which a transnational business class with links to the Chinese state-society complex may be forming:

1. *Economically*, the pull of China as a trading partner, source of finance and development aid, if sustained, will almost inevitably drive elites from other countries to shift their general orientation towards Chinese economic engagements (Schwartz, 2018). This is tied up with impressive infrastructural project delivery capabilities. As Blanchette (2020) notes,

'"CCP Inc." possesses an unrivalled ability to deliver the complete value package when entering overseas investment deals: it can buy, build, and finance on a scale and speed that is unmatched.'

2. *(Geo)politically*, China has become deeply embedded within institutions of global governance (Xu, 2016) while establishing important new organizations over which it has substantial leverage: the Asian Infrastructure Investment Bank (AIIB), Shanghai Cooperation Organisation, Regional Comprehensive Economic Partnership, Belt and Road Forums, and a swathe of bilateral trade agreements and memorandums of understanding (Beeson and Wilson, 2015). The overwhelming majority of investment and loan capital are disbursed through the major state-owned policy banks – China Development Bank (CDB) and The Export–Import Bank of China (EXIM), rather than multilaterally through the AIIB.

3. *Ideologically*, Beijing has cultivated a narrative of 'win-win' forms of 'non-interventionist' engagement, while it explicitly sanctions statist development strategies and tolerates a variety of political regimes.

These three axes establish the terms of and space for the potential integration of national bourgeoisies and China's increasingly self-confident capitalist class as it expands overseas. Three brief case studies explore the extent to which a Sino-centric TCC is emerging in Venezuela, Serbia, and New Zealand.

## Venezuela

Venezuela was firmly rooted in the US sphere of influence for much of the 20th century, receiving substantial development aid during the Cold War and becoming a major oil supplier to the US during the Persian Gulf embargo of 1973. The country possesses a quarter of the world's known oil reserves and, under the Pérez administration's import substitution industrialization in the 1970s, the nationalization of the Petróleos de Venezuela, S.A. (PDVSA) – the principal oil company – was tacitly sanctioned by the US. This company fuelled the rise of a state rentier class structure centred around the capture of oil revenue and its redistribution to workers (through social programmes) and capitalists (through industrial subsidies and industrialization drives) (Coronil, 1997; Chiasson-LeBel, 2016). During the 1980s, the state pursued partial privatization of national industries and the introduction of 'strategic associations' with resource extraction MNCs in exchange for access to an IMF credit line. As part of the 'Pink Tide' of left-wing governments which came to power during the late 1990s and early 2000s, Hugo Chávez's presidency reversed neoliberal reforms and re-nationalized the PDVSA. However, Chávez encountered significant opposition from the increasingly powerful domestic bourgeoisie which had become deeply connected with US finance capital. This class mounted a coup attempt in 2002 which was staved off by popular resistance.

During Chávez's presidency, China and Venezuela established close commercial and political relations. Chinese capital inflows focused overwhelmingly on loan-for-oil deals and financed extractive infrastructure. This coincided with a re-nationalization of the PDVSA and termination of cooperation agreements with foreign MNCs, thus re-centring the state class and sidelining private capital. Chinese influence grew substantially after 2005, when Venezuela broke off military cooperation with the US and Chávez announced the pursuit of 'socialism in the 21st century' at the World Social Forum. Chinese oil and infrastructure funds were directed through the CDB, which contributed US$30 billion to a 'Great Volume Fund' by 2011 in exchange for a decades guarantee of oil exports (Giacalone and Ruiz, 2013). A bilateral China–Venezuela Joint High Level Committee was established in 2010 to manage the relationship (Ferchen, 2020), and CDB funds were mobilized for social programmes like low-income housing construction. Overall Chinese capital inflows amounted to over US$40 billion during the Chávez presidency.

Maduro's presidency (2013–) witnessed a downward economic spiral due to the collapse in oil prices and US sanctions. A further US$20 billion in Chinese investment from 2013–20 did not stem the recessionary tide, and Venezuelan GDP contracted persistently (shrinking by a devastating 35 per cent in 2019 and 30 per cent in 2020). Economic crisis precipitated large-scale outmigration, collapsing infrastructure, spiralling inflation, and chronic food shortages. As a result, the private capitalist class re-emerged as an organized political force, most evident in anti-government protests and presidential hopeful Juan Guaidó's attempted coup.

Isolated from the US both geopolitically and now economically, with a state class struggling to retain power, Venezuela demonstrates both an advanced case of Sino-centric TCC formation, and the limits of such a process. Should the private capitalist class be able to overthrow statist elites represented by Maduro, it would most likely distance Venezuela from China and join a growing majority of Latin American states pivoting toward the US in the wake of the Pink Tide (the name of the political turn away from neoliberal ideology in Latin America, towards left-leaning governance, see Chodor, 2015). China has done little to prevent this beyond continuing to recognize the existing government. Chinese aid, finance, and investment flows have largely dried up – abandoning the internationally influential state class that emerged under Chávez.

## Serbia

Serbia's economy suffered a deep recession during the break-up of the former Yugoslavia and the Balkans conflicts. Recovery has been slow, with some estimates putting GDP per capita at a lower level today than in 1990 (Uvalic, 2010). During the 2000s, GDP growth fluctuated around 5 per cent and was driven by European investments in the automotive sector, manufacturing, and formerly state-owned financial intermediation and telecoms. The 2008 crisis significantly weakened but did not transform this export-oriented model, and left annual growth averaging a much lower 1–2 per cent in the decade to 2020. Economically, Serbia remains largely dependent upon EU-based FDI, which represented 73 per cent of cumulative foreign investment from 2010–17 (Chinese FDI constituted a mere 2 per cent). Beyond the productive, export-oriented sectors, much of Serbia's economic activity has taken the form of what Bedo et al (2011) term 'wild capitalism'. This (non-)variety of capitalism is animated by 'clientelism and reliance on personal networks' that shape a system in which 'business activity is characterized by insider dealing as privatized state assets are bought up by individuals who are well connected with (sometimes) competing elites'. This largely deregulated and predatory economic system exists beyond the EU's regulatory scrutiny (despite an ongoing accession application). For these reasons, Serbia has proven fertile ground for a Chinese investment and presence in Europe in recent years.

Serbian President Aleksandar Vučić's administration has sought to strategically play Chinese investors against the EU and its member states. Chinese investment activities in Serbia amount to US$29 billion in infrastructure alone, and are broadly balanced across strategic sectors, with 31 per cent in transportation, 13 per cent in energy, 20 per cent in information and communication technologies and 20 per cent in manufacturing (Conley et al, 2020). Investment in sectors like steel smelting and copper mining may bolster the power of Serbian state elites. For example, the state-owned Zelezara Smederevo steel plant was rescued by Chinese SOE HeSteel in 2016 for US$56 million (Hopkins, 2021), and is now Serbia's largest exporter. On the other hand, greenfield investment in frontier sectors such as high-speed rail may serve to catalyse broad-based economic growth. The Budapest–Belgrade railway, which forms one leg of the larger network intended to link China's flagship port of Piraeus in Athens with Central and Eastern Europe, is a key Chinese infrastructure project (Kynge et al, 2017). The Serbian component, valued at US$1.1 billion, is financed by China EXIM Bank loans, while construction is undertaken by a consortium involving China Communications Construction Company, China Railway International, and Russia's RZD International.

Perhaps more significant is the depth of technological integration between Chinese digital companies and Serbian elites. Serbia and China have signed a Digital Silk Road memorandum of understanding, while in 2019 Huawei introduced smart city technologies to Belgrade. This presaged a deeper entanglement in the sphere of digital technologies, and Huawei subsequently partnered with the state-owned Telekom Srbija to roll out a US$177 million high-speed broadband network. Meanwhile, the Chinese tech giant opened an Innovations and Development Centre in Belgrade in September 2020 (Dragojlo, 2020). Geopolitically, Serbia has become one of China's most robust allies in the 17+1, a diplomatic forum connecting Central and Eastern European states with China. While relations with other regional states have soured, Serbia was one of just three 17+1 participants in the Belt and Road videoconference forum in the summer of 2021, along with Greece and Hungary. The resilience of this relationship is in part due to Beijing's refusal to recognize Kosovan independence, a substantial barrier to its potential for achieving UN recognition – a major geopolitical goal of the Serbian government (Milic, 2020). However, in the US-mediated normalization deal signed with Kosovo in September 2020, Serbia signed on to the 'Clean Network' initiative, a global telecommunications initiative established late in the Trump administration that apparently commits Serbia to strictly limit the incorporation of Huawei-developed components and services in its telecommunication networks.

There is considerable Western interest surrounding Chinese influence in Serbia. A Washington, DC-based think tank argued that Serbia is turning into a Chinese 'client state' (Conley et al, 2020). However, in addition to Serbia's participation in the Clean Network initiative, the US International Development Finance Corporation opened its first overseas office in Belgrade (Baptista, 2020). Thus, rather than establishing an economic, political, or ideological commitment to China, Serbian elites have largely welcomed American overtures. In addition, they remain deeply economically and culturally integrated with the EU and anticipate ratification of Serbia's ongoing membership application in the coming decade. Rather than furthering processes of Sino-centric class formation, Serbian elites seem content to exploit the country's unique geographical and political position to balance competing interests.

## New Zealand

While Australia's relationship with China has exploded into acrimony since 2018, neighbouring New Zealand has maintained amicable relations with China. Like its larger neighbour, New Zealand became profoundly economically dependent upon China during the commodities supercycle during the decade to 2014, while it also remained ideologically committed to the US-led liberal world order. China replaced Australia as New Zealand's major export market in 2012, and it absorbed 27.6 per cent of New Zealand's total exports in 2019 (OEC, 2021).

Despite concerns about agricultural export dependency, China has made prominent investments in New Zealand firms and become its second-largest source of FDI (9 per cent in 2019, versus 50 per cent from Australia). For instance, Pengxin New Zealand Farm Group is a major shareholder in Fonterra, New Zealand's largest agribusiness corporation. Pengxin is owned by Milk NZ Holdings which, in turn, has been majority-owned (57 per cent) by Chinese retail giant Alibaba since 2017. Such investments are typical. Similar acquisitions of large or controlling stakes, often worth hundreds of millions of dollars, have been made by Chinese firms in Silver Fern Farms (a meat production firm), Westland Milk Products (a dairy product maker), and Oceania Dairy, among others (Watanabe and Shin, 2020).

China's role as a local landowner has grown considerably, despite attempts since 2018 to restrict foreign landownership. China Forestry Group New Zealand Company – itself owned by China's State Forestry Administration – was, for instance, New Zealand's 19th largest landholder in 2019 (Newton, 2019). Although it owned less land than certain Australian, Malaysian, and Japanese multinationals, this contributed to growing concern surrounding Chinese acquisition of agricultural land and real estate. Growing migration (the number of ethnic Chinese in New Zealand grew by 60 per cent from 2006 to 2018) and capital flight from China have also contributed to the ballooning value of New Zealand's housing stock, which has increased at more than twice the rate of the OECD average since 1990 – though again, Chinese capital is only a minor contributor to this price boom. Such concerns are nevertheless wedded to nebulous anxieties about Chinese interference in politics, business, and academia through its ostensible use of 'United Front'[1] influence activities (Smyth, 2020a, 2020b). In response, New Zealand strengthened legislation in its Overseas Investment (Urgent Measures) Amendment Bill, which is intended principally to scrutinize Chinese investments in New Zealand firms (Walters, 2020).

Agricultural and urban land values are leveraged as collateral by New Zealand's oversized financial sector, a source of potential economic instability (Kelsey, 2015). China helps entrench this growth model by absorbing agricultural exports, boosting the value of farmland, and purchasing the financialized land and corporate assets such exports enable. The expansion of this cycle has deepened New Zealand's dependence upon China, while awareness of this dependence has also strengthened the ideological commitment of elites to the US liberal order. Yet, the business community remains highly sensitive to critiques of China, and New Zealand upgraded its 2008 FTA with China in 2021. Political elites have sought to minimize escalating tensions with China and maintain US-oriented foreign policy. Thus, despite testing its boundaries, New Zealand remains embedded within its historic US-oriented geopolitical alignment. New Zealand has expanded military cooperation with the US during recent years, hosting naval warships operating in the region, alongside two US military bases. Huawei's role in 5G provision was shelved in favour of Samsung in 2019, due in large part to US pressure. Similarly, two important New Zealand technology companies – Rocos (a robotics maker) and Icehouse Ventures (a business incubator) were forced to cut ties with US-blacklisted Chinese technology company iFlytek in early 2021 (Penfold, 2021).

In sum, New Zealand's capitalist class depend on relations with Chinese firms yet remain politically and ideologically committed to the US-led international order. The emergence of a large, influential Sino-centric TCC in New Zealand is unlikely in the near term – and no major pro-China political formation (either formal or informal) has yet surfaced.

## Conclusion

The internationalization of China's economy has a dramatic impact on societies far beyond its boundaries. Nevertheless, the cases presented here demonstrate the severe limitations of Sino-centric TCC formation. In New Zealand, where economic dependence on China is profound, most striking is how little appetite there appears among any major representatives of the capitalist class or the state to challenge the 'China threat' narrative now pervasive in Oceania. In Serbia, despite rhetoric and concerted Chinese efforts to integrate the economy through prestige projects, there is little immediate prospect of reversing the European orientation of state and private sector elites. Finally, in Venezuela, a state class with powerful economic, political, and ideological orientation towards China did cohere during the commodities supercycle but appears to be disintegrating under the pressures of a collapse in oil prices and the downturn in Chinese lending, alongside US sanctions. This is not to say that a Sino-centric TCC could not emerge in the future, but such examples do undermine narratives surrounding Chinese influence on countries' domestic politics and economies – as many countries and their elites appear to hedge their bets rather than align with one hegemon.

This chapter challenges economistic and networked-based accounts by pointing to political, ideological, and structural dimensions of transnational class formation and geopolitical influence. Such an orientation emphasizes historical openness of the present situation. Will the complex supply chains, centred on the US TCC, prove impossible to unwind, and impede the further extension of Chinese overseas influence? Or, could the international political economic system fragment into autarchic blocs and give rise to rival 'empires within the world-economy' (Wallerstein, 1984)? The result would be a patchwork of partially overlapping but discrete regional systems, shaped – absent from outright conflict – by the dynamics of inter-systemic competition. A third possibility is that smaller states avoid substantial alignment with either great power, instead playing them off against each other (see Kuik's analysis of host-country agency, this volume). This would likely lead to the revival of 'national bourgeoisies' that enjoy significant autonomy from both the US and China and greater developmental policy space. But in sum, despite weaknesses in and the increasingly disciplinary nature of US power, a Sino-centric transnational capitalist class remains embryonic for the time being.

## Note

[1] The United Front – 统一战线 (Tǒngyī Zhànxiàn) – is a political strategy of advancing CCP interests abroad through the mobilization of influential individuals and groups sympathetic to party interests.

## References

Arrighi, G. (1994) *The Long Twentieth Century: Money, Power, and the Origins of Our Times*, London: Verso.

Babic, M., Garcia-Bernardo, J., and Heemskerk, E.M. (2020) 'The rise of transnational state capital: state-led foreign investment in the 21st century', *Review of International Political Economy*, 27(3): 433–75.

Babones, S., Åberg, J.H., and Hodzi, O. (2020) 'China's role in global development finance: China challenge or business as usual?', *Global Policy*, 11(3): 326–35.

Baptista, E. (2020) 'Serbia under Washington's gaze as China builds influence in Balkan states', *South China Morning Post*, 4 November.

Bedo, Z., Upchurch, M., and Marinković, D. (2011) 'Wild capitalism, privatisation and employment relations in Serbia', *Employee Relations*, 33(4): 316–33.

Beeson, M. and Wilson, J.D. (2015) 'Coming to terms with China: managing complications in the Sino-Australian economic relationship', *Security Challenges*, 11(2): 21–38.

Blanchette, J. (2020) 'From "China Inc." to "CCP Inc.": a new paradigm for Chinese state capitalism', *China Leadership Monitor*. Available from: www.prcleader.org/blanchette

Bond, P. and Garcia, A. (2020) 'BRICS from above, commoning from below', in S.A.H. Hosseini, J. Goodman, S.C. Motta, and B.K. Gills (eds) *The Routledge Handbook of Transformative Global Studies*, Abingdon: Routledge.

Bräutigam, D. (2020) 'A critical look at Chinese 'debt-trap diplomacy': the rise of a meme', *Area Development and Policy*, 5(1): 1–14.

Breslin, S. (2021) *China Risen? Studying Chinese Global Power*, Bristol: Policy Press.

Carroll, W.K. (2013) *The Making of a Transnational Capitalist Class: Corporate Power in the 21st Century*, London: Zed Books.

Chiasson-LeBel, T. (2016) 'Neo-extractivism in Venezuela and Ecuador: a weapon of class conflict', *The Extractive Industries and Society*, 3(4): 888–901.

Chodor, T. (2015) *Neoliberal Hegemony and the Pink Tide in Latin America*, London: Palgrave Macmillan.

Conley, H., Hillman, J.E., McCalpin, M., and Ruy, D. (2020) Becoming a Chinese client state: The case of Serbia, CSIS. Available from: www.csis.org/analysis/becoming-chinese-client-state-case-serbia

Coronil, F. (1997) *The Magical State: Nature, Money, and Modernity in Venezuela*, Chicago, IL: University of Chicago Press.

Dragojlo, S. (2020) 'China's Huawei opens tech centre, consolidating presence in Serbia'. *Balkan Insight*, 15 September.

Fennema, M. and Heemskerk, E.M. (2018) 'When theory meets methods: the naissance of computer assisted corporate interlock research', *Global Networks*, 18(1): 81–104.

Ferchen, M. (2020) *China-Venezuela Relations in the Twenty-First Century: From Overconfidence to Uncertainty*, US Institute of Peace. Available from: www.jstor.org/stable/pdf/resrep26639.pdf

Gallagher, J. and Robinson, R. (1953) 'The imperialism of free trade', *The Economic History Review*, 6(1): 1–15.

Giacalone, R. and Ruiz, J.B. (2013) 'The Chinese–Venezuelan oil agreements: material and nonmaterial goals', *Latin American Policy*, 4(1): 76–92.

Gowan, P. (1999) *The Global Gamble: Washington's Faustian Bid for World Dominance*, London: Verso.

Harris, J. (2014) 'Transnational capitalism and class formation', *Science and Society*, 78(3): 312–33.

Heilmann, S. (2018) *Red Swan: How Unorthodox Policy-Making Facilitated China's Rise*, Hong Kong: The Chinese University of Hong Kong Press.

Hopkins, V. (2021) 'Serbs fret over environmental costs of Chinese investment', *Financial Times*, 27 April.

Hymer, S. (1972) 'The internationalization of capital', *Journal of Economic Issues*, 6(1): 91–111.

Jepson, N. (2020) *In China's Wake: How the Commodity Boom Transformed Development Strategies in the Global South*, New York: Columbia University Press.

Jessop, B. and Overbeek, H. (2018) *Transnational Capital and Class Fractions: The Amsterdam School Perspective Reconsidered*, Abingdon: Routledge.

Jones, L. (2020) 'Beyond China, Inc: understanding Chinese companies', Amsterdam: TNI Institute. Available from: https://longreads.tni.org/stateofpower/understanding-chinese-companies-beyond-china-inc.

Jones, L. and Zeng, J. (2019) 'Understanding China's "Belt and Road Initiative": beyond "grand strategy" to a state transformation analysis', *Third World Quarterly*, 40(8): 1415–39.

Kelsey, J. (2015) *The Fire Economy: New Zealand's Reckoning*, Wellington: Bridget Williams Books.

Kynge, J., Beesley, A., and Byrne, A. (2017) 'EU sets collision course with China over "Silk Road" rail project', *Financial Times*, 20 February.

Lee, C.K. (2017) *The Specter of Global China: Politics, Labor, and Foreign Investment in Africa*, Chicago, IL: University of Chicago Press.

Li, R. and Cheong, K.C. (2019) '"Going out", going global, and the Belt and Road', in *China's State Enterprises*, London: Springer.

Milic, J. (2020) 'China is not replacing the West in Serbia', *The Diplomat*, 3 April. Available from: https://thediplomat.com/2020/04/china-is-not-replacing-the-west-in-serbia/.

Murphy, A., Haverstock, E., Gara, A., Helman, C., and Vardi, N. (2021) 'Global 2000: how the world's biggest public companies endured the pandemic', *Forbes*, 13 May. Available from: forbes.com/lists/global2000/#2f0741295ac0.

Newton, K. (2019) 'New Zealand's biggest 50 landowners revealed', *Stuff*, 17 October. Available from: www.stuff.co.nz/business/farming/116661441/new-zealands-biggest-50-landowners-revealed.

OEC (2021) 'Country Profile: New Zealand'. Available from: https://oec.world/en/profile/country/nzl.

Panitch, L. and Gindin, S. (2012) *The Making of Global Capitalism*, New York: Verso Books.

Penfold, P.C.L. (2021) 'Icehouse axes partnership with blacklisted Chinese company iFlytek', *Stuff*, 15 April.

Poulantzas, N. (1975) *Classes in Contemporary Capitalism*, London: New Left Books.

Robinson, W.I. (2004) *A Theory of Global Capitalism: Production, Class, and State in a Transnational World*, Baltimore, MD: JHU Press.

Rolf, S. (2020) *China's Uneven and Combined Development*, London: Springer.

Rude, C. (2009) 'The role of financial discipline in imperial strategy', in *American Empire and the Political Economy of Global Finance*, London: Springer.

Saad-Filho, A. and Ayers, A.J. (2008) 'Production, class, and power in the neoliberal transition', in A.J. Ayers (ed) *Gramsci, Political Economy, and International Relations Theory*, London: Springer.

Schwartz, H.M. (2018) *States Versus Markets: Understanding the Global Economy*, New York: Macmillan International Higher Education.

Smyth, J. (2020a) 'New Zealand finance minister warns on foreign influence', *Financial Times*, 23 February.

Smyth, J. (2020b) 'New Zealand university dismisses complaints against China expert', *Financial Times*, 11 December.

Starrs, S.K. (2017) 'The Global Capitalism School tested in Asia: transnational capitalist class vs taking the state seriously', *Journal of Contemporary Asia*, 47(4): 641–58.

Starrs, S.K. (2018) 'Can China unmake the American making of global capitalism', *Socialist Register*, 55(2019): 173–200.

Stuenkel, O. (2017) *Post-Western World: How Emerging Powers Are Remaking Global Order*, Chichester: John Wiley & Sons.

Sum, N.-L. (2019) 'The intertwined geopolitics and geoeconomics of hopes/ fears: China's triple economic bubbles and the "One Belt One Road" imaginary', *Territory, Politics, Governance*, 7(4): 528–52.

UNCTAD (2020) 'World Investment Report 2020'. Available from: https:// unctad.org/webflyer/world-investment-report-2020.

Uvalic, M. (2010) *Serbia's Transition: Towards a Better Future*, London: Springer.

Van Apeldoorn, B., de Graaff, N., and Overbeek, H. (2012) 'The reconfiguration of the global state–capital nexus', *Globalizations*, 9(4): 471–86.

Van Der Pijl, K. and Yurchenko, Y. (2015) 'Neoliberal entrenchment of North Atlantic capital. From corporate self-regulation to state capture', *New Political Economy*, 20(4): 495–517.

Wallerstein, I. (1984) *The Politics of the World-Economy: The States, the Movements and the Civilizations*, Cambridge: Cambridge University Press.

Walters, L. (2020) 'Protecting against China's post-COVID buying spree', Newsroom. Available from: www.newsroom.co.nz/page/ protecting-against-chinas-post-covid-buying-spree.

Watanabe, F. and Shin, M. (2020) 'China's hunger for farms and food alarms Australia and New Zealand', Nikkei Asia, 16 January.

Xu, J. (2016) *Beyond US Hegemony in International Development: The Contest for Influence at the World Bank*, Cambridge: Cambridge University Press.

PART II

# Infrastructural Governance and State Restructuring

# 6

# Contradictory Infrastructures and Military (D)Alliance: Philippine Elite Coalitions and Their Response to US–China Competition

*Alvin Camba, Jerik Cruz, and Guanie Lim*

The Belt and Road Initiative (BRI) and the Philippines' 'pivot to China' have driven shifts in the country's infrastructure governance apparatus. This chapter discusses how elites in the administration of Rodrigo Duterte (2016–) have carved out new state spaces amid intensified US–China geostrategic competition and the expansion of the BRI in the Philippines.[1] While observers argue that Duterte represents a distinct 'pro-China' faction (Parameswaran, 2016; de Castro, 2019; Chao, 2021), we instead suggest that the country's recent shifts in foreign economic policy are the result of competing political, economic, and military coalitions that collectively underpin a convoluted geopolitical approach towards US and China. Beyond Duterte's immediate role, our account draws attention to a broader constellation of actors and conflicts behind the country's management of the BRI and geopolitics in general.

In the Philippines, BRI implementation has intersected with major policy shifts and coalitional cleavages in two key areas. First, we examine how business elites and key economic managers embarked on a state-led infrastructure drive by tapping the BRI and modifying the Philippines' market-liberal infrastructure regime, spawning a contradictory infrastructure build-up. Second, we show that Filipino military elites simultaneously increased their institutional engagements with US and China to expand their assets and capabilities. While soldiers affiliated with the Armed Forces of the Philippines (AFP) – the Army, Navy, and Airforce – have been sceptical of Chinese support due to prominent maritime disputes, actors in the Coast Guard have demonstrated openness to cooperation with the Peoples' Republic of China (PRC) due to the political and economic opportunities presented by the government's 'China pivot' (Tarriela, 2019), possibly supporting a redrawing of political and economic geographies in the South China Sea (SCS) in line with PRC norms.

In the context of US–China competition, Philippine elites are pursuing longstanding political, economic, and spatial objectives through state restructuring. In our first case, we illustrate how Philippine economic managers, particularly those in the Duterte administration, shifted their infrastructure strategy from a market-oriented approach leaning heavily on public–private partnerships (PPPs) to a hybridized usage of PPPs and foreign funding. Yet following an initial period of uncertainty, business elites have been able to take advantage of the situation by expanding the use of unsolicited proposals. In addition, the availability of Chinese aid has supported efforts by Philippine state elites to restructure its domestic public finance and infrastructure apparatus towards this hybrid model, thus tapping multiple multinational development banks and countries to fund infrastructure, while enhancing the internal territorial reach of the state. In our second case, elites within the Philippine military, particularly those in the Coast Guard, leveraged Duterte's (d)alliance with China to expand their jurisdiction and capacity. Though the AFP has remained resolutely against Chinese expansion, the Coast Guard's latent objectives – the expansion of its jurisdiction vis-à-vis the navy – has benefitted from Duterte's direct interaction with Chinese largesse, particularly via China's development assistance. This chapter accounts for the pursuit of interests of economic, technocratic, and military elites, contributing to scholarship on host country agency (Liu and Lim, 2019; Camba, 2020; Camba, 2021b; Lim et al, 2021).

## The political economic foundations of the Philippine infrastructure state

Among Southeast Asian countries, the recalibration of the Philippines as an infrastructure state under Rodrigo Duterte can be seen as a bellwether of broader trends in the region, given longstanding legacies of the country's economic liberalization. In the aftermath of a severe economic recession at the tail-end of the Marcos dictatorship – propelled by the regime's debt-driven surge in infrastructure spending in the 1970s and early 1980s – the country witnessed one of the most dramatic shifts to a market-oriented policy regime in the Asia-Pacific region (Montes and Cruz, 2020). While initiated by the Philippines' inclusion among the guinea pigs of World Bank–International Monetary Fund structural adjustment programmes in the early 1980s (Broad, 1981), the 'neoliberal revolution' in the country garnered substantial internal momentum during the Fidel Ramos administration (1992–98) in the subsequent decade (Bello et al, 2005; Bello, 2009).

Practically all domains of economic policy (trade, investment, taxation, and debt servicing) were upended by neoliberal reforms. Yet from vantage point of the country's economy of space, these changes were particularly prominent in the planning, financing, and operation of infrastructure-intensive public utilities and other forms of public goods. In the 1990s, the country embarked on some of the biggest privatizations in the region (including Manila's water system); it was likewise the first in the world to permit privately developed special economic zones. Even the provision of social infrastructure was impacted, with responsibility for public housing delegated to the private sector by means of yearly mandatory quotas among property developers for socialized housing projects (Bello et al, 2005; McKay, 2006). Thus, even as the country underwent some strategic decoupling from the US with the Senate's non-renewal of the US military bases agreement in 1991, it was also drawn decisively into the international liberal economic order, epitomized by its adoption of the 'Washington Consensus' purveyed by international financial institutions, including the World Bank and the Japan-aligned Asian Development Bank (ADB), both globally and regionally (Harvey, 2005; Dent, 2008). Apart from the US and other liberal powers, Japan played an especially critical role in the economic transition, having served as the Philippines' largest development donor since the 1990s (Setsuho and Yu-Jose, 2003).

While meant to pre-empt the governance failures that proliferated during the dictatorship period, the turn towards private-led urban and infrastructure development has oftentimes not met expectations. For one, 'cronyism' and corruption dynamics have persisted across administrations in both public and private-led projects, with a wide variety of privatizations, PPPs, and joint ventures bearing the stamp of particularistic political interference, with kickbacks from road constructions reportedly reaching heights of 40 per cent of allotted budgets during Gloria Macapagal Arroyo's presidency (2001–10) (De Dios and Ferrer, 2001; Co, 2010). Only during the reformist administration of Benigno Aquino III (2010–16) were credible commitments made to align the Philippine infrastructural governance regime with global regulatory, accountability, and good governance norms (Raquiza, 2014; Montes and Cruz 2020). This was most visibly evidenced by a revamp of regulations surrounding PPPs (Aquino, 2010; Patalinghug, 2017; World Bank, 2018).[2]

Likewise, in the context of decentralization of urban governance and fragmented infrastructure development across agencies, the increased prominence of private developers has translated into the de facto 'privatization' of urban and infrastructural planning. This is the case in terms of the prominence of unsolicited public–private partnership (PPP) ventures and the unprecedented influence of private conglomerates in urban development and administration processes (Shatkin, 2008; Camba, 2017; Mendoza and Cruz, 2020; Mouton and Shatkin, 2020). Not only has this relegated investment for 'public city' zones to residual status, but it also bred a series of market and coordination failures across most areas of urban governance. Generally, market liberalization has intensified politically-mediated competition among domestic business factions for sources of urban rents, evidenced by the rapid and dramatic diversification of practically all major conglomerates into the real estate, construction, utilities, and retail sectors (Mendoza et al, 2018).

Traffic congestion is emblematic of these long-running dysfunctions. Transport planning has been neglected by successive administrations, private bus services are practically unregulated, and investment in transport infrastructure (especially non-car mobility) has been anaemic in the past decades (Cervero, 2000; Hansen, 2017). Not surprisingly, the country's pre-COVID-19 economic boom drove a stark increase of road traffic levels, largely due to increasing private car sales and usage in urban areas. Unfortunately, the adaptation of the infrastructure bureaucracy to the low-fiscal resource environment engendered by decades of debt servicing has worked at cross-purposes with the growing needs and demands for broad-based public infrastructure investment. Indeed, recent efforts to accelerate public capital expenditures, both in the Aquino and Duterte governments, have been frustrated by a recalcitrant dilemma of public sector 'underspending', with a majority of budget allocations to government infrastructure development agencies remaining unspent (Monsod, 2016; Cuenca, 2020).[3]

## Duterte's China pivot and the Build, Build, Build programme

With the start of Duterte's presidency in 2016, there were tectonic shifts in infrastructure governance. Clamouring to plug worsening infrastructure deficits, the Duterte government sought to aggressively accelerate infrastructure spending to attain a 'Golden Age of Infrastructure'. Yet in contradiction of the liberal regime of large-scale infrastructure development, Duterte's Build, Build, Build (BBB) programme instead hiked domestic funding for infrastructure and public works, and increasingly relied on official development assistance (ODA) and foreign loans to finance the most capital-intensive flagship projects (Camba, 2017). Ostensibly, this was due to growing dissatisfaction with the inefficiency of PPP-led procedures, despite the past administration's stronger safeguards (Ito, 2018). But there is no doubt, at least early in Duterte's tenure, that this state-oriented approach was also shaped by the administration's much-ballyhooed 'pivot to China' that Duterte cemented during his visit to China in October 2016. Filipino cabinet officials and their Chinese counterparts signed at least 27 cooperation memorandums (Cardenas, 2017) led to the prevailing perception that BBB privileged Chinese ODA at the expense of traditional bilateral partners, including multilateral agencies (such as the World Bank and the ADB), Western governments, as well as other longstanding donors within the region (Japan, Australia, and South Korea).

However, the picture is more complex: though the allocation of proposed ODA sources shifted, Japan remains behind many of the most expensive foreign-funded projects (including the Metro Manila Subway Project, and the North-South Commuter Railway). In the National Economic and Development Authority's (NEDA) 2019 ODA review, Japanese financing accounted for 39 per cent of active loans and grants (see Figure 6.1), followed by the ADB (26.4 per cent), and the World Bank (19.9 per cent) – all traditional development partners. Chinese ODA accounted for only around 2.7 per cent of the total ODA portfolio, reflecting the low number of China-funded projects (two grants and two loans) active by the end of 2019. In addition, the country had one active loan that year with the multilateral, but China-linked, Asian Infrastructure Investment Bank (for the Metro Manila Flood Management project). Among the administration's planned flagship projects (including projects not yet actively implemented), Japanese-funded projects comprised a 35.6 per cent of all ventures as of 2021, whereas Chinese-assisted ones accounted for 14.6 per cent of the total (Cruz and Juliano, 2021).

Arguably a more important repercussion of the BBB programme than the increase in Chinese aid has been the legitimation of major fiscal and institutional reforms that reoriented the country's infrastructure governance regime. To fund its ambitious infrastructure drive, for instance, the Duterte government legislated the most comprehensive series of tax reforms in the country since the ratification of the 1987 Constitution (Reside, Jr. and Burns, 2016; Manasan, 2018). Similarly, to address institutional constraints in the country's infrastructure bureaucracy, administration technocrats have, since 2017, solicited technical assistance from the ADB through an Infrastructure Preparation and Innovation Facility for improving project preparation, feasibility analysis, approval, and implementation procedures among Department of Public Works and Highways and Department of Transportation projects (Marquez, 2017). In early 2020, this was supplemented by a PhP3.8 billion (US$78.5 million) Philippine–South Korea Project Preparation Facility, which was established to support preparatory activities and analyses, especially for water, flood control, irrigation, bridge, and road projects (de Vera, 2020). For Chinese-funded projects, the government also established more stringent evaluation procedures and guidelines in pre-project and project implementation stages, which were approved in November 2016 (DOF, 2016).

**Figure 6.1:** Philippines' top ten official development assistance partners by share, 2019 (%)

Total ODA (USD million)

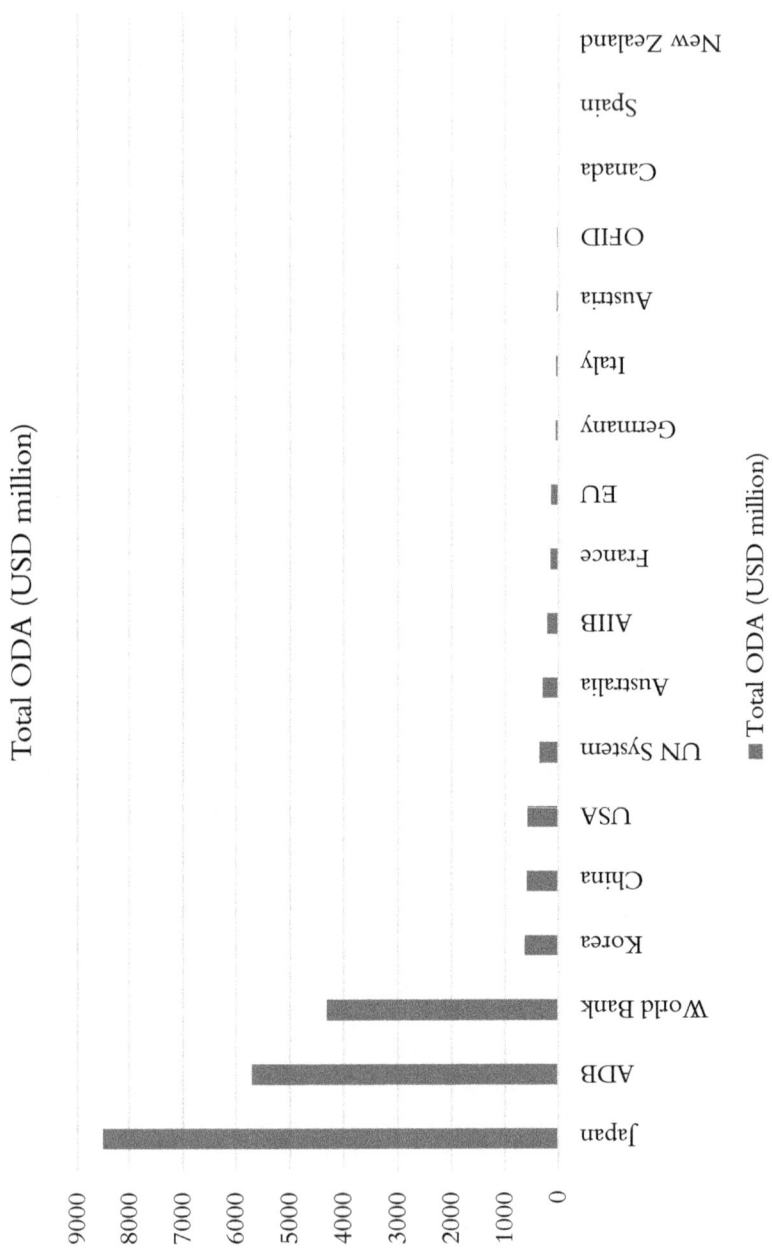

■ Total ODA (USD million)

Source: NEDA ODA Portfolio Review (2019)

However, pushback from conglomerates with substantial capabilities in infrastructure development, as well as gaps in the capacity of government agencies point to the limits of such governance shifts. For one, although Duterte technocrats claimed to curb the scourge of infrastructure 'underspending' during the Aquino presidency, high levels of underutilization of infrastructure budgets persisted well into the present administration (Habito, 2019; Cuenca, 2020; Patalinghug, 2020). In the same vein, there was a major revision of the BBB list of now-100 flagship projects in 2019, in which the government substantially modified its state-heavy infrastructure drive. The administration's economic managers dropped 29 of the original 75 projects due to cost and infeasibility, and added a cavalcade of unsolicited PPP projects (Rivas, 2019). As of August 2020, such PPP ventures championed by the most influential business groups comprise more than one-sixth of BBB projects (19 of 104) (NEDA, 2020).

The revision of the BBB list indicates that 'project selection and evaluation were haphazardly done' (Patalinghug, 2020) at earlier stages. The newfound reliance on unsolicited PPPs essentially overturned governance reforms advanced in the Aquino administration, which aimed to limit the scope of unsolicited infrastructure ventures. Though the longer-term repercussions of private conglomerates' infrastructure clout remain unclear, their presence, along with the staying power of Japanese ODA for large-scale projects, already attests to the limits of Duterte's so-called pivot, as well of leader-centred accounts in explaining resulting dynamics.

## *Military elites and the (geo)politics of infrastructural agendas*

Developments in the Duterte administration also sparked commentary surrounding security and military capabilities. Under Aquino, the Philippine government famously filed, and in 2016 won, a petition with the Permanent Court of Arbitration (PCA) in the Hague disputing China's territorial claims in the SCS and Spratly Islands. Yet, shortly thereafter, the Duterte government shelved the PCA's ruling, continuing bilateral exercises and military cooperation with the US (de Castro, 2016, 2017, 2018). Though much ink has been spilled on senior officials' efforts to 'rebalance' foreign policy with the US and China, a major part of the incoherence in the administration's policy has stemmed from divisions within the ranks of the Philippine military itself (Parameswaran, 2016; Jones and Jenne, 2021).

The AFP comprises the Army, Navy, and Air Force. During the Cold War, the AFP maintained internal stability by fighting the Hukbalahap, a paramilitary that fought against the Japanese during World War II. Working with the American Central Intelligence Agency, the AFP effectively eliminated the Hukbalahap by the late 1950s (Greenberg, 1987). The Communist Party of the Philippines (CPP), created in the 1968, emerged as the main threat against the Philippine state and it aimed to overhaul the entire state structure. At the start of his tenure in 1965, which he extended by declaring of Martial Law in 1972, Ferdinand Marcos (1965–86) established a military-supported dictatorship and declared war against the CPP. Marcos engaged in a protracted anti-communist struggle that culminated in widespread human rights violations against the civilian populace (Overholt, 1986). The Philippine Constabulary – part of the AFP during Marcos' time – was guilty of most of these violations. Nonetheless, after Marcos' rule, elites undertook key reforms, such as the professionalization of the AFP and budgetary reforms.

Since the return to democracy in 1986, the CPP has been relegated to a marginal, yet persistent, threat to the Philippine state. Nonetheless, the New People's Army (NPA), the Chinese Communist Party's (CCP) military arm, became a threat to local elites who rely on local private armies and police to safeguard their regional suzerainties (Kowalewski, 1992; Camba, 2015). Though barely a revolutionary movement, the NPA has been a persistent challenge in some provincial areas of the Philippines, partly due to the underfunding of military and police forces, uneven national development plans, and rampant abuses by subnational elites that perpetuated local-level CPP support (Holden, 2013, 2014).

Crucially, the post-Marcos institutional environment curtailed the budgetary allotment of the AFP (Ross, 1990; Chambers, 2012). Indeed, Philippine political elites limited financial resources to the AFP and the Department of National Defence (DND), effectively tilting the balance of power between the political and military elites in favour of the former. Requesting annual funding for the AFP and the DND is a straightforward process that all departments follow (see Department of Budget and Management, 2021). Both institutions jointly determine what they need for operations, the payment of personnel, and new projects. Afterwards, this budget is proposed to Congress, and a bargaining ensues between members of Congress and the AFP to determine the final request. The Philippine president also intervenes, particularly by proposing and accepting the final annual budget.

There are three conflicts of interest in this budgetary process. First, members of Congress draw from the same pile of money, generating a clear conflict for the budget. A higher budget for the AFP decreases the amount of overall budget, particularly for congressional pork barrel or development projects, which were available until Aquino III's term. Pork or arbitrary congressional funding is often used for the congressional patronage activities that are crucial to re-election or expansion of influence. Second, AFP funding tends to disproportionately redound towards conflict-affected areas. Congressional representatives of conflict areas (particularly with the NPA and the terrorist groups in Mindanao) are incentivized to fund the

AFP or win members of the DND. However, those outside of conflict areas typically have less motivation to do so. Finally, there is a general wariness toward strengthening the military. As the AFP was a key cog of the Marcos regime, there is some consensus among Philippine elites about restricting the AFP from partisan politics to pre-empt the rise of a would-be dictator. An unintended effect of this fear is that it keeps the AFP disempowered as an institution. Hence, to garner the support of the AFP, Philippine political elites usually resort to giving positions to generals to co-opt the military, rather than strengthening the institution in a programmatic fashion.

This political economy generates divisions within the AFP, where political objectives lead each branch to pursue alliances with various political elites, outside forces, and resource partners. During the Cold War, for instance, US military bases provided AFP personnel training and spillover military modernization. When the American military base agreement was not renewed in 1991, the Visiting Forces Agreement (VFA) became the substitute for US military transfers to the Philippines. However, the VFA has been expanded and constrained by various administrations. Aquino III pursued the expanded defence cooperation agreement with the US (Rappler, 2016), while in recent years, Duterte threatened to cancel it on multiple occasions (Reuters, 2021). As such, AFP branches have pursued diverse alliances, with the interest and governance of infrastructure circumscribed by the agendas of various AFP agencies. The Navy is primarily concerned with protecting the maritime boundaries from external actors. Similarly, the Air Force has both internal and external disputes, defending against potential enemy naval assets and secessionist movements. The Army is the largest of the three AFP branches and primarily concerned with internal conflict (Camba, 2021a). With respect to the People's Republic of China, there is a consensus among all branches that the country is a clear and present threat.

## China and the Philippine Coast Guard

The Philippine Coast Guard best exemplify state restructuring around infrastructure within the broader umbrella of US–China competition, rescaling the political and economic geographies of the SCS. Hence, the Coast Guard case, while not entailing the usual hard infrastructure, can be seen as entailing (a) the 'soft infrastructure' of PRC laws/dispute mechanisms/tactics/normative frameworks, and offering (b) a critical case of the broader spatial transformation of the region, complementing connectivity infrastructure investments elsewhere.

During the Cold War, the Coast Guard was under the Navy and guarded ports, overseeing customs inspection and safety regulations (Burdeos, 2008). While the Navy was tasked to protect maritime borders and instigate offensives, the Coast Guard dealt with illegal poachers, fishers, and trespassers.

However, the two branches functionally overlap. As the three services divided the congressional budget, the Navy essentially decided on the Coast Guard budgetary allotments, limiting the Coast Guard's finances. As such, various elites, from Ferdinand Marcos to Fidel Ramos advocated for the Coast Guard to be separated from the Navy. The Navy, however, fought doggedly against this until Ramos legislated their separation (Tarriela, 2020), placing the Coast Guard under the Department of Transportation. While free from the naval control, they continued to receive insufficient budget allocation. Between the DND and the Department of Transportation, the former is among the five most well-endowed departments while transportation lingers in the top ten (Department of Budget and Management, 2021).

The Navy–Coast Guard rivalry escalated during Arroyo's presidency. Arroyo prioritized the Coast Guard as part of her development plan, turning the service into a combination of transportation, traffic enforcers, and police in the maritime areas (Saavedra, 2019). During the Aquino III era, the Navy was central in the stand against China, but the Coast Guard still mobilized for modernization and received more resources from the US, Japan, and Australia (Official Gazette, 2013). However, high-profile events, particularly the Scarborough shoal episode, placed the Coast Guard at the centre of policing and empowered it vis-à-vis the Navy (Tarriela, 2019). During the incident, Aquino III moved the Navy to support a Philippine research vessel that had an encounter with the Chinese coast guard (Rosen, 2014). The People's Liberation Army responded by dispatching naval vessels to the region, threatening regional peace. The US brokered a deal whereby the Philippines and China agreed to a mutual retreat, but the People's Liberation Army remained and militarized the area.

The Philippine Coast Guard has been pivotal to Duterte's foreign policy strategy. Given the Coast Guard's more neutral, policing function, in contrast to the Navy's role in security and defence, Duterte deployed it to deal with Chinese fishers and ships in the contested maritime boundaries (Tarriela, 2019). Deploying the Navy would have risked provoking China, jeopardizing Duterte's political strategy of appeasing the Chinese (Godbold, 2019). Additionally, Duterte channelled funding from the US, Australia, and Japan towards the Coast Guard for more ships and training (Tarriela, 2018).

The Coast Guard has used its newfound political and strategic place in the Duterte administration to expand its jurisdictional reach. It acquired ships from South Korea, sent hundreds of officers to train in the US, and works with Japan to upgrade ships (Philippine Coast Guard, 2019; US Indo-Pacific Command, 2021). At the same time, it has taken a more direct role dealing with Chinese vessels and established a working relationship with members of the Chinese Coast Guard. However, the AFP views the Coast Guard as 'co-opted' by the Chinese government, particularly as it began to use Chinese maritime laws to interpret fishing disputes between Chinese and Filipino fishers. Naval officers accused members of the Coast Guard of transferring the Chinese way of interpreting maritime disputes to the Philippines through trainings with China. An intelligence officer noted that Coast Guard officers regularly join sponsored trips to China. While relationships between the Coast Guard and their Chinese counterparts to some account for the co-optation of the Coast Guard, according to a commander of the Coast Guard, working with the Chinese Navy comprises a single aspect of their external partnerships. The Coast Guard's official training list shows that 95 per cent of their officers were sent to the US. The Navy has similarly pursued cooperation with the US and Japan, exploiting US–China competition. The Navy relies on the US military for the acquisition of new ships and vehicles, such as the transfer of two corvette class ships, unmanned aerial system, and a maritime research vessel (Gotinga, 2020). As the Coast Guard expands its jurisdictional and political capacity, the Navy, through the VFA and other agreements, uses its longstanding links with the US government to pursue infrastructure modernization (including pre-existing and next generation assets) outside the reliance on the Philippine budget (ABS-CBN News, 2020).

## Conclusion

This chapter analysed the Philippines' 'pivot to China' strategy, complicating the oft-mentioned Sino-US (and to a smaller extent, Japanese) competitive dynamics and their role in reconfiguring regional development. It took stock of the heterogeneous set of stakeholders and unpacked how their actions have transformed state (physical and non-physical) spaces and the resulting infrastructure governance apparatus. Although Duterte is clearly influential, a host of other interest groups and socio-political forces shape Filipino overtures to US and China. In contrast to political-realist assumptions that rational actors use US–China competition in pursuit of state goals, our chapter illustrated that host states restructure in the context of US–China competition in accordance with the parochial interests of elite coalitions, illustrating the heterogeneity of powers and interests in the host country.

More prosaically, we have shown through two cases how the infrastructure state manifests in the Philippines, amid the increasing levels of US–China rivalry. In the first case, increasing Chinese largess has allowed Filipino elites to tweak a hitherto market-oriented approaches to infrastructure development (relying primarily on PPPs) to one that is more multimodal (combining PPPs and foreign funding). In the second case, key AFP actors draw on US–China competition to exploit opportunities in activities ranging from training, equipment acquisition, to resolving fishing disputes. Elites, especially those in the Coast Guard, have cultivated clout following Duterte's warming ties with China, notwithstanding the AFP's general stance against Chinese territorial expansion in the South China Sea and historical ties with the US. This, in effect, feeds directly into broader inter-state efforts to redraw the political and economic geographies of the South China region. Each case can be understood as the outcome of negotiations among political, economic, and military coalitions strategically engaging with ongoing processes of spatial transformation.

## Notes

[1]  On the notion of 'new state spaces' see Brenner (2004).

[2]  Finding investors has not been a major problem, though there were concerns with the speed of the Aquino-era PPP due to the focus on anti-corruption that resulted in a more cumbersome procedure for infrastructure projects. For non-PPP public expenditure, underspending challenges were rooted in how sustained austerity led to institutional measures to conserve fiscal resources, which worked at odds with accelerated spending once those resources became available.

[3]  Despite China's low-level contribution, direct comparison with Japan should come with two caveats. First, some Chinese-funded projects remain outside the ODA model, a point Deborah Bräutigam makes in her analysis of Chinese financing. In the Philippines, these include successfully-built engineering procurement and construction projects, such as two drug rehabilitation facilities in Mindanao, and a management consultancy contract for the South Long-Haul Project. Second, Japan has a long history of funding ODA projects in the country, making the Chinese target projects that may be less 'bankable' riskier than Japanese ones. Some projects are prone to more risk than others, and it is necessary to place China's performance in a proper context.

## References

ABS-CBN News (2020) 'Amid fears of VFA scrapping, PH Navy to sustain training with US', 29 January. Available from: https://news.abs-cbn.com/news/01/29/20/amid-fears-of-vfa-scrapping-ph-navy-to-sustain-train ing-with-us.

Aquino, B. (2010) 'Speech of President Aquino during the PPP Conference, Infrastructure Philippines'. Available from: https://ppp.gov.ph/speeches/speech-of-president-aquino-during-the-ppp-conference-infrastructure-philippines-2010/.

Bello, W. (2009) 'Neoliberalism as hegemonic ideology in the Philippines: rise, apogee, and crisis', *Philippine Sociological Review*, 57: 9–19.

Bello, W., de Guzman, M., Malig, M.L., and Docena, H. (2005) *The Anti-Development State: The Political Economy of Permanent Crisis in the Philippines*, New York: Zed Books.

Brenner, N. (2004) *New State Spaces: Urban Governance and the Rescaling of Statehood*, Oxford: Oxford University Press.

Broad, R. (1981) 'New directions at World Bank: Philippines as guinea pig', *Economic and Political Weekly*, 16(47): 1919–22.

Burdeos, R. (2008) *Filipinos in the US Navy and Coast Guard during the Vietnam War*, Bloomington, IN: Author House.

Camba, A. (2015) 'From colonialism to neoliberalism: critical reflections on Philippine mining in the "long twentieth century"', *The Extractive Industries and Society*, 2(2): 287–301.

Camba, A. (2017) 'Inter-state relations and state capacity: the rise and fall of Chinese foreign direct investment in the Philippines', *Palgrave Communications*, 3(1): 41.

Camba, A. (2020) 'The Sino-centric capital export regime: state-backed and flexible capital in the Philippines', *Development and Change*, 51(4): 970–97.

Camba, A. (2021a) 'How Duterte strong-armed Chinese dam-builders but weakened Philippine institutions', Carnegie Endowment for International Peace. Available from: www.carnegieendowment.org/2021/06/15/how-duterte-strong-armed-chinese-dam-builders-but-weakened-philippine-institutions-pub-84764.

Camba, A. (2021b) 'How Chinese firms approach investment risk: strong leaders, cancellation, and pushback', *Review of International Political Economy*. doi: 10.1080/09692290.2021.1947345.

Cardenas, K. (2017) 'Duterte's China deals, dissected', Philippine Center for Investigative Journalism, 8 May. Available from: https://pcij.org/article/1705/dutertes-china-deals-dissected.

Cervero, R. (2000) *Informal Transport in the Developing World*, Nairobi: United Nations Centre for Human Settlements.

Chambers, P. (2012) 'A precarious path: the evolution of civil–military relations in the Philippines', *Asian Security*, 8(2): 138–63.

Chao, W.-C. (2021) 'The Philippines' perception and strategy for China's Belt and Road Initiative expansion: hedging with balancing', *The Chinese Economy*, 54(1): 48–55.

Co, E. (2010) 'The long and winding road to infrastructure development and reform', *Philippine Journal of Public Administration*, 54(2): 153–66.

Cruz, J. and Juliano, H. (2021) *Assessing Duterte's China Projects: Governance, White Elephants, and COVID-19 in the Build, Build, Build Program*, Manila: Asia Pacific Pathways to Progress Foundation.

Cuenca, J. (2020) 'Review of the "Build, Build, Build" program: implications on the Philippine Development Plan 2017–2022', PIDS Discussion Paper Series, 54.

de Castro, R.C. (2016) 'The Duterte administration's foreign policy: unravelling the Aquino administration's balancing agenda on an emergent China', *Journal of Current Southeast Asian Affairs*, 35(3): 139–59.

de Castro, R.C. (2017) 'The Duterte administration's appeasement policy on China and the crisis in the Philippine–US alliance', *Philippine Political Science Journal*, 38(3): 159–81.

de Castro, R.C. (2018) 'Explaining the Duterte administration's appeasement policy on China: the power of fear', *Asian Affairs: An American Review*, 45(3–4): 165–91.

de Castro, R.C. (2019) 'China's Belt and Road Initiative (BRI) and the Duterte administration's appeasement policy: examining the connection between the two national strategies', *East Asia*, 36(3): 205–27.

De Dios, E. and Ferrer, R. (2001) 'Corruption in the Philippines: framework and context', *Public Policy*, 5(1): 1–42.

de Vera, B. (2020) 'PH gets $50-M Korean loan for Infra Project preparation', Inquirer.net, 29 January. Available from: https://business.inquirer.net/289 208/ph-gets-50-m-korean-loan-for-infra-project-preparation.

Dent, C.M. (2008) 'The Asian Development Bank and developmental regionalism in East Asia', *Third World Quarterly*, 29(4): 767–86.

Department of Budget and Management (2021) *People's Proposed Budget: Reset, Rebound, and Recover: Investing for Resiliency and Sustainability*, Manila: Philippines Department of Budget and Management. Available from: www.dbm.gov.ph/wp-content/uploads/2012/03/PGB-B2.pdf.

Department of Finance (DOF) (2016) 'NEDA Board approves availment guidelines for Chinese support for proposed PHL investments', 15 November. Available from: www.dof.gov.ph/neda-board-approves-availment-guidelines-for-chinese-support-for-proposed-phl-investments/.

Godbold, T. (2019) 'Rise of the white hulls: Indo-Pacific coast guards become diplomatic tools', *Stars and Stripes*, 25 April. Available from: www.stripes.com/news/pacific/rise-of-the-white-hulls-indo-pacific-coast-guards-become-diplomatic-tools-1.578358.

Gotinga, J. (2020) 'New navy chief wants "modern systems, mindsets" as new assets replace ailing ships', Rappler. Available from: www.rappler.com/nation/250901-navy-chief-wants-modern-systems-mindsets-february-2020/.

Greenberg, L. (1987) *The Hukbalahap Insurrection: A Case Study of a Successful Anti-Insurgency Operation in the Philippines: 1946–1955*, Washington, DC: Analysis Branch, US Army Center of Military History.

Habito, C. (2019) 'Disturbing slowdown', Inquirer.net, 16 August. Available from: https://opinion.inquirer.net/123329/disturbing-slowdown.

Hansen, A. (2017) 'Hanoi on wheels: emerging automobility in the land of the motorbike', *Mobilities*, 12(5): 628–45.

Harvey, D. (2005) *The New Imperialism*, Oxford: Oxford University Press.

Holden, W. (2013) 'The never-ending war in the wounded land: the New People's Army on Samar', *Journal of Geography and Geology*, 5(4): 29–49.

Holden, W. (2014) 'The New People's Army and neoliberal mining in the Philippines: a struggle against primitive accumulation', *Capitalism Nature Socialism*, 25(3): 61–83.

Ito, S. (2018) 'PPP vs ODA revisited: key issues for PPP infrastructure development in the Philippines', *Philippine Review of Economics*, 55(1–2): 56–86.

Jones, D.M. and Jenne, N. (2021) 'Hedging and grand strategy in Southeast Asian foreign policy', *International Relations of the Asia-Pacific*. Available from: https://doi.org/10.1093/irap/lcab003.

Kowalewski, D. (1992) 'Counterinsurgent paramilitarism: a Philippine case study', *Journal of Peace Research*, 29(1):71–84.

Lim, G., Li, C. and Syailendra, E.A. (2021) 'Why is it so hard to push Chinese railway projects in Southeast Asia? The role of domestic politics in Malaysia and Indonesia', *World Development*, 138: 105272.

Liu, H. and Lim, G. (2019). 'The political economy of a rising China in Southeast Asia: Malaysia's response to the Belt and Road Initiative', *Journal of Contemporary China*, 28(116):216–231.

Manasan, R. (2018) 'Assessment of Republic Act 10963: The 2017 Tax Reform for Acceleration and Inclusion', PIDS Discussion Paper Series, 27.

Marquez, R. (2017) 'Making "Build, Build, Build" work in the Philippines', Asian Development Bank, 30 October. Available from: www.adb.org/news/features/making-build-build-build-work-philippines.

McKay, S. (2006) 'The scalar strategies of capital, state and labor in evolving Philippine economic zones', *Journal of Comparative Asian Development*, 5(2): 203–27.

Mendoza, R. and Cruz, J.P. (2020) 'Governing the "golden age of infrastructure": assessing transparency innovations in Philippine infrastructure development', *Asian Politics and Policy*, 12(2): 175–204.

Mendoza, R., Arbo, M.D.G., and Cruz, J.P. (2018) 'In search of Philippine chaebols', ASOG Working Paper, 19–009.

Monsod, T. (2016) 'Government "underspending" in perspective: incompetence, inertia, or indigestion?', UP School of Economics Discussion Papers, 12.

Montes, M. and Cruz, J. (2020) 'The political economy of foreign investment and industrial development: the Philippines, Malaysia and Thailand in comparative perspective', *Journal of the Asia Pacific Economy*, 25(1): 16–39.

Mouton, M. and Shatkin, G. (2020) 'Strategizing the for-profit city: the state, developers, and urban production in mega Manila', *Environment and Planning A: Economy and Space*, 52(2): 403–22.

NEDA (2019) 'Official development assistance portfolio review 2019'. Available from: www.neda.gov.ph/oda-portfolio-review-2019/.

NEDA (2020) 'Revised list of infrastructure flagship projects'. Available from: https://www.neda.gov.ph/wp-content/uploads/2020/09/Revised-List-of-IFPs-as-of-19-August-2020.pdf.

Official Gazette (2013) 'Message of President Aquino to the Philippine Coast Guard on the occasion of their 112th Anniversary'. Available from: www.officialgazette.gov.ph/2013/10/10/message-of-president-aquino-to-the-philippine-coast-guard-on-the-occasion-of-their-112th-anniversary-october-10-2013/.

Overholt, W. (1986) 'The rise and fall of Ferdinand Marcos', *Far Eastern Survey*, 26(11): 1137–63.

Parameswaran, P. (2016) 'The limits of Duterte's US–China rebalance', *The Diplomat*, 24 October. Available from: https://thediplomat.com/2016/10/the-limits-of-dutertes-us-china-rebalance/.

Patalinghug, E. (2017) 'Assessment of planning and programming for capital projects at the national and agency levels', PIDS Discussion Paper Series, 37: 371–80.

Patalinghug, E. (2020) *The Build, Build, Build Program: Will It Live Up to the Expectations?* Manila: Stratbase ADRI Publications.

PCA (2016) 'PCA press release: the South China Sea Arbitration (The Republic of the Philippines v. The People's Republic of China)'. Available from: https://pca-cpa.org/en/news/pca-press-release-the-south-china-sea-arbitration-the-republic-of-the-philippines-v-the-peoples-republic-of-china/.

Philippine Coast Guard (2019) 'Coast Guard presents newly acquired rescue assets, braces up for rainy season', 31 May. Available from: https://coastguard.gov.ph/index.php/news/news-2020/145-news-2019/3343-coast-guard-acquires-additional-p31-million-worth-rescue-assets-braces-up-for-rainy-season.

*Rappler* (2016) 'Fact check: was EDCA not signed by Aquino?'. Available from: www.rappler.com/newsbreak/iq/148124-edca-aquino-signature/.

Raquiza, A. (2014) 'Changing configuration of Philippine capitalism', *Philippine Political Science Journal*, 35(2): 225–50.

Reside, Jr., R. and Burns, L. (2016) 'Comprehensive tax reform in the Philippines: principles, history and recommendations', UP School of Economics Discussion Papers, 10.

Reuters (2021) 'Philippines' Duterte tells US 'You have to pay' if it wants to keep troop deal', 12 February. Available from: www.reuters.com/article/us-philippines-usa-defence/philippines-duterte-tells-u-s-you-have-to-pay-if-it-wants-to-keep-troop-deal-idUSKBN2AC1K2.

Rivas, R. (2019) 'Expensive, too much work: Duterte's team rethinks Infra Projects', Rappler, 23 October. Available from: www.rappler.com/business/243206-duterte-team-rethink-infrastructure-projects/.

Rosen, M. (2014) *Philippine Claims in the South China Sea: A Legal Analysis*, Arlington, VA: CNA Corporation.

Ross, J. (1990) 'Militia abuses in the Philippines', *Third World Legal Studies*, 9: 171–203.

Saavedra, J.R. (2019) 'GMA thanks Cebuanos for backing RoRo during "sentimental tour"', Philippine News Agency, 17 June. Available from: www.pna.gov.ph/articles/1072553.

Setsuho, I. and Yu-Jose, L. (2003) *Philippines–Japan Relations*, Quezon City: Ateneo De Manila University Press.

Shatkin, G. (2008) 'Decentralization and the struggle for participation in local politics and planning: lessons from Naga City, the Philippines', in V. Beard, F. Miraftab, and C. Silver (eds) *Planning and Decentralization: Contested Spaces for Public Action in the Global South*, New York: Routledge.

Tarriela, J.T. (2018) 'Is the Philippine Coast Guard being militarized?', *The Diplomat*, 14 April. Available from: https://thediplomat.com/2018/04/is-the-philippine-coast-guard-being-militarized/.

Tarriela, J.T. (2019) 'Duterte's Coast Guard diplomacy', *The Diplomat*, 23 December. Available from: https://thediplomat.com/2019/12/dutertes-coast-guard-diplomacy/.

Tarriela, J.T. (2020) 'Why the Philippines needs to revise its National Defense Act', *The Diplomat*, 24 September. Available from: https://thediplomat.com/2020/09/why-the-philippines-needs-to-revise-its-national-defense-act/.

US Indo-Pacific Command (2021) 'US Military and Philippine Coast Guard conduct tactical combat casualty care training in Palawan', 2 February. Available from: www.pacom.mil/Media/News/News-Article-View/Article/2490584/us-military-and-philippine-coast-guard-conduct-tactical-combat-casualty-care-tr/.

World Bank (2018) *Philippines Economic Update: Staying in the Course Amid Global Uncertainty*, Washington, DC: World Bank.

# 7

# Infrastructure-Led Development with Post-Neoliberal Characteristics: *Buen Vivir*, China, and Extractivism in Ecuador

*Nicholas Jepson*

After reintroducing democratic elections in 1979, Ecuador experienced a pattern of debt crises, austerity, and structural adjustment that was punctuated by waves of protest and occasional retrenchment (Hey and Klak, 1999). During this time the country remained firmly within the US orbit – even more so after a banking collapse in the late 1990s precipitated full dollarization. A succession of presidents campaigned on reversing neoliberalization, then shifted towards orthodoxy once in office, often prompting popular pressure and their eventual ouster. Given these precedents, when Rafael Correa won the presidential elections in 2006 on a left-populist platform, observers predicted another neoliberal turn and/or political instability once Correa assumed office (Kraul, 2006). Instead, Correa's tenure lasted ten years and three terms. His administration was one of Ecuador's most consequential, as it moved away from Washington and laid out an ambitious path of state-led developmentalism, infrastructural upgrading, and redistribution. Correa benefitted greatly from rising oil revenues, driven by the China-driven commodity boom of the early 21st century (Jepson, 2020). However, Correa's infrastructure drive also required external financing. After the government selectively defaulted on its sovereign debt in 2008 – effectively severing relations with international financial institutions (IFIs) and capital markets – China became the only realistic alternative.

Ecuador's story during this period demonstrates how Chinese finance enabled greater national policy sovereignty, serving as an alternative to the (post-)Washington Consensus. It is also illustrative of the pitfalls and limitations of Chinese funding. An examination of Correa's tenure indicates that developmental agendas – whether heterodox or orthodox, China- or IFI-backed – are converging towards a model of infrastructure-led growth (Schindler and Kanai, 2019). Yet there remains significant variation within this convergence. States may choose Chinese finance as the primary external partner in these endeavours, or else turn to the US-centred complex of investors and IFIs that comprise an emerging Wall Street Consensus (WSC; Gabor, 2021). This WSC encourages states to engage in various forms of 'de-risking' in order to attract global institutional investors as a primary source of infrastructure finance, rather than traditional public debt financing. The choice between WSC and Chinese finance is not a simple binary and many countries in the Global South have borrowed heavily from Chinese policy banks in recent years while maintaining programmes with IFIs. However, Ecuador under Correa represents a case where the WSC–China choice *was* effectively treated as a binary, and as such it provides an opportunity to explore the political and economic consequences of reliance on Chinese credit as an alternative to the WSC. Three different financing modalities have implications for the types of projects and national development policies that are feasible with each type of partner. More than this, the WSC (following traditional approaches of multilateral development banks) applies disciplinary mechanisms that would today rule out a broader heterodox economic and social agenda of the type pursued by Correa.

As this chapter shows, Correa's development agenda was flawed and contradictory in both conception and execution. It nevertheless brought relative stability and economic growth, while its reduction of poverty and inequality are unmatched in recent Ecuadorian history. To a significant extent, Correa's government was successful in its efforts to reverse three decades of neoliberal economic reform. In 2009 the government declared US$3.2 billion of external debt illegitimate, defaulting on these bonds and then buying them back at a fraction of the price. This constituted a rupture with global finance that cut Ecuador off from international capital markets for several years. Correa also ended IMF surveillance of the country's economy, expelled the World Bank's representative in Quito, and moved to restrict the power of the domestic financial sector. The size and reach of the state grew, with public spending as a percentage of GDP rising from 24 per cent in 2007 to 44 per cent in 2014 (IMF, 2020a). Social spending doubled, conditional cash transfer payments roughly tripled, and the minimum wage increased well above inflation.

While Ecuador's economy remained dollarized and continued to depend on the US as a primary export market, Correa framed his 'Citizens' Revolution' as a counterhegemonic project, a stance that hardened over time. An American base at Manta on the Pacific Coast was closed in 2009 and in 2014 US military staff working on counter-narcotics were expelled from Ecuador (Alvaro et al, 2014). Asylum was offered to whistleblower Edward Snowden and granted to Wikileaks founder Julian Assange (in Ecuador's London embassy). Correa was also active in various regional forums designed as alternatives to associations accused of being too US-centric. For example, Ecuador joined Venezuela, Bolivia, and Cuba in ALBA (Bolivarian Alliance for the Peoples of Our America) and agreed to host the new organization Unasur (Union of South American Nations), meant to foster regional integration.

Fundamentally, as I have argued in previous work, early 21st-century post-neoliberal turns such as that seen in Ecuador were facilitated by a China-driven commodity boom, in conjunction with particular domestic state-society dynamics (Jepson, 2020; compare with Kaplan, 2016). As an oil exporter Ecuador benefitted from crude prices which more than quadrupled in real terms between 2002 and 2008, maintaining roughly those levels until 2014 (except for a dip in 2009; World Bank, 2020). Along with an aggressive renegotiation of oil contracts and a drive to improve tax collection, the oil boom increased state revenues that could be drawn upon to pay for much of the new social spending and state building.

But Correa's aims ran beyond redistribution and state reconstruction, to a developmental agenda of structural transformation in service of long-term economic autonomy. This involved a major public investment programme focused on infrastructure, with 8,500 km in new or improved roads, eight large dam projects as well as new schools, hospitals, and housing, especially in the Amazon. The most ambitious was Yachay, a planned city, science park, industrial zone, and research university which, it was hoped, would engender new comparative advantages for Ecuador, in part by leveraging the country's remarkable biodiversity as a resource for new technologies. Plans for Yachay incorporated state planning, industrial upgrading, respect for indigenous knowledge, and harmony with nature – emblematic of the often laudable, often confused ideology of *buen vivir*.

## *Buen vivir* and its discontents

Correa's 2006 victory was part of the 'pink tide' that swept across Latin America during the first decade of the 21st century and resulted the election of left and centre-left parties. Some commentators collapse pink tide political tendencies into a moderate/radical binary or conflate radicalism and populism in analytically clumsy ways. Nevertheless, there is resemblance between the Alianza Pais government in Ecuador and its Andean counterparts in Venezuela and Bolivia in terms of agendas and populist political style. Correa, a Keynesian economist who in 2005 briefly served as finance minister in a caretaker government, ran in 2006 on a platform of overhauling the country's institutional and economic structures. Once in office, near constant popular mobilization and frequent referenda were used to bypass existing institutional barriers, and with them, the power of the country's established political and economic elite (derided by Correa as *pelucones*, a reference to the powdered wigs worn by the early 19th-century aristocracy).

The state was 're-founded' via a new constitution and overhaul of judicial and political institutions. This 2008 Constitution attracted international attention for being the first in the world to officially recognize the rights of nature and call for social and ecological harmony, summed up as a philosophy of *buen vivir* (loosely translated as *living well*). Though somewhat slippery in both definition and application, *buen vivir* seeks to combine indigenous traditions of reciprocity and community (the Kichwa phrase *Sumak Kawsay* is often seen as synonymous)[1] with varied strands of leftist thought – including developmentalist, Marxist, and post-developmentalist tendencies. Tensions within this unwieldy whole can be seen in many of the contradictions of the Correa era, reflecting his government's rather awkward and incomplete embrace of popular movements, not least regarding indigenous struggles (Gonzalez-Vicente, 2017).

The creation of the powerful National Secretariat of Planning and Development (SENPLADES) facilitated a conscious return to planning. SENPLADES was tasked with elaborating the concept of *buen vivir* via a series of three- and four-year national development plans, with all but the first of these bearing the name *Plan Nacional de Buen Vivir*. The state's role grew significantly, with 24 new or reorganized state-owned enterprises and 53 new public institutions (Sanchez and Polga-Hecimovich, 2019). Social spending as a percentage of GDP doubled between 2006 and 2016, with increases in education, health, and housing (Weisbrot et al, 2017). An existing cash transfer programme, the Bono de Desarrollo Humano that paid US$11.50–15.00 per month to poor families, increased to $50 per month by 2013. In 2010, it covered 44.3 per cent of the population (Rinehart and McGuire, 2017). During Correa's tenure poverty rates declined by 38 per cent and extreme poverty by 48 per cent (Weisbrot et al, 2017). Average GDP growth was solid but unspectacular at 3.5 per cent per year, weighed down by the 2008 global financial crisis, a 2014 commodity downturn, and the impact of a 2016 earthquake on the Pacific coast. Thus, in terms of headline numbers, then, results were fairly impressive, even if, as discussed later, some of the government's grander designs proved harder to execute.

Alongside macroeconomic policy shifts, the government forged ahead with state-led investment for transport infrastructure and hydroelectric projects. Public investment spending rose from 4 per cent of GDP in 2006 to a peak of 14.8 per cent by 2013 (Weisbrot et al, 2017). A portion of this was financed by growing oil and tax revenues. But without the possibility of borrowing from international capital markets, meeting the full costs of Correa's agenda involved obtaining credit from a variety of sources, including

Brazil, Venezuela, the Andean Development Corporation, and the Inter-American Development Bank. By far the biggest source of external finance, however, was China, with an estimated US$18.4 billion in loans agreed between 2010 and 2018 (Gallagher and Myers, 2021).[2]

Extractive industries were the most obvious sources of contradictions in the new development model. A government ostensibly committed to a policy of harmony with nature actually relied on intensification and expansion of the extractive frontier in the Amazon. This included extracting oil from protected areas and in the face of dwindling oil reserves (and an uncertain future for fossil fuels in general), attracting investment to exploit the country's largely untapped mineral resources was a priority. However, taxation and royalty rates demanded by the government proved unattractive to most northern investors, and China became the key partner again. Most notably, a consortium of Tongling Nonferrous Metals Group and the China Railway Construction Corporation developed the Mirador copper mine, sparking major protests in the southern Amazon (van Teijilngen and Hogenboom, 2016).

The 2008 Constitution had, in theory, established new safeguards around indigenous rights and environmental damage. In practice, though, they were often overlooked, and Correa, for his part, railed against environmental and indigenous protesters, calling them 'infantile leftists' and stressing the need to exploit resources in service of development. In this understanding, extraction was a necessary evil, and a means to an end. Here the developmentalist aspects of Alianza Pais' ruling coalition dominated. SENPLADES documents reflect the influence of neo-structuralist ideas, which had at the time become strongly associated with CEPAL (United Nations Economic Commission for Latin America and the Caribbean; Leiva, 2008; see Dutt, 2019 on neostructuralism). Most notable was the idea of using resource revenues as a means of generating new comparative advantages in knowledge-based activities. The following section explores these efforts to 'transform the productive matrix' (SENPLADES, 2013).

## Dams and developmentalism

Ecuador's economic problems, according to neo-structuralist diagnoses, stemmed from a reliance on primary commodity exports (especially oil), which generated few well-paid or skilled jobs, thus hindering the growth of internal demand and the diversification of the domestic economy (SENPLADES, 2009). The proposed solution involved a mix of targeted import substitution with a drive for competitiveness in higher value-added manufacturing and services. Resource revenues, managed by the state, were supposed to be invested in these new sectors, resulting in the temporary intensification of extraction in order to move beyond oil and mining. In addition to import substitution industrialization designed to increase local production of value-added goods and efforts to augment 'immaterial' sectors such as biotechnology, the government invested heavily in infrastructure. Often not conceived of in direct terms as sources of value-addition and knowledge-based production,[3] roads, bridges, airports and, above all, dams, would, it was hoped, provide the necessary foundations for broad structural transformation. Phrased in SENPLADES' terms, 'transforming the energy matrix' was a necessary step in 'transforming the productive matrix' (Warner et al, 2017).

The government put in motion plans to build eight new dams in 2010, of which six were majority Chinese-financed and built (another was bankrolled by EXIM Bank of Russia and built by a Chinese firm, and the last was constructed by Odebrecht and financed by the Brazilian BNDES). Ecuador borrowed approximately US$3 billion from China Development Bank and China EXIM Bank for these projects (Garzón and Castro, 2020: 33–6), which have been controversial and beset by social and environmental problems (26 workers died during construction). All but the Odebrecht-built Manduriacu dam suffered delays and cost overruns (Lozano, 2019) and as of 2019, four of the seven projects that involved Chinese firms or finance were reportedly completed, producing a combined 2422MW at peak output (Lozano, 2019). If judged purely in terms of shifting Ecuador's energy mix, the programme was relatively successful, as hydroelectric power grew from 7.13TWh in 2006 (47 per cent of Ecuador's annual electricity production) to 24.66TWh by 2019 (75 per cent of the total) (BP, 2020).

Selective import substitution industrialization plans were consciously linked to this hydropower build-out. Most prominently, the government attempted to stimulate local production of electric induction cooking units in hopes of creating a new domestic industry and eliminating costly gas subsidies. However, this hinged on the notion that a reliable and cheaper energy supply would drive demand for induction cookers. To Correa's frustration, local business failed to increase production despite subsidies. Eventually the government removed barriers to importing the cookers from China (Purcell et al, 2017).

More ambitious, but still linked to a neo-structuralist agenda, was the Yachay City of Knowledge project (see Hernandez and Gaybor, 2018). Constructed on a 47 km² site in the Urcuquí Valley 120 km north of Quito, Yachay was partly financed by a China EXIM Bank loan of US$197 million. It was simultaneously envisaged as a sustainable city, research university, and science park, and according to Correa it was the most important project in Ecuador's history (*El Telégrafo*, 2015). It is fitting, then, that Yachay neatly sums up the advances and contradictions of Correa's vision, as a 'post-neoliberal version of systemic competitiveness' (González et al, 2018, 336; author's translation), while seeking to blend high modernism with communitarian principles and indigenous notions of harmonious living

(*Yachay* in Kichwa roughly translates to 'wisdom'). Yachay seems to have become a totem for the hope that Ecuador could not only transcend its peripheral status but rapidly catapult itself to a position at the technological frontier, as a knowledge rather than commodity producer. Yachay-Tech University opened in 2014, but by the time Correa left office in 2017 only 40 per cent of a planned US$1 billion in investments had materialized and other elements of the project (an industrial park and zones for biotechnology and agrotourism) remained partially constructed or at the planning stage.

## Opting out of the Wall Street Consensus

Part of the significance of Correa's infrastructure drive lies in how Chinese capital enabled such programmes to be combined with ambitious state-led developmentalist and redistributive agendas (albeit one which brought further marginalization to some communities at the extractive frontier). Dam and road building in themselves did not require higher oil prices or Chinese finance and construction firms, since Correa's years in power coincided with the re-emergence of infrastructure as a major focus for development institutions worldwide (World Bank, 2009; Bhattacharyay et al, 2012). Funding for infrastructure in the Global South has become increasingly accessible via the OECD's 'blended' model, whereby some multilateral concessional financing is provided as a means of 'de-risking' projects and making them attractive to global investors (Gabor and Sylla, 2020; Musthaq, 2021).

This imperative to make projects 'bankable' for global finance, however, implies a much broader agenda of de-risking, with the onus falling largely upon the state (Mawdsley, 2018; Gabor, 2021). Gabor argues that the power of institutional investors within the global economy after 2008 drove the emergence of a Wall Street Consensus (WSC) framework of financialized development. While large infrastructure projects as instruments of development may have experienced a resurgence, the WSC inherits a focus on fiscal discipline from the Washington Consensus, whereby external borrowing to fund infrastructure is discouraged. Instead, public–private partnerships (PPPs) are favoured, along with raising finance on local bond markets, since both options ostensibly reduce debt sustainability risks. Rather than public debt, PPPs are often financed by repaying investors via a project's eventual revenue, while local bonds are denominated in domestic currency, and so, from the view of the issuing state, not prone to foreign exchange risks.[4] However, such benefits may be illusory since the state is usually required to assume most risks in the last instance (Gabor, 2021). International investors participating in southern domestic bond markets may insist that issuing states backstop foreign exchange risk. Similarly, if a PPP-funded project does not generate sufficient user fees to meet repayments, states are generally required to step in. Multilateral development banks may themselves provide guarantees for some of these risks, but even so, activated guarantees are often converted into loans, making the state ultimately liable. Finally, blended finance packages tend to come with demands for further neoliberal reform. This may be sectoral, as in Gabor's example of a US$1 billion World Bank loan to Nigeria conditioned on privatization of the country's energy sector. More broadly, the WSC emphasizes the development of local capital markets and the removal of controls on flows of portfolio investment.

A version of Correa's infrastructure plan attempted without Chinese finance, therefore, would have looked rather different. It would have been difficult to reverse the liberalization of energy production and distribution, as Correa's government did, and also attract global private finance and multilateral development bank funds. In contrast to the turnkey projects built by Chinese firms, greater private sector involvement in terms of equity stakes or operational responsibilities would have been likely. And while roads may still have been constructed, the WSC model of generating returns directly via user fees implies a toll system rather than the free to use public highways that were built in Ecuador. Thus, Ecuador's experience under Correa can be seen as a Chinese-backed (rather than WSC) variation on the emergent infrastructure-led development regime. There is a contradictory element here, as in many aspects of Ecuador's developmental efforts during this period. On the one hand, as detailed earlier, development planning was oriented around ending dependence on primary commodity exports and establishing economic sovereignty. On the other, however, many of these policies carried more conventional readings of infrastructure-led development that stress value-added production, upgrading within global value chains, systemic competitiveness, and transnational connectivity. For example, new roads, bridges, and schools in Amazonian 'strategic zones' were built in part to foster the sense of *patria* (nation). But many of these were also part of larger, transborder ventures such as the Manta–Manaus multimodal transport corridor, designed primarily to facilitate Brazilian soy exports to China and connect Chinese factories with Brazilian assembly plants (Valdivia, 2017; Wilson and Bayón, 2017).

Many of the more radical elements of the Correa government's agenda could not feasibly have been pursued *in combination* with an infrastructure drive, without both the China-driven oil boom and loans from Chinese policy banks. High oil prices and the state's increased bargaining power in the sector meant the government could afford to do without IFI and bond market financing, while turning to Chinese loans to cover much of the costs of a public investment push. Quito emphatically turned its back on global capital markets with its 2009 bond default, after which IFIs struggled to discipline Ecuador and influence its domestic policy choices (Jepson, 2020). This afforded Correa the opportunity to introduce financial sector reforms that rank among the most radical enacted in recent years. First, central bank independence was abolished, moving monetary policy back under the purview of the executive branch. Banks were required to hold 45 per cent of liquid assets domestically (this was later increased to 60 per cent) and capital leaving the country was taxed (which raised US$1 billion annually). These reforms, among others, seem to have ameliorated periods of economic instability around both the global financial crisis and the end of the commodity boom (Weisbrot et al, 2017). Such policies, of course, run counter to both neoliberal orthodoxy and the principles of the emergent WSC.

## Moreno and the neoliberal restoration

Lenín Moreno was Correa's vice president from 2007 to 2013 and became his successor after a narrow win in 2017. Initially expected to extend, if perhaps moderate, Correa's agenda, Moreno instead quickly distanced himself from Correa's legacy. This precipitated a split in the Alianza País party, leading Moreno to shore up support among Ecuador's conservative political and economic elites and shift rightward on policy and rhetoric. In terms of foreign policy Moreno closed the Unasur (Union of South American Nations) headquarters in Quito and took the country out of the organization. Julian Assange's London embassy asylum was revoked, and Moreno lined up behind the US in recognizing Juan Guaidó as President of Venezuela.

Moreno's domestic reforms included shrinking the state bureaucracy, cutting social spending, introducing a series of tax breaks for large firms, and removing capital controls. Ecuador also returned to the IMF, but a US$4.2 billion loan deal agreed with the Fund was met by mass protests that forced the government to back down on a planned elimination of fuel subsidies. In 2020 the COVID-19 pandemic hit Ecuador particularly hard, presenting twin public health and economic crises. This dire situation prompted the government to default on some of its Eurobond holdings (albeit with investor consent for restructuring) and sign another IMF agreement for US$6.5 billion. While infrastructure spending slowed during Moreno's tenure, several new large projects were initiated under a PPP model, reflecting a turn towards a more WSC-friendly model of development.

In the context of an internal party power struggle, Moreno leveraged allegations of corruption and claims of wasteful spending on infrastructure against Correa and his allies.[5] The scandal around Brazilian construction firm Odebrecht, which has reverberated throughout Latin America, was central to these struggles. Jorge Glas, who served as vice president under both Correa and Moreno, was convicted in late 2017 of taking US$13.5 million in bribes from Odebrecht and sentenced to six years in prison (*El Comercio*,

2013). During his time in government Glas was responsible for many dam projects and the Chinese policy bank loans that financed them. A dedication to Glas stands at the entrance to the largest of these projects, the Coca Codo dam (Casey and Klauss, 2018). Because Glas was arrested shortly after writing a letter denouncing Moreno and sent to serve his sentence at a maximum-security prison, Correa supporters claimed that the prosecution was politically motivated. Though it seems highly unlikely that all recent corruption cases have been baseless, they have fallen disproportionately onto Correa loyalists. This, together with procedural irregularities, has given the impression of a 'lawfare' campaign to sideline political opponents.

Moreno employed a 'discourse of crisis' (King and Samaniego, 2019; von Schoettler, 2020) in order to justify his shift back to neoliberalism. His tenure began with Ecuador in recession, struggling with the oil downturn and costs of the 2016 earthquake. Nevertheless, once in office Moreno exaggerated the severity of economic problems inherited from Correa in order to justify policy shifts. As Franklin Ramirez Gallegos puts it, '[g]overnment and media have taken up the new Latin American argument that left-wing redistribution results in corruption, and austerity is therefore a moral imperative' (Ramírez Gallegos, 2018). China has loomed large in this imaginary. Nearly every official involved in the Coca Codo dam deal has been imprisoned, and Chinese loans have been blamed as enabling graft and prolificacy (Kraul, 2018). Beyond this, though, ministers began to discuss Chinese finance in terms that resonate with accusations of debt-trap diplomacy commonly levelled by Washington. According to Moreno's Energy Minister Carlos Perez, for example, 'the strategy of China is clear. They take economic control of countries' (Casey and Klauss, 2018). Amid a deteriorating economic situation, Moreno's government negotiated various restructurings, payment deferrals, and new loans with Chinese policy banks, but these focused on meeting immediate financing needs rather than being tied to new projects (Reuters, 2018; IMF, 2020b). While China remains a significant economic actor in the country, the US is intent on facilitating a break between Quito and Beijing. In the last days of the Trump administration, the International Development Finance Corporation (DFC) negotiated a deal whereby it will pay off some of Ecuador's debts to China and provide new investment in exchange for a commitment from Quito to bar Chinese firms from Ecuadorian telecoms networks (*Financial Times*, 2021).

Ecuador held elections in early 2021, with Moreno declining to run for a second term. Conservative Guillermo Lasso narrowly beat out *correista* candidate Andres Arauz, who had promised to end IMF agreements and return to China for greater financial support (Stott and Long, 2020). While Lasso is likely to have warmer relations with the IMF, his background suggests a more antagonistic approach to China. Lasso's think tank Ecuador

Libre is a member of the US-based Atlas Network of libertarian and anti-communist organizations, whose funders have included ExxonMobil and the Koch brothers (The Intercept, 2017; Público, 2021). Nevertheless, former Ecuador Libre head and now presidential advisor Aparicio Caicedo spoke of a desire to 'maintain a very good relationship' with China, even while ending 'this ideological vision of preferring China to the United States' (*Wall Street Journal*, 2021). Ecuador's perilous economic situation means that the country may have little choice but to request further rescheduling and relief on its Chinese debts, so some efforts at conciliation on Lasso's part are expected. The impact of Moreno's DFC agreement on Ecuador–China relations may be significant, however, assuming that new administrations in Quito and Washington implement a deal that some are now urging they cancel (Gallagher and Myers, 2021).

Elsewhere in Latin America, left of centre governments in Bolivia and Argentina face similarly tough decisions around a COVID-19-related debt crisis and where to turn for external help. Meanwhile, recent falls in Chinese bilateral finance could signal a new wariness on the part of PRC policy banks which, burned by their experiences in Venezuela, may be concerned about the political baggage that increasingly seems to come with lending in Latin America (Myers and Gallagher, 2020). The infrastructure-led template for development looks set to stay, albeit with two distinct funding models: a China-backed version funded by policy bank loans; or a WSC variant entailing a PPP model and much lower immediate fiscal costs, balanced by for the requirement for receiving states to assume risk on behalf of investors. So far, these have not been entirely mutually exclusive – a state contracting Chinese loans can still adopt a WSC approach, and vice versa, and a number of governments in recent years have sought external finance from multiple sources at once. Should deals like the one between the DFC and Ecuador become more widely adopted, however, states may face more of a binary choice between models and partners. While various forms of spatial planning and industrial policy may well be possible under either model, closing off one major source of external financing or the other seems likely to constrain developmental plans to at least some degree and may rule out the sort of ambitious efforts seen in Correa's Ecuador.

## Notes

[1] *Sumak Kawsay* is often translated instead as 'harmonious' or 'beautiful life'; considerable controversy exists over whether the term's use alongside *buen vivir* represents a misappropriation of the concept.

[2] Much of this lending fits the by now familiar pattern of credits tied to specific infrastructure projects contracted to Chinese construction and engineering firms. Somewhat unusually, however, around US$8.4 billion of loans seem to have been at least partially discretionary and used to finance a mix of government priority development projects and programs.

Government estimates show outstanding public debt to China of US$6 billion as of 2020 (Ecuador Ministry of Finance, 2020).

[3] Exceptions here being the Yachay City of Knowledge, discussed later, and the planned Esmeraldas Refinery, meant to facilitate upgrading within the petroleum value chain, which as of 2021 has only been partially constructed.

[4] Since Ecuador still uses the US dollar as its currency, such benefits from domestic bond issuance would not apply in this case.

[5] Correa has resided in Belgium since the end of his presidency, and he was convicted in absentia and sentenced to eight years' imprisonment over a bribery scandal. The case hinged that appears to have been fabricated, and the current government went to considerable lengths to prevent Correa from joining Andres Arauz's ticket as the vice-presidential nominee in 2021 (Long, 2020).

# References

Alvaro, M., Kzak, R., and Dube, R. (2014) 'Ecuador expels Pentagon group attached to US embassy', *Wall Street Journal*, 25 April. Available from: www.wsj.com/articles/SB10001424052702304788404579524111277765526.

Bhattacharyay, B.N., Kawai, M., and Nag, R.M. (2012) *Infrastructure for Asian Connectivity*, Cheltenham: Asian Development Bank and Edward Elgar.

BP (2020) *Statistical Review of World Energy*. Available from: www.bp.com/en/global/corporate/energy-economics/statistical-review-of-world-energy.html.

Casey, N. and Klauss, C. (2018) 'It doesn't matter if Ecuador can afford this dam. China still gets paid', *New York Times*, 24 December. Available from: www.nytimes.com/2018/12/24/world/americas/ecuador-china-dam.html.

Dube, R. (2021) 'Ecuador's new leader needs help from US, but China will stay close', *Wall Street Journal*, 12 April. Available from: www.wsj.com/articles/ecuadors-new-leader-needs-help-from-u-s-but-china-will-remain-close-11618266593.

Dutt, A.K. (2019) 'Structuralists, structures, and economic development', in M. Nissanke and K. Ocampo (eds) *The Palgrave Handbook of Development Economics*, Cambridge: Palgrave Macmillan, pp 109–41.

Ecuador Ministry of Finance (2020) 'Boletin de deuda publica interna y externa'. Available from: www.finanzas.gob.ec/https-wwwdeuda-publica-nueva-metodologia/.

*El Comercio* (2013) 'El hombre de los contratos millionarios', 23 January. Available from: https://web.archive.org/web/20130125205313/http://www.elcomercio.com/politica/hombre-contratos-millonarios-eleccion es_2013-Jorge_Glas_Espinel_0_852514810.html.

*El Telegrafo* (2015) 'Correa: Yachay es el proyecto más importante de la historia del país', 5 August. Available from: www.eltelegrafo.com.ec/noticias/sociedad/6/correa-22.

*Financial Times* (2021) 'US development bank strikes deal to help Ecuador pay China loans', 14 January. Available from: www.ft.com/content/affcc432-03c4-459d-a6b8-922ca8346c14.

Gabor, D. (2021) 'The Wall Street Consensus', *Development and Change*, 52(3): 429–59.

Gabor, D. and Sylla, N.S. (2020) 'Planting budgetary time bombs in Africa: the Macron Doctrine En Marche', *Groupe d'études géopolitiques*. Available from: https://geopolitique.eu/en/2020/12/23/planting-budgetary-time-bombs-in-africa-the-macron-doctrine-en-marche/

Gallagher, K. and Myers, M. (2021) *China-Latin America Finance Database*, Washington, DC: Inter-American Dialogue.

Garzón, P. and Castro, D. (2018) 'China-Ecuador relations and the development of the hydro sector', in E. Dussel Peters, A.C. Armony, and S. Cui (eds) *Building Development for a New Era: China's Infrastructure Projects in Latin America and the Caribbean*, Mexico City: Asian Studies Center, Center for International Studies, University of Pittsburgh, and Red Académica de América Latina y el Caribe sobre China, pp 23–57.

González, M.F., Álvarez, M.C., and Purcell, T. (2018) 'Urbanismo utópico, realidades distópicas: una etnografía (im) posible en Yachay, ciudad del conocimiento', *Etnográfica*, 22(2): 335–60.

Gonzalez-Vincente, R. (2017) 'South–South relations under world market capitalism: the state and the elusive promise of national development in the China–Ecuador resource-development nexus', *Review of International Political Economy*, 24(5): 881–903.

Hey, J.A. and Klak, T. (1999) 'From protectionism towards neoliberalism: Ecuador across four administrations (1981–1996)', *Studies in Comparative International Development*, 34(3): 66–97.

IMF (2020a) 'World Economic Outlook Database, October 2020'. International Monetary Fund. Available from: www.imf.org/en/Publications/WEO/weo-database/2020/October.

IMF (2020b) 'Ecuador: request for an extended arrangement under the Extended Fund Facility', International Monetary Fund. Available from: www.imf.org/-/media/Files/Publications/CR/2020/English/1ECUEA2020002.ashx.

Jepson, N. (2020) *In China's Wake: How the Commodity Boom Transformed Development Strategies in the Global South*, New York: Columbia University Press.

Kaplan, S.B. (2016) 'Banking unconditionally: the political economy of Chinese finance in Latin America', *Review of International Political Economy*, 23(4): 643–76.

King, K. and Samaniego, P. (2019) 'The crisis narrative of Ecuador's Lenín Moreno has obscured the real winners and losers of recent economic policy', LSE Blog: Latin America and The Caribbean. Available from: https://blogs.lse.ac.uk/latamcaribbean/2019/09/12/the-crisis-narrative-of-ecuadors-lenin-moreno-has-obscured-the-real-winners-and-losers-of-recent-economic-policy/.

Kraul, C. (2006) 'Leftist rolling to victory in Ecuador's presidential race', *Los Angeles Times*, 27 November. Available from: www.latimes.com/archives/la-xpm-2006-nov-27-fg-ecuador27-story.html.

Kraul, C. (2018) 'Por los créditos negociados con China, Ecuador se enfrenta a un enorme déficit presupuestario', *Los Angeles Times*, 10 December. Available from: www.latimes.com/espanol/internacional/la-es-por-los-creditos-negociados-con-china-ecuador-se-enfrenta-a-un-enorme-deficit-presupuestario-20181210-story.html.

Leiva, F.I. (2008) *Latin American Neostructuralism: The Contradictions of Post-Neoliberal Development*, Minneapolis, MN: University of Minnesota Press.

Long, G. (2020) 'In Lenín Moreno's Ecuador, democracy is in danger', Jacobin, 21 September. Available from: https://jacobinmag.com/2020/09/equador-lenin-moreno-election-2021-democracy.

Lozano, G. (2019) 'Ecuador's China-backed hydropower revolution', *Dialogo Chino*, 25 July. Available from: https://dialogochino.net/en/climate-energy/29133-ecuadors-china-backed-hydropower-revolution/.

Mawdsley, E. (2018) '"From billions to trillions": financing the SDGs in a world "beyond aid"', *Dialogues in Human Geography*, 8(2): 191–5.

Musthaq, F. (2021) 'Development finance or financial accumulation for asset managers? The perils of the global shadow banking system in developing countries', *New Political Economy*, 26(4): 554–73.

Myers, M. and Gallagher, K. (2020) *Scaling Back: Chinese Development Finance in LAC, 2019*, China–Latin America Report, March, The Dialogue and Boston University Global Development Policy Center. Available from: www.thedialogue.org/analysis/scaling-back-chinese-development-finance-in-lac-2019/.

Público (2021) 'El lobby anticomunista 'Atlas Network' recurre a Casado y Ayuso para amplificar su discurso en América Latina', 24 April. Available from: www.publico.es/politica/lobby-anticomunista-atlas-network-recurre-casado-ayuso-amplificar-discurso-america-latina.html.

Purcell, T.F., Fernandez, N., and Martinez, E. (2017) 'Rents, knowledge and neo-structuralism: transforming the productive matrix in Ecuador', *Third World Quarterly*, 38(4): 918–38.

Ramírez Gallego, F (2018) 'Ecuador veers to neoliberalism. December', *Le Monde Diplomatique*. Available from: https://mondediplo.com/2018/12/06ramirez.

Reuters (2018) 'Ecuador clinches $900 mln loan from China – Moreno', 12 December. Available from: www.reuters.com/article/ecuador-debt-china/ecuador-clinches-900-mln-loan-from-china-moreno-idUSL1N1YH0YJ.

Rinehart, C.S. and McGuire, J.W. (2017) 'Obstacles to takeup: Ecuador's conditional cash transfer program, the bono de desarollo humano', *World Development*, 97(September): 165–77.

Sánchez, F. and Polga-Hecimovich, J. (2019) 'The tools of institutional change under post-neoliberalism: Rafael Correa's Ecuador', *Journal of Latin American Studies*, 51(2): 379–409.

Schindler, S. and Kanai, J.M. (2019) 'Getting the territory right: infrastructure-led development and the re-emergence of spatial planning strategies', *Regional Studies*, 55(1): 1–12.

SENPLADES (2009) *Plan Nacional para el Buen Vivir 2009–2013*, Quito: Secretaría Nacional de Planificación y Desarrollo. Available from: www.ecuadorencifras.gob.ec/wp-content/descargas/%20Informacion-Legal/Normas-de-Regulacion/Plan-Nacional-para-el-Buen-Vivir/Plan+Nacional+del+Buen+Vivir+2009-2013.pdf.

SENPLADES (2013) *Plan Nacional para el Buen Vivir 2013–2017*, Quito: Secretaría Nacional de Planificación y Desarrollo. Available from: https://observatorioplanificacion.cepal.org/es/planes/plan-nacional-del-buen-vivir-2013-2017-de-ecuador

Stott, M. and Long, G. (2020) 'Ecuador presidential contender hits at IMF deal and pledges reforms', *Financial Times*, 23 December. Available from: www.ft.com/content/276b1718-0707-4308-a2e6-77ae5ebd3f12.

The Intercept (2017) 'Sphere of influence: how American libertarians are remaking Latin American politics', 9 August. Available from: https://theintercept.com/2017/08/09/atlas-network-alejandro-chafuen-libertarian-think-tank-latin-america-brazil/.

Valdivia, G. (2017) 'Oil citizens of the revolution: is Esmeraldas, Ecuador, spotlighting the contradictions of Rafael Correa's Citizens' Revolution?', *NACLA Report on the Americas*, 49(4): 429–35.

van Teijlingen, K. and Hogenboom, B. (2016) 'Debating alternative development at the mining frontier: buen vivir and the conflict around El Mirador mine in Ecuador', *Journal of Developing Societies*, 32(4): 382–420.

von Schoettler, W.V. (2020) 'Análisis del discurso de los Informes a la Nación 2017–2019 del presidente Lenín Moreno. Del progresismo al neoliberalismo autoritario en el Ecuador', *Revista Ibérica de Sistemas e Tecnologias de Informação*, 26: 376–87.

Warner, J.F., Hoogesteger, J., and Hidalgo, J.P. (2017) 'Old wine in new bottles: the adaptive capacity of the hydraulic mission in Ecuador', *Water Alternatives,* 10(2): 322–40.

Weisbrot, M., Johnston, J., and Merling, L. (2017) *Decade of Reform: Ecuador's Macroeconomic Policies, Institutional Changes and Results*, Washington, DC: Center for Economic and Policy Research.

Wilson, J. and Bayón, M. (2017) 'Fantastical materializations: interoceanic infrastructures in the Ecuadorian Amazon', *Environment and Planning D: Society and Space*, 35(5): 836–54.

World Bank (2009) *World Development Report 2009: Reshaping Economic Geography*, Washington, DC: World Bank. Available from: https://openknowledge.worldbank.org/handle/10986/5991.

World Bank (2020) 'World Bank commodity price data'. Available from: www.worldbank.org/en/research/commodity-markets.

# 8

# Centralizing Infrastructure in a Fragmenting Polity: China and Ethiopia's 'Infrastructure State'

*Zhengli Huang and Tom Goodfellow*

## Introduction

During the Belt and Road Initiative (BRI) Summit in April 2019, Prime Minister of Ethiopia Abiy Ahmed secured US$1.8 billion in funding from Beijing for the country's electricity transmission and distribution network. Following the summit, the Chinese Ambassador noted that Ethiopia would continue to be a leading partner in Sino-African relations, serving as a model of cooperation with the BRI. In March 2020, the US signalled its willingness to invest US$5 billion in Ethiopia through its new International Development Finance Corporation (IDFC), in sectors with the potential to 'crowd in' private investment such as telecommunications, geothermal energy, logistics, and sugar production (Pilling, 2020). As with many IDFC investments, this has been widely interpreted as an attempt by the US government to counter the influence of China in Ethiopia (and across the world more broadly).

These developments reflect growing competition between the US and China and illustrate the sectors and spaces in which it is playing out through their respective African policies. Both countries are drawn to Africa by the huge potential of some of the world's fastest growing economies, significant reserves of natural resources, and emerging investment opportunities. Interestingly, however, it is in Ethiopia – a country with very few natural resources – that current superpower competition has become particularly evident.

Ethiopia has had firm and enduring political ties with the Chinese government and has received more Chinese development finance than any other non-resource-rich country in Africa. This is not only because of opportunities presented by the country's growing economy and its geopolitical position as an entryway to Sub-Saharan Africa, but also because of the relative 'affinities' between Chinese and Ethiopian governance processes (Cabestan, 2012). Ethiopia's previous regime under the Ethiopia People's Revolutionary Democratic Front (EPRDF), which ruled the country from the 1990s to 2018, advocated a 'developmental state' model influenced by the East Asian experience. Chinese finance in Ethiopia during this time grew steadily, with infrastructure as one of the key areas of investment and way of pursuing spatial objectives. The most significant example of Chinese finance in Ethiopia's infrastructure sector is the 756 km Addis Ababa–Djibouti Railway. Planned as a turnkey project to facilitate Ethiopia's structural transformation, the railway was financed by the Chinese government and built by Chinese enterprises.

At the same time, Ethiopia has long had close ties with a range of OECD donors, including the US. Despite the EPRDF's resistance to liberalization and affinity for an 'East Asian' development model, the US viewed Ethiopia as a key regional partner in relation to both the 'War on Terror' and its apparently effective use of international aid (McVety, 2012). Aid to Ethiopia rose substantially under the Obama administration and remained high under Trump, making it one of the top three recipients of US aid in Sub-Saharan Africa since 2001.[1] However, tensions between the Ethiopian government and Western donors regarding the ideological direction of the EPRDF and its resistance to deeper economic and political liberalization were common (Feyissa, 2011).

While Sino-Ethiopia ties have historically been strong, they have been challenged by the radical political changes in Ethiopia since 2018, which initially appeared to suggest closer alignment with US agendas. When Abiy Ahmed emerged as Ethiopia's new prime minister following several years of unrest in the country, he initiated a series of political and economic reforms including the liberalization of key parts of the economy and the reversal of industrialization policies promoted by the EPRDF regime. This was accompanied by an intentional turning away from 'developmental state' rhetoric, challenging China's role in Ethiopia. Abiy's ideological reorientation and pursuit of other sources of foreign investment, combined with shifts in Chinese lending strategy and concerns about Ethiopian debt, led to widespread speculation in 2018 that China–Ethiopia ties would weaken (Aglionby and Feng, 2018). The increasing financial commitments from the US government also raised the question of how US–China rivalry would affect the development of Chinese-financed infrastructure projects, and Ethiopia's commitment to infrastructure-led development more generally.

However, the political upheavals and ideological reorientations in Ethiopia since 2018 have not reduced the government's famed enthusiasm for large-scale infrastructure projects. Continuity in infrastructure-led development is epitomized by the high-profile La Gare project, which relies on private investment from the United Arab Emirates (UAE) to 'facelift' the centre of Addis Ababa,[2] and the 'Beautifying Sheger' project to revitalize the city's riverside areas, stimulate urban tourism and promote flood resilience (Terrefe, 2020). Meanwhile, the US$5 billion Grand Ethiopian Renaissance Dam is rarely out of the news, Ethiopia remains central to the BRI in Africa, and China is vital for realizing many of Abiy's goals. In 2019, the Chinese government and aerospace companies facilitated the launch of Ethiopia's first satellite.[3] The US government, meanwhile, has made increasing commitments to Ethiopia's infrastructure development, though concrete results are yet to be seen. This suggests that far from a shift away from infrastructure, Ethiopia's new government has instead thrown it open to a wider range of financial players, and is seeking to capitalize on its unique qualities as a site of infrastructure investment to maintain and promote a narrative of growth and development under conditions of increasing crisis. This involves capitalizing on the growing US–China rivalry to maximize the range of infrastructure finance available, while also building on legacies of state centralization that pre-date (but have been further strengthened by) Chinese support for infrastructure construction and finance in recent decades.

In this chapter, we argue that Ethiopia has received so much finance from China (both before and after the announcement of the BRI) partly *because* of Ethiopia's potentiality as an 'infrastructure state', due to its relatively centralized state structures and hierarchical governance processes. This chapter therefore begins by examining why and how Ethiopia became such an important partner for China in Africa, and how Sino-Ethiopian infrastructure relations thrived partly due to the relative affinity between Chinese and Ethiopian governance processes and their shared spatial objectives. It then examines how Chinese infrastructure finance has facilitated the restructuring of state institutions to deliver major transport infrastructure projects and Ethiopia's industrial parks strategy. Here we show that Ethiopia has drawn on China's own experience of infrastructure governance and territorial integration, but also argue that it is far from being a powerless partner in its dealings with Beijing and with Chinese State-Owned Enterprises (SOEs). Finally, we turn to the recent period of political crisis, particularly since 2018, and show that US engagement in the Ethiopian infrastructure sector offers potential opportunities and risks. The future of the 'infrastructure state', developed partly through Chinese assistance, remains uncertain as the Ethiopian government struggles to consolidate territorial integration and control, and political fragmentation threatens to unravel a centuries-long project of centralization. Moreover, it is not clear whether the finance provided by a new range of actors with an interest in Ethiopian infrastructure will contribute to centralization or undermine it.

## Centralization and Sino-Ethiopia ties

Modern Ethiopian history has seen the building of the central authority across the nation's complex and highly fragmented territory. Ethiopia is often characterized in three distinct territorial zones: the 'highland core' in the Northern and Central Highlands (historically known as Abyssinia), the 'highland periphery' to the South (including Oromia), and 'lowland periphery' stretching out to the Northeast and East, particularly the Somali region and area towards Djibouti (Markakis, 2011; Clapham, 2017). The attempt by the 'highland core' to integrate the other territories through imperial conquest into a modern nation state was the dominant agenda of the emperors ruling Ethiopia from the mid-19th century (Donham and James, 1986). In contrast with African states where borders were drawn by the colonial powers, extending and deepening control over territory has been a gradual and continuous project in Ethiopia for at least 150 years.

Such centralizing processes were extended under Emperor Haile Selassie in the mid-20th century and the communist Derg regime from 1974–91 (James et al, 2002). The resulting legacies of social hierarchy, state land ownership and state control are unusual in comparison to many other African countries, which influences Ethiopia's relationship with China. While it is often noted that Ethiopia is an important strategic partner for China due to is diplomatic importance and market size (Adem, 2012; Hackenesch, 2013), the nature of Ethiopia's governance – alongside the Marxist-Leninist origins of the EPRDF regime – contribute to the strength of Sino-Ethiopian ties in recent decades (Cabestan, 2012), and also partly explain the exponential rise of Chinese finance in Ethiopia's infrastructure development over recent decades. From 2000 to 2018, Chinese loans to Ethiopia have financed 52 projects amounting to US$13.7 billion, in Africa second only to those in Angola. A major proportion of these loans, around US$4.8 billion, flowed into the transport sector.[4] The largest transport infrastructure project is the US$3.4 billion Addis-Djibouti railway, 70 per cent of which was financed through a loan from China EXIM Bank. The second most significant Chinese-financed Ethiopian transport project is Addis Ababa Light Rail Transit (LRT), with a cost of US$475 million, 85 per cent of which was financed by China EXIM Bank. Chinese firms and state-owned enterprises (SOEs) also play a significant role in constructing large-scale infrastructures in Ethiopia, whether financed by Chinese capital or not (Huang and Chen, 2016).

Our interviews with Chinese SOEs operating in Ethiopia show that they generally hold a positive view of the Ethiopian government in comparison to other African states, with several SOE representatives emphasizing their confidence in the federal government; one noted that "the work efficiency of Ethiopian Government, especially the Federal Government, is remarkable".

Another pointed out that government efficiency was "rooted in its modern history". They also cited a "strong will to learn from China". This is not to say that relations are always smooth; in particular, tensions among the federal, state and, municipal governments have generated a range of challenges and delays for Chinese-financed projects. However, the sense of a shared commitment to a state-driven model of development with massive investment in physical capital at its core contribute to the intensity and success of China–Ethiopia ties by the 2010s.

In this regard, the most significant phase in the evolution of Sino-Ethiopian cooperation on infrastructure occurred well before the unveiling of the BRI. The tendency to refer to BRI as a driving force of infrastructural integration in contexts such as Ethiopia is therefore misleading. More generally, the popular narrative of BRI that sees it as 'an ambitious Chinese-constructed infrastructure network that seeks to link the economies of participating countries to that of China's' (Taylor and Zajontz 2020) obscures the complex interactive dynamics of Chinese-financed infrastructure development in Africa. There is no evidence that the BRI itself is the driving force of infrastructure development in Africa (Sun, 2014). In fact, the biggest transport infrastructure projects financed by China in Africa, including the Addis–Djibouti railway, were designed by African states and construction began years before Xi Jinping introduced the BRI in 2013.

Chinese infrastructure finance is animated by policy banks such as China's EXIM Bank as well as construction companies, and especially the SOEs. Policy banks enable projects, but their role in planning is very limited. They tend to rely heavily on information provided by the project contractors themselves, whose influence in shaping Chinese development assistance has generally been under-emphasized (Zhang and Smith, 2017; Goodfellow and Huang, 2021a). Most China EXIM-funded development projects are directed by the SOE contractors, and project initiation is not necessarily a consequence of bilateral negotiations between the bank and the country government, as is normal for OECD aid. Instead, Chinese SOEs play the role of mediator, and often initiate and carry out projects, and then seek financial support from China EXIM Bank. Their advantages stem from their familiarity with host country politics, often developed over years or decades of working in a particular country, and their relationship with Chinese lenders (Zhang and Smith, 2017, 2340; Soule, 2019).

Dealing with Chinese construction companies confidently, and enabling them to proceed with project implementation effectively, is therefore a significant challenge for any state receiving large amounts of Chinese finance. In the Ethiopian case, this finance was of such a scale that it necessitated state restructuring. In order to overcome some of the obstacles and conflicts built into its complex political system, which could delay its ambitious infrastructure vision, the Ethiopian government rapidly created several semi-autonomous state agencies to manage the inflow of Chinese resources and channel them towards new territorial projects. These agencies were conceived in order to bypass problems associated with Ethiopia's federal governance structure, and they channelled national authority across new scales of governance in ways similar to those conceptualized by Brenner (2004). Two of the most important agencies in this regard, the Ethiopian Railway Corporation (ERC) and the Industrial Parks Development Corporation (IPDC), have enabled the Ethiopian state to consolidate central control over projects motivated in part by the desire to enhance territorial integration.

## Chinese infrastructure finance and Ethiopian state restructuring

*The Ethiopian Railway Corporation*

As we argue, relatively centralized bureaucratic control has long been a feature of Ethiopian governance, which has in part lubricated Chinese engagement in Ethiopia – yet it is also increasingly evident that Chinese finance in Ethiopian infrastructures has precipitated state restructuring. The establishment and evolution of the ERC during the implementation of two major railway projects financed by China EXIM Bank in Ethiopia: The Addis Ababa–Djibouti Railway (ADR) and the Light Rail Transit (LRT) in Addis Ababa demonstrate this point.

The 759-km long ADR is often depicted as a 'Chinese transportation project in Ethiopia' by the international media, yet the plan was initiated years before Chinese finance was secured. The railway route itself is more than a century old; there was a railway between Ethiopia and Djibouti financed by the French and completed in 1917, though it had deteriorated significantly and was almost abandoned by the 2000s. ERC was established in 2007 by the Ethiopian government through Regulation Number 141/2007 of the Council of Ministers, to facilitate the building of a modern nationwide railway network (Woldemariam, 2020). It was also commissioned to plan for the National Railway Network of Ethiopia, which along with the modern ADR included a number of auxiliary railways linking the main line with different regions in the country, and also the city-scale LRT.

Chinese funding for the ADR and LRT was not secured until 2011. Though this finance originates from China EXIM Bank, it was the profit-seeking Chinese contractors, mainly represented by the China Railway Engineering Corporation (CREC) in these two cases, that were driving the projects' implementation. The SOE approached China EXIM Bank after securing the contract with ERC. Despite this, ERC was identified as the 'owner' and commissioner of the projects, and the railways were built following its plans and designs. The Ethiopian government thus vested a great deal of authority in this new agency, as it began consolidating the country as an 'infrastructure state'. At the time ERC was formed, there was a huge capacity gap in building and managing national and city-scale railway projects. This gap narrowed as Chinese contractors began work on both the ADR and LRT, so while China was not the fundamental cause of this particular infrastructure-driven reorganization within the state, it has clearly been an important catalyst. As the following cases show, negotiation with powerful contractor partners from China contributed to the capacity building of ERC, thus further enhancing centralization in infrastructure governance.

One of the major disagreements between the contractors and ERC was whether to apply an electrified system. The Chinese operators complained that ERC insisted on building an electrified railway despite the consensus that unstable electricity supply would haunt the project for years after its completion. The ambition of the Ethiopian state to build a modern and electrified railway over-rode the cost-benefit analysis of international contractors, demonstrating the relatively strong bargaining power of ERC in its negotiations with the SOEs. In addition, the ERC's authority to overcome opposition from other Ethiopian governmental institutions and engage SOEs at a high level was cemented in just a few years. In one case, a Chinese SOE found itself confronted by a hostile local government authority regarding the removal of ballast along the route. The local officials claimed that the Chinese SOE had illegally used land and materials within their jurisdiction, while the latter insisted that they followed procedures established for ADR construction. These issues were exacerbated by growing tensions between the federal government and Ethiopia's regional and local governments from 2016 onwards. As one SOE representative noted, "the political struggles in this country are too complicated for us to understand. We are not able to intervene with any of these issues. That's why we have to go through ERC when it comes to governmental issues".

Both the ADR and LRT were designated as 'turnkey' projects, meaning that the obligations of the contractor would end after a fixed period of time and the project would be handed over to ERC. However, the Ethiopian government managed to extend the services of the Chinese contractors as operators. The Chinese firms are also responsible for technical training of ERC personnel, which is an important part of capacity transfer. Yet the Chinese companies complain that ERC constantly intervenes with staffing arrangements, diminishing training effectiveness and thus slowing down their exit plan. According to the SOEs, ERC rarely adopts their suggestions about who to appoint to key positions based on training results, and managerial positions within ERC are clearly based on 'political preference' rather than competency.

Many similar dynamics, in terms of difficult negotiations between Chinese SOEs and the ERC in which the latter ultimately prevailed, are evident in the implementation of the Addis Ababa LRT project. The LRT is similarly depicted as an embodiment of modernity and the Ethiopian 'renaissance', often hailed as the first light rail system in Sub-Saharan Africa and a model for the continent (Boudet et al, 2015; Nallet, 2018). Again, the idea of a light rail system predates the availability of Chinese finance, and some route designs had been made before CREC became involved, but CREC's design affiliation proposed an alternative LRT network. According to CREC, its proposal would integrate the city's zoning plan and expansion process more effectively, but the proposal was rejected by the Ethiopian government because the cost was much higher, and it the implementation would take longer. ERC also insisted on the inclusion of a number of updates to the LRT system, but these were not included in the original proposal and not factored in to CREC's initial bid (see Goodfellow and Huang, 2021a). Representatives of CREC estimate the excess cost of these alterations at somewhere around US$60–70 million. ERC rejected this estimate and negotiations were ongoing at the time of writing.

The conception, procurement, and construction of the LRT was thus fraught with conflicts. The reality is far from one of domineering Chinese influence or the 'debt-trap' caricature through which Beijing is said to exploit a highly unbalanced power relationship to engineer Ethiopia's indebtedness and dependency. While the LRT project certainly had difficulties, and the prospects for repaying the loan appear improbable, these are less about Chinese influence or the terms of Chinese finance that the institutional conflicts and obstacles on the Ethiopian side. ERC – established in 2007 with the aim of facilitating national and international rail development – took charge of planning for the LRT city-level project in 2010 with very little experience. Meanwhile, many city government institutions were sidelined, and city-level authorities have largely disowned the LRT (Rode et al, 2020; Goodfellow and Huang, 2021a). Through this new agency, the central government refashioned as an infrastructure state was able to take control of more local scales of governance and planning in order to implement a major territorial political project in the capital city – despite ERC's lack of expertise in city-level projects and its poor relationship with municipal authorities.

## The Industrial Parks Development Corporation

The restructuring of the state to transcend the scalar constraints of Ethiopia's federal system is also evident in the country's industrialization strategy. The cornerstone of this strategy is the construction of industrial parks (IPs) focused on light manufacturing, of which Chinese companies and institutions play important roles in their construction (Goodfellow and Huang, 2021b). The Industrial Parks Development Corporation (IPDC) was established in 2014 to realize this strategy and drive industrialization (Oqubay, 2015; Staritz and Whitfield, 2017; Sun, 2017), with Hawassa Industrial Park as its flagship project. Hawassa has attracted interest from academic and policy communities because it is the biggest industrial park in Sub-Saharan Africa, is government owned and has demonstrated considerable productivity and employment potential (Mihretu and Llobet, 2017; Oqubay and Lin, 2020). By the end of 2019, the park employed around 30,000 workers, and is projected to house 90,000 workers at its peak – which could mean employing as many as 270,000 people over three 8-hour shifts.

There are more than 20 such IPs planned within Ethiopia, at different stages of completion and involving various types of manufacturing companies, almost all of which are being built by Chinese companies. Although Chinese finance is limited in these IPs developed by IPDC, which are domestically resourced unlike most of the private IPs emerging in Ethiopia (Goodfellow and Huang 2021b), Chinese contractors have played a central role in the IPDC's learning process. After Chinese contractors finished the construction of the flagship Hawassa IP, its management was entrusted to experts from Kunshan Industrial Park Company, an experienced special economic zone (SEZ) management company that facilitated the growth of Kunshan, one of the most significant SEZs in China. They introduced a 'one-stop service centre' within each park and a local desk for each of the corresponding departments from the federal government.

As well as SOEs, private investors from China influence industrial development in Ethiopia and in so doing shape the IPDC as a key architecture of the infrastructure state. Ethiopia's first industrial zone – the Eastern Industrial Zone (EIZ) established in 2007 – is owned and managed by a private Chinese company, Qiyuan Group. The planning, construction, and management of EIZ has been a story of the Chinese company's efforts to overcome institutional and legal barriers in order to turn the zone into a leasable, manageable, serviced, and productive piece of land (Goodfellow and Huang, 2021b). As well as facing obstacles relating to the sub-leasing of land plots, EIZ experienced complications surrounding taxation and customs clearance arrangements, all of which underlined the importance of having an effective national agency to streamline industrial park development, implement key decisions and oversee new legislation. Drawing significantly on the experiences of EIZ's establishment, the Ethiopian government introduced new legislation for IP development and management aiming to overcome some of these challenges. The IPDC was also born out of this process.

The role of IPDC is especially significant in terms of the Ethiopian state's long-term project of territorial integration. It represents a new level of centralization of state power in the sphere of infrastructure development, and Ethiopia's industrial strategy of establishing parks that are strategically dispersed in the country's regional states. The spatial aspect of Ethiopia's

**Figure 8.1:** Location of government-owned industrial parks in Ethiopia (operational or under construction in 2020) and the Addis Ababa–Djibouti Railway

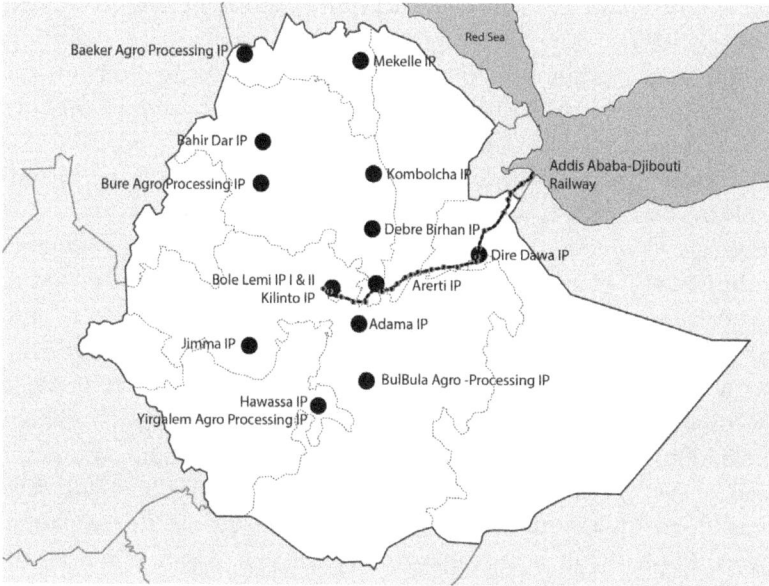

Source: Authors' drawing based on information from the Ethiopian Investment Commission.

industrial strategy was designed to diffuse rising political tensions that were already evident within the Ethiopian federation by 2014–15, as policy makers sought to foster balanced regional growth and pre-empt the narrative that certain areas were disproportionately benefitting from public investment. In the context of a highly fragmented nation state, some IPs were designated in places with no access to basic resources and infrastructure for production, thus making no economic sense for industrial development (see Figure 8.1). Yet, IPDC was tasked with integrating all regions into its new economic agenda. Although this spatial strategy was designed to empower regions, it was implemented in a top-down manner that, as we will show in the next section, sidelined local officials, and exacerbated political tensions.

## The infrastructure state in crisis

Since 2016, Ethiopia has been in a state of near-permanent crisis. Ethnic conflicts and confrontations among political forces have continued despite the attempt by Abiy to calm tensions by emphasizing Ethiopian unity and releasing political prisoners. In November 2020, armed conflict broke out between the federal government and Tigrayan People's Liberation Front, the leading party with the previous EPRDF coalition. Coming in the middle of a global pandemic that further destabilized the economy, the conflict has led to a devastating humanitarian crisis in Tigray, and consequently damaged the international reputation of the Abiy government.

Facing both global and domestic challenges and having turned away from the 'developmental state' ideology that lent some coherence to Ethiopia's economic strategy, Ethiopia's 'infrastructure state' too is experiencing crisis. Abiy's political reform led to leadership restructuring in parastatal firms, including replacement of executive staff within IPDC and ERC. This raised concerns over whether signature infrastructure projects, including the railways and the IPs, would be completed.

However, as noted in the introduction, the Abiy government maintained its enthusiasm for large infrastructure projects. Abiy also happened to come to power at a time when international infrastructure finance was rapidly evolving and increasingly available. Growing interest in financing African infrastructure by international banks and private investors (Goodfellow, 2020) coincided with the Trump administration's agenda countering Chinese influence on the continent. Meanwhile Abiy's stated commitment to democracy and peace with Eritrea solidified partnerships with several Western donors, while simultaneously capitalizing on the renewed interest of several Persian Gulf states eager to establish allegiances in the Horn of Africa (Kibsgard, 2020; Miller and Verhoeven, 2020). In December 2019, the IMF pledged US$2.9 billion to support Abiy's proposed 'Homegrown Economic Reform' plan – one of the biggest programmes in the IMF's history in Africa (Pilling, 2020).

As international development finance evolves, clearly fuelled in part by China–US rivalry in Africa, there is scope for heightened US influence over infrastructural development in Ethiopia. For example, the Trump administration froze aid to Ethiopia due to controversy surrounding the Grand Ethiopian Renaissance Dam, but this policy was recently reversed by the Biden administration (Solomon, 2021) The decision of the US government to channel almost 10 per cent of the IDFC's entire lending capacity into Ethiopia reflects both the relative significance of China in Ethiopia and the extent to which Abiy has aligned himself with US economic agendas. During a visit to Addis Ababa, US Secretary of State Mike Pompeo made a 'veiled swipe' at China by noting that Ethiopia and other countries 'should be wary of authoritarian regimes and their empty promises' (Paravicini, 2020).

These new financing channels available within Ethiopia should not be interpreted as a decisive turn away from China, but rather are consistent with 'Ethiopia's historic pattern of balancing external powers against each other to reaffirm its own independence' (Kibsgard, 2020). The question that remains, therefore, is whether promises of direct investment from Western sources, especially the US government, in specific sectors can deliver the same gains as Chinese loans that operated through the creation of strong, centralized, new semi-autonomous state agencies. Far from being 'empty promises', Chinese finance has delivered concrete results. Even if the benefits of these infrastructures on the ground remain in question, the combination of Chinese SOE expertise and Ethiopian state restructuring together has been a powerful driver of change in Ethiopia.

As noted in the introduction to this chapter, the Chinese government has made significant new pledges since Abiy took the helm, most notably in relation to energy infrastructure. This is consistent with the infrastructural support that China offered the previous regime: one of the main obstacles for the Addis–Djibouti railway and the national industrial parks is the lack of consistent power. In a public response to the commitment of the Chinese government to invest in Ethiopia's energy sector, Abiy stressed that the main purpose of this finance was to improve electricity supply along the rail line and increase the supply to the country's industrial parks. China's pledges on electricity include the transmission of 6,000MW of hydroelectric power from the Ethiopian Grand Renaissance Dam (Yalew and Changgang, 2020, 179). Meanwhile, as Abiy's agenda pivots towards maximizing foreign investment and the liberalization of the economy, real estate is likely to become another key area of state restructuring, with early Chinese investors exerting significant influence due to their 'first mover' role in this sector. The question is whether the new and increasingly fragmented and uncoordinated forms of investment in infrastructure can build on the legacy of the 'infrastructure state' or will undermine it.

## Conclusion

Ethiopia's long history of state centralization and hierarchical control made it a particularly suitable partner for Chinese financial institutions, with a relatively strong capacity to implement infrastructure projects on a massive scale. However, the Ethiopian political system is also rife with tensions and institutional challenges due to the complexities of ethnic federalism and fragmented central, state, and local relations. This led the central government to develop new and powerful quasi-autonomous central agencies with the authority to manage Chinese finance and projects. In this respect, the restructuring of the Ethiopian state in pursuit of the territorial rollout of infrastructure cannot be separated from the presence of large amounts Chinese finance.

These dynamics have, however, been further complicated by the 2018 regime change in the context of political crisis. This crisis has underscored the urgency of projects of territorial integration, as the Ethiopian polity threatens to fragment at every turn, but also reflects the uncomfortable relationship between Abiy's agenda of political and economic liberalization and the ongoing effort to exert control. As China–US competition continues to play out across the continent, the increasing commitment from the US government provides new opportunities for the Ethiopian government to venture into new models of state-led infrastructure development. But in the context of the country's political unravelling, whether US direct investment can provide a force for both economic development and political integration remains to be seen. As a new range of donors and investors funnel resources into Ethiopian infrastructure, the nature of the infrastructure state that evolved through Sino-Ethiopian relations – in which single agencies directed huge amounts of finance with unprecedented levers of political authority – seems unlikely to survive unscathed.

## Notes

1   See https://explorer.usaid.gov/cd/ETH.
2   See www.lagare.com/newsroom/eagle-hills-abu-dhabi-expands-to-ethiopia-with-the-launch-of-la-gare/.
3   See *Space in Africa*, 2019.
4   See https://chinaafricaloandata.org/.

## References

Adem, S. (2012) 'China in Ethiopia: diplomacy and economics of Sino-optimism', *African Studies Review*, 55(1): 143–60.

Aglionby, J. and Feng, E. (2018) 'China scales back investment in Ethiopia', *Financial Times*, 3 June. Available from: www.ft.com/content/06b69c2e-63e9-11e8-90c2-9563a0613e56.

Boudet, L., Gendreau, L., and Marchand, Q. (2015) 'Les transports à Addis Abeba', Université de Rennes 1, France.

Brenner, N. (2004) *New State Spaces: Urban Governance and the Rescaling of Statehood*, Oxford: Oxford University Press.

Cabestan, J.P. (2012) 'China and Ethiopia: Authoritarian affinities and economic cooperation', *China Perspectives*, 4: 53–62.

Clapham, C.S. (2017) *The Horn of Africa: State Formation and Decay*, Oxford: Oxford University Press.

Donham, D. and James, W. (1986) *The Southern Marches of Imperial Ethiopia*, Cambridge: Cambridge University Press.

Feyissa, D. (2011) 'Aid negotiation: the uneasy "partnership" between EPRDF and the donors', *Journal of Eastern African Studies*, 5(4): 788–817.

Goodfellow, T. (2020) 'Finance, infrastructure and urban capital: the political economy of African "gap-filling"', *Review of African Political Economy*, 47(164): 256–74.

Goodfellow, T. and Huang, Z. (2021a) 'Contingent infrastructure and the dilution of 'Chineseness': reframing roads and rail in Kampala and Addis Ababa', *Environment and Planning A: Economy and Space*, 53(4): 655–74.

Goodfellow, T. and Huang, Z. (2021b) 'Manufacturing urbanism: improvising the urban–industrial nexus through Chinese economic zones in Africa', *Urban Studies*. doi:10.1177/00420980211007800.

Hackenesch, C. (2013) 'Aid donor meets strategic partner? The European Union's and China's relations with Ethiopia', *Journal of Current Chinese Affairs*, 42(1): 7–36.

Huang, Z. and Chen, X. (2016) 'Is China building Africa?', *European Financial Review*, 7.

James, W., Donham, D., Kurimoto, E., and Triulzi, A. (eds) (2002) *Remapping Ethiopia: Socialism and After*, Oxford: James Currey.

Kibsgard, D. (2020) 'Sino-Ethiopian relations from Meles Zenawi to Abiy Ahmed: the political economy of a strategic partnership', *China Currents*, 19: 2.

Markakis, J. (2011) *Ethiopia: The Last Two Frontiers*, Suffolk: Boydell & Brewer.

McVety, A.K. (2012) *Enlightened Aid: US Development as Foreign Policy in Ethiopia*, Oxford: Oxford University Press.

Mihretu, M. and Llobet, G. (2017) *Looking Beyond the Horizon: A Case Study of PVH's Commitment in Ethiopia's Hawassa Industrial Park*, Washington, DC: World Bank.

Miller, R. and Verhoeven, H. (2020) 'Overcoming smallness: Qatar, the United Arab Emirates and strategic realignment in the Gulf', *International Politics*, 57(1): 1–20.

Nallet, C. (2018) 'The challenge of urban mobility: a case study of Addis Ababa light rail', Paris: Institut Français des Relations Internationales.

Oqubay, A. (2015) *Made in Africa: Industrial Policy in Ethiopia*, Oxford: Oxford University Press.

Oqubay, A. and Lin, J.Y. (2020) *The Oxford Handbook of Industrial Hubs and Economic Development*, Oxford: Oxford University Press.

Paravicini, G. (2020). 'Pompeo takes veiled swipe and China on final leg of Africa trip', Reuters, 19 February. Available from: www.reuters.com/article/uk-usa-pompeo-ethiopia/pompeo-takes-veiled-swipe-at-china-on-final-leg-of-africa-trip-idUKKBN20D1PW.

Pilling, D. (2020) 'US ready to back Ethiopian reform with $5bn investment', *Financial Times*, 5 March. Available from: www.ft.com/content/b0c2963c-5e1a-11ea-8033-fa40a0d65a98.

Rode, P., Terrefe, B., and da Cruz, N.F. (2020) 'Cities and the governance of transport interfaces: Ethiopia's new rail systems', *Transport Policy*, 91: 76–94.

Solomon, S. (2021) 'US restoration of foreign aid to Ethiopia signals new course', VoA News. Available from: voanews.com/a/africa_us-restoration-foreign-aid-ethiopia-signals-new-course/6202535.html.

Soule, F. (2019) 'How to negotiate infrastructure deals with China: four things African governments need to get right', *The Conversation*. Available from: www.theconversation.com/how-to-negotiate-infrastructure-deals-with-china-four-things-african-governments-need-to-get-right-109116.

*Space in Africa* (2019) 'Chinese Foreign Minister visits Addis Ababa to strengthen Sino-Ethiopia cooperation', *Space in Africa*. Available from: https://africanews.space/chinese-foreign-minister-visits-addis-ababa-to-strengthen-sino-ethiopia-cooperation/.

Staritz, C. and Whitfield, L. (2017) 'Made in Ethiopia: the emergence and evolution of the Ethiopian apparel export sector', CAE Working Paper 2017:3, Roskilde.

Sun, I.Y. (2017) *The Next Factory of the World: How Chinese Investment Is Reshaping Africa*, Cambridge, MA: Harvard Business Press.

Sun, Y. (2014) *Africa in China's Foreign Policy*, Washington, DC: Brookings Institution. Available from: www.brookings.edu/wp-content/uploads/2016/06/africa-in-china-web_cmg7.pdf.

Taylor, I. and Zajontz, T. (2020) 'In a fix: Africa's place in the Belt and Road Initiative and the reproduction of dependency', *South African Journal of International Affairs*, 27(3): 277–95.

Terrefe, B. (2020) 'Urban layers of political rupture: the "new" politics of Addis Ababa's megaprojects', *Journal of Eastern African Studies*, 14(3): 375–95.

Woldemariam, W.L. (2020) 'Assessment of organizational culture practice: the case of Ethiopian railways cooperation', Master's thesis, St. Mary's University.

Yalew, M.T. and Changgang, G. (2020) 'China's 'Belt and Road Initiative': implication for land locked Ethiopia', *Insight on Africa*, 12(2): 175–93.

Zhang, D. and Smith, G. (2017) 'China's foreign aid system: structure, agencies, and identities', *Third World Quarterly*, 38(10): 2330–46.

# 9

# Radioactive Strategies: Geopolitical Rivalries, African Agency, and the *Longue Durée* of Nuclear Infrastructures in Namibia

*Meredith J. DeBoom*

In December 2018, then US National Security Advisor John Bolton introduced Prosper Africa, the Trump administration's new Africa strategy, in a speech at the Heritage Foundation. Bolton asserted that the new policy reflected the US government's desire for 'African nations to succeed, flourish, and remain independent in fact and not just in theory' (Bolton, 2018, 5). His speech, however, sounded more like a 'China policy' than a thoughtful proposal for relations with a continent of 54 sovereign states. Rather than focusing on priorities identified by Africans, Bolton warned his audience that African leaders were succumbing to the geopolitical machinations of China, Russia, and terrorist organizations. The former, he emphasized, was 'aggressively' investing in infrastructure to 'gain a competitive advantage over the United States' (Bolton, 2018, 4). In total, Bolton's 21-minute speech mentioned 'China' or 'Chinese' 20 times.

Although ostensibly a 'new' policy, Prosper Africa's China-centric narrative has deep historical roots. Great power policymakers, analysts, and media outlets have long characterized Africa as a 'chessboard' for geopolitical competition rather than a continent whose leaders pursue their own spatial objectives. These characterizations are most prominent during times of intense geopolitical rivalry. Diplomatic correspondences from the late 1800s and early 1900s European 'Scramble for Africa' (Koponen, 1993) and the US–Soviet Cold War, for example, are replete with 'chessboard' metaphors. Scholars have identified similar themes in contemporary debates over the 'new Cold War' between the US and

China (DiCarlo and Schindler, 2020). While 'chessboard' narratives are excellent fodder for selling foreign affairs magazines, they risk overlooking how non-great power actors adapt to and influence the on-the-ground dynamics of geopolitical rivalries and, moreover, leverage those rivalries to pursue their own spatial objectives.

This chapter challenges 'chessboard' interpretations of infrastructural geopolitics in a realm often assumed to be the sole purview of great powers: nuclear energy. More specifically, it analyses how Namibian actors have employed *radioactive strategies* – a term used here to refer to the tactics through which actors use the geopolitical significance of nuclear infrastructures to advance their spatial objectives – to pursue their goals across temporalities ranging from German colonialism to apartheid South Africa's quest for 'the bomb' and the Cold War to the war on climate change. Drawing inspiration from Enns and Bersaglio's (2020) *longue durée* (Braudel, 2009) analysis of infrastructural megaprojects in East Africa and Hecht's (2012) research on the historical technopolitics of uranium, the chapter uses archival data and interviews to explain how the South West Africa People's Organization (SWAPO), Namibia's liberation-movement-turned-ruling-party, has transformed nuclear infrastructures from symbols of colonial and apartheid exploitation into core components of its quest for state-led extractive development. The chapter begins with an overview of the colonial and apartheid origins of Namibian uranium before embarking on a more detailed analysis of SWAPO's radioactive strategies during the Cold War, the US-led War on Terror, and, finally, the current moment of US–China rivalry. It concludes with a discussion of how attending to historical and geographic contexts and host country agency can shed new light on infrastructural geopolitics across multiple scales.

## Infrastructures of colonial violence: the origins of Namibian uranium

The origins of uranium mining in Namibia date to German colonization. In 1884, Chancellor Otto Von Bismarck claimed German South West Africa[1] as a German protectorate. Reflecting Bismarck's preference for financing colonial infrastructures through private rather than public funds, the protectorate's mineral rights were awarded to the private German Colonial Society for South West Africa in 1885 (Dierks, 2002). Bismarck hoped that commercial investors would extend Germany's territorial control over Namibia by constructing new infrastructures for the efficient extraction of its natural resources, but German investors failed to provide the necessary capital to facilitate territorial consolidation or mineral extraction. The German government was forced to resort to British capital instead. In 1892, the South West Africa Company, a German–British shareholder venture whose operations would outlast German colonialism, took over most mining operations (Voeltz, 1988).

From the very beginning, local resistance constrained the infrastructural ambitions of European capitalists, political leaders, and settlers. Nama and Herero leaders resisted the expropriation of their herding, farming, and hunting lands for mining operations. They also refused to recognize German military authority, on which private companies often relied for security during prospecting and surveying expeditions. In 1892, Herero forces halted a new mining project by refusing to allow the German military to escort labourers working for the South West Africa Company (Gewald, 1999). Resistance intensified in the early 1900s, culminating in the siege of a German fort by Herero forces in 1904. German commander Lothar von Trotha initiated Germany's four-year genocidal war against the Herero and Nama shortly thereafter. Survivors of German military attacks were forced into starvation in the Kalahari Desert or sent to concentration camps, where they were subjected to slave labour and medical experiments (Steinmetz, 2007; Zimmerer and Zeller, 2008; Erichsen and Olusoga, 2010). Although exact numbers are disputed, German forces killed an estimated 70–80 per cent of the Herero and 50 per cent of the Nama during this time.

Uranium mining in Namibia can be traced to the Herero and Nama genocide, which facilitated subsequent land expropriation and settler colonialism. Between 1908 and the 1914 outbreak of World War I, mining operators in Namibia experienced some of the most profitable operating conditions in German colonial history (Gewald, 1999). They also benefitted from African forced or coerced labour, which facilitated the cheap extraction of raw materials for European industrialization. Although German surveyors identified uranium traces near Rössing Mountain, investors at the time considered uranium to be an unremarkable element of little commercial value; it was used primarily as a glass colouring agent and a sepia tint for photography (Goldschmidt, 1989). Even after the discovery of radioactivity by Henri Becquerel and Marie Curie, uranium was valued primarily as a source of radium, a rarer element that could be extracted from uranium ore and used for x-rays and cancer treatments (Burke, 2017). The geopolitical significance of Rössing's uranium deposits would remain opaque until the discovery of nuclear fission in 1938 – over two decades after the end of German colonial rule.

## Infrastructures of occupation: early South African rule

The Union of South Africa captured German South West Africa in 1915 as part of a World War I military campaign endorsed by the British government. Sensing an opportunity for territorial expansion, South African authorities encouraged Afrikaners to settle in the occupied territory. After World War I, the League of Nations rewarded South African expansionism by granting it a Class C Mandate over South West Africa, a designation reserved for territories considered to be the 'least developed'. It quickly became clear that South African officials had little intention of guiding their new mandate toward independence. In 1922, South African General Jan Smuts laid the rhetorical groundwork for permanent occupation in a letter to the League of Nations. The Permanent Mandates Commission, he argued, 'must recognise the fact that C mandates are in effect not far removed from annexation' (League of Nations, 1922, 91). South Africa would continue to occupy Namibia for nearly 70 more years.

Mining infrastructures played a key role in the first objective of the newly independent South African state: to consolidate its territorial control over Namibia and, through industrialization-led development, over South Africa itself. Although nuclear fission was still a decade away, German geologist Peter Louw began uranium exploration at Rössing Mountain in 1928 with the encouragement of South African leaders (Louw, 2018). He confirmed the uranium deposits identified during German colonialism and attempted to market Rössing for the mining of radium, which remained a higher-value commodity than uranium (Louw, 2018). Unfortunately for Louw, his efforts were overshadowed by the Great Depression.

For the next 25 years, South African leaders struggled to attract investors to Namibia's uranium deposits. Rössing's low grades made it poorly suited for the labour-intensive practices of early uranium extraction. Systematic exploration only began in the 1950s, shortly after South Africa commenced uranium exports from its own mines to the US (UNCN, 1980). By that time, both the uranium market and the territorial politics of Namibia had been transformed. In 1938, uranium's confirmed nuclear fission potential catalysed a global search for the commodity that had previously been treated as a waste product of radium (Burke, 2017). Then, in 1946, South Africa rejected the newly established United Nations' (UN) plan to convert South West Africa into a UN trust territory as part of the dissolution of the League of Nations. Instead, South Africa announced that it would absorb Namibia as a new province – thus avoiding the prospect of an independent, black majority-ruled state on the border of white minority-ruled South Africa. This promised provincial status never came to fruition. Instead, South Africa became an apartheid state with the National Party's election in 1948, and Namibia became its occupied territory.

## Infrastructures of apartheid nuclear ambition: the Cold War

By the 1960s, the combination of the Cold War race for nuclear weapons and growing concerns about US dominance in the uranium supply chain meant that Rössing's mediocre geology was no longer an obstacle for investors. The apartheid regime took advantage of these twin dynamics – increased demand and limited supply – to pursue its nuclear ambitions. In 1965, South African mining representatives met with British and Canadian mining companies to discuss alternative uranium sources to the US (Hecht, 2012). That same year, the Canadian government limited uranium exports from Canadian mines to peaceful purposes (Hunt, 1977). Canada's decision was a problem for the UK, as it meant that Canadian uranium could no longer be used to fuel the British nuclear weapons programme. The same was not true, however, for uranium from South Africa, which 'seemed happy to waive "end-use" restrictions for the UK as long as Namibia remained governed by South Africa and its atomic energy legislation' (Hecht, 2012, 97). British mining company Rio Tinto purchased the licence for Rössing the following year.

By the time Rio Tinto began commercial-scale mining at Rössing in 1976, the mine had become a geopolitical liability for the UK. In October 1966, the same year that Rio Tinto purchased Rössing, the UN General Assembly voted to end South Africa's mandate over Namibia. This decision was the cumulation of years of activism by African leaders, including the governments of Ethiopia and Liberia, which had initiated formal UN proceedings against South Africa in 1960. In 1969, the UN reinforced its position by calling for immediate South African withdrawal from Namibia. Then, just as it seemed that the situation could not worsen for British diplomats, South African Minister of Mines Piet Koornhof acknowledged that South Africa had constructed a full-scale uranium enrichment plant (UN, 1974). Koornhof's announcement fuelled suspicions that South Africa would pursue its own nuclear weapons – suspicions that were confirmed in 1993 when President F.W. de Klerk acknowledged and destroyed six nuclear warheads on the eve of South Africa's 1994 democratic transition.

Yet Rio Tinto pressed on with mining operations at Rössing, which expanded its export market to France, Japan, the US, and West Germany (Cooper, 1988). The continued commercial appeal of Namibian uranium despite South Africa's imminent pariah status and the mine's unremarkable geology can be explained in large part by South Africa's apartheid labour system (Hecht, 2009). With uranium producers like the US, Australia, and Canada facing tightened labour regulations – and hence production costs – at home, the South African government and Rio Tinto framed Rössing's apartheid operations as 'low-cost' and thus commercially advantageous. This 'Rössing advantage' was reinforced by South African police officers, whose on-site presence was designed to deter strikes (Hecht, 2012). Rössing's apartheid labour regime gave the mine a commercial edge within the capitalist world.

Some Western governments also attempted to portray Rössing as a force for global good in the Cold War context, framing South Africa's integration into the capitalist world via its nuclear infrastructures as essential to the longer-term political reform of the apartheid state. Hecht (2012) chronicles how apartheid leaders amplified this message for their own ends, leveraging US–Soviet rivalry to argue that Rössing was essential to Western nuclear and capitalist supremacy. In a 1970 statement to the Parliament of South Africa – delivered in English to be accessible to Western audiences – Prime Minister B.J. Vorster argued that:

> under South African conditions, a large scale [uranium enrichment] plant can be competitive with existing plants in the West ... South Africa does not intend to withhold the considerable advantages inherent in this development from the world community. We are therefore prepared to collaborate in the exploitation of this process with any non-communist country(ies) desiring to do so, but subject to the conclusion of an agreement safeguarding our interests. (Joint Committee on Atomic Energy, 1970: 240–1)

Vorster's allusion to 'South African conditions' can be interpreted as a coded reference to South Africa's nuclear infrastructures, which depended on 'cheap coal mined by cheap black labor' (Hecht, 2012, 91). Through its uranium mines and enrichment plants, the apartheid regime framed itself as offering a geopolitical and geoeconomic service to the capitalist world.

## Infrastructures of independence: Namibia's liberation struggle

As anti-apartheid sentiments grew in the late 1970s, the Namibian origins of Western nuclear weapons engendered political outrage. In the US, the UK, and Western Europe, the intertwined infrastructures of Western uranium supply chains and the apartheid state provided ample fodder for both anti-apartheid and anti-nuclear activists to castigate Western governments (Roberts, 1980; Saunders, 2009). These campaigns also provided a new option for African leaders to undermine apartheid, reinforcing the diplomatic strategies pursued at the UN. In 1979, for example, the UK-based Anti-Apartheid Movement founded the World Campaign Against Military and Nuclear Collaboration with South Africa with the support of Tanzanian President and pan-Africanist leader Julius Nyerere (Reddy, 1994). Similar campaigns emerged in Sweden, the Netherlands, and the US.

Meanwhile, Namibia's liberation movements leveraged growing activist opposition to Rössing to pursue their own goal: Namibian independence. Although several movements were involved, the South West African People's Organization (SWAPO) emerged as the lead actor in 1964 when the Organization of African Unity (predecessor to the African Union) decided to support SWAPO over its primary rival, the South West Africa National Union (Saunders, 2018). The UN followed suit in the early 1970s, designating SWAPO as the authentic representative of the Namibian people in a series of decisions that rendered South Africa's continued occupation of Namibia – and extraction of its resources – illegal (UN, 1974). This UN designation provided SWAPO with significant moral authority to wage a campaign against the apartheid state and its Western government enablers.

SWAPO complemented moral support from Western activists and the UN with the cultivation of military support from the communist bloc, which shared SWAPO's goal of defeating the South African regime and its Portuguese colonial allies in Angola and Mozambique. Namibia's liberation struggle was long and violent, and Western anti-nuclear activists could not provide the weapons and training needed to counter South African military and paramilitary forces. By the 1970s, SWAPO had secured financial and military support for its guerrilla operations from Cuba, the Soviet Union, China, North Korea, and East Germany (Saunders, 2019). This multi-actor strategy – military support from a variety of communist states, moral support from activists and the UN – enabled SWAPO to play multiple sides of the Cold War in its pursuit of Namibian independence.

National independence was not SWAPO's only goal during this period. As momentum for independence grew, SWAPO leaders drew upon the affective power of nuclear infrastructures to cultivate international and domestic support for SWAPO as the assumed ruling-party-to-be of an independent Namibia. As one SWAPO veteran of the liberation struggle explained during an interview:

> 'Uranium was key to SWAPO those days. ... It took us out of the apartheid shadow and made Namibian independence a cause of its own. Here we were, the African colony still fighting for freedom! ... We [SWAPO] could not go without it [uranium] in terms of economics, but the same was true for politics.'

SWAPO's radioactive strategy to win support for its political future was remarkably successful. Activist groups like the Anti-Apartheid Movement began to call not only for Namibia's independence, but also for 'solidarity with SWAPO' (Anti-Apartheid Movement, 1982). The moral authority granted to SWAPO, to the exclusion of other Namibian liberation organizations, enhanced its domestic legitimacy and its global political economic standing. It also laid the groundwork for SWAPO's post-independence dominance of Namibian politics (Melber, 2014).

## Infrastructures of state structuring: neoliberal Namibia

Calls to nationalize Rössing were a cornerstone of SWAPO's rhetoric in the 1980s (UNCN, 1980; SWAPO, 1981). Some SWAPO leaders, however, privately recognized that nationalization might not be in the best interests of the soon-to-be-independent state or its ruling party (Hecht, 2012). Uranium mining is capital-intensive and legally complicated. It relies on long-term contracts and requires significant organizational expertise. Accounting for 10 per cent of national GDP and 40 per cent of export earnings, Rössing was a leading contributor to the emergent Namibian economy on the eve of independence – and those revenues appeared to be in jeopardy. Due to the approaching end of the Cold War nuclear arms race, uranium spot prices in the late 1980s were less than half what they had been when Rössing opened, and the mine's late 1980s production was less than half its peak levels. These declines, combined with a shift toward more capital-intensive production strategies, had prompted Rössing to cut its workforce roughly in half by the time SWAPO won independent Namibia's first election. Wary of alienating any actors that could prove helpful to securing its ruling party status, SWAPO navigated a careful line between cultivating anti-apartheid outrage and positioning itself for future political economic success.

Broader geopolitical and geoeconomic dynamics further discouraged any nationalization ambitions still harboured among SWAPO leaders. With the end of the Cold War at hand, the ideological winds had begun to shift. Neoliberal approaches were gaining traction among international financial institutions as support for state-led approaches to resource extraction dwindled. Recognizing these changing conditions, SWAPO shifted its political strategy from nationalization to neoliberalism. While other governments were commencing early 'state *re*structuring' (Brenner, 2003, emphasis added), SWAPO was 'structuring' the newly independent Namibian state to align with the emerging geopolitical and geoeconomic landscape.

SWAPO's neoliberal shift was nowhere clearer than in its relationship with Rössing. By Namibia's independence in 1990, it was clear that SWAPO would not pursue nationalization (Bauer, 1998). Instead, SWAPO leaders established a 'typical neoliberal state' (Harvey, 2005, 70) of the postcolonial variety, with a foreign investment and extraction-centric economy characterized by low mining royalties and a minimal state role. By the mid-2000s, SWAPO's neoliberal transformation was so incontrovertible that the revolutionary movement formerly aligned politically with the Soviet Union was praised by Nicholas Kristof (2006) in the *New York Times* as a 'pioneer ... stable, pleasant, and safe, and its government has tried hard to entice foreign investors'.

Yet despite their public neoliberal strategy, many SWAPO leaders remained privately determined to break Rössing's uranium monopoly. The opportunity to do so would not arise until 20 years after independence, when the Chinese government's pursuit of nuclear energy would catalyse a new 'uranium rush' (Conde and Kallis, 2012) and, with it, the possibility of a new radioactive strategy. In the meantime, SWAPO leaders catered to

Rio Tinto – the very company they had previously accused of plundering Namibian uranium (UNCN, 1980) – and its Western government allies. SWAPO's dependence on Rössing, while far from a point of pride, was a strategic necessity.

## Infrastructures of US surveillance: the 'War on Terror'

After 11 September 2001, the close ties that SWAPO leaders had forged with Rio Tinto encountered a new challenge: US surveillance. In his 2003 State of the Union address, US President George W. Bush cast the geopolitical spotlight on African uranium producers when he claimed that 'Saddam Hussein recently sought significant quantities of uranium from Africa' (Bush, 2003). The increased attention to African uranium that followed did not come with increased geopolitical power for African states. Instead, Western governments channelled concerns that terrorist groups and 'rogue' states would acquire African uranium into extra-territorial efforts to surveil uranium production networks across the continent. In Namibia, tensions peaked in 2009 after two Rössing employees and a member of the Namibian Defense Force stole yellowcake (unenriched uranium concentrate powder) from the mine (Hartman, 2009). Following the theft, the US Department of Energy's National Nuclear Security Administration audited Rössing, which in turn requested on-site assistance from the Department of Energy's Global Threat Reduction Initiative to secure its operations.

Many of the Namibian government officials I interviewed bristled at this exercise of extra-territorial sovereignty, but they had little choice but to comply. At the time, the US accounted for roughly 30 per cent of Rössing's exports. The Namibian government's reliance on US foreign aid further reduced its manoeuvrability. Although Namibian officials publicly deferred to the US, the incident stoked frustration and embarrassment within SWAPO. When asked to identify factors that facilitated subsequent Chinese investments in Namibian uranium, my informants often described the stationing of US security forces at Rössing, combined with the perceived paternalism of US officials, as influential in cultivating support for the 'China model'. Others emphasized the willingness of Chinese investors and officials to contribute to Namibian government priorities, including a state ownership role in the infrastructures of uranium extraction. The longer the War on Terror persisted, the more appealing non-Western investment sources became.

Intensified US surveillance also came at a moment when the post-independence support Rössing had garnered within SWAPO was beginning to wane. In 2003, Rössing announced plans to end its operations by 2007 due to falling uranium prices. By 2004, it had reduced its workforce to just 833 people. Finally, as the proverbial *coup de grâce*, Rio Tinto announced its intent to challenge the Namibian government's already-low mining royalties in court. At the time, this legal challenge appeared likely to prompt a renegotiation of the mining tax in Rio Tinto's favour. Within a year, however, a new geopolitical power had arrived, this time pursuing uranium not for nuclear weapons, but rather for 'green' nuclear energy.

## Infrastructures of climate change mitigation: China's nuclear rise

Due to uranium's limited uses, its spot prices are heavily dependent on nuclear energy generation forecasts. In the early 2000s, concerns about climate change prompted speculation that existing uranium supplies were insufficient to meet the ambitious plans of emerging nuclear energy producers like China and India. After three decades of hovering around US$10 per pound, spot prices increased more than 13-fold between 2003 and early 2007. Uranium exploration in Namibia surged accordingly, with investments nearly quadrupling (DeBoom, 2017). Investment sources initially ranged from Chinese, Canadian, and Australian private companies to French and Russian state-owned enterprises (SOEs), but Chinese SOEs soon emerged as frontrunners. In 2004, Rössing became the first Western-owned mine to export uranium directly to the Chinese government via Shanghai Power Utility. Four years later, the state-owned Aluminum Corporation of China (Chinalco) purchased a 12 per cent ownership stake in Rio Tinto. Chinese SOEs also began to invest in uranium exploration, most notably via China Nuclear International Uranium Corporation (SinoU), a 2006 spin-off of China National Nuclear Corp (CNNC).

Namibia's uranium boom came to an abrupt halt in March 2011 following the Fukushima nuclear disaster in Japan. By then the world's fifth-largest uranium producer, Namibia was hit particularly hard by Fukushima; Japan's 50 nuclear power plants represented a quarter of Namibia's uranium export market when they went idle. Namibia's entire mining sector contracted 10 per cent the following year as several European countries abandoned nuclear power and global uranium price stagnation set in. By 2014, prices had declined to US$30 per pound, less than a quarter of their 2007 high. When the price is below US$40 per pound, more than half of the world's uranium mines, including all Namibian mines, operate below the break-even point. Prices remained low through late 2020, halting exploration projects around the world.

At first glance, this market decline appeared to bode poorly for SWAPO's ambition to reduce Namibia's dependency on Rössing. Not all uranium producers, however, make investment decisions based on spot prices alone. Reflecting the logics of Chinese state capitalism (Lee, 2017) and reminiscent of Cold War era investments that defied the logics of commercial profitability, most Chinese investments in Namibian uranium are driven by state-based incentives: namely the Chinese Communist Party's (CCP) plan to reduce carbon emissions and pursue 'Ecological Civilization' through the rapid intensification of nuclear energy generation (DeBoom, 2021). Chinese SOEs – most notably China General Nuclear Power Corporation (CGN) – were *the* catalyst in Namibia's record-breaking US$1 billion in uranium investments in 2014. As one industry informant explained, "[t]here was and is a great suspicion [in the uranium industry] of the Chinese, but the fact remains that we would have collapsed without them". The persistence of uranium mining in Namibia despite market conditions that shuttered mines elsewhere is inconceivable without Chinese state-based investments.

## Infrastructures of state ownership: the Husab mine

China's nuclear rise has facilitated fundamental changes in the ownership structure and scale of uranium mining in Namibia. In 2000, Namibia hosted zero Chinese uranium investments; Rössing was the only investment actor. By 2020, Chinese SOEs held majority stakes in all of Namibia's operating uranium mines. Chinese SOE ownership includes Rössing itself, in which CNNC purchased a majority stake (68.62 per cent) from Rio Tinto in 2019. It also includes the CGN majority-owned Husab mine, which is the world's second-largest operating uranium mine and, at US$5 billion, the single-largest investment in independent Namibia. The same country that was on the brink of losing its only uranium mine in 2003 was the world's fourth-largest uranium producer in 2020.

In addition to forestalling the demise of Namibian uranium, Chinese investments provided an opening for the SWAPO-led Namibian state to achieve two goals that many Namibians presumed impossible: an end to Rössing's dominance and Namibian state ownership in the infrastructures of uranium extraction. When CGN began production at Husab in 2015, SWAPO finally realized its goal of ending Rössing's nearly 40-year dominance. Husab also (re)introduced an alternative economic model to Namibia's post-independence neoliberalism: state ownership. Unlike previously operating mines, Husab's ownership is entirely state-based: CGN owns 90 per cent, while Epangelo, the Namibian government's mining SOE, owns the remaining 10 per cent. Epangelo, which means 'government' in Namibia's Oshiwambo language, was created in 2008 to increase Namibian state ownership in the mining sector in the name of national development. Husab is its first major direct investment in uranium.

The Namibian state's ownership stake in Husab was not a mere gift or historical accident. Recognizing a geopolitical opening that had not existed since the Cold War, Namibian officials responded to China's rise by pursing a new radioactive strategy: state capitalism. They negotiated partial ownership for Epangelo during Husab's mining licence approval process in 2014 and secured a Chinese loan to fund Epangelo's stake. This loan raises the spectre of 'debt-trap diplomacy' (Bräutigam, 2020; Carmody, 2020; DeBoom, 2020) – CGN could gain full control of the mine if Epangelo is unable to repay its loan through production dividends – but such a scenario is unlikely given the strong diplomatic ties between the CCP and SWAPO, which were forged through China's support during Namibia's independence struggle (Taylor, 1997; Dobler, 2008; Melber, 2013; DeBoom, 2017). In geopolitical terms, Husab represents the intersection of China's infrastructure-centric approach to strategic integration and SWAPO's own approach to 'getting the territory right' (Schindler and Kanai, 2019), this time through a Namibian interpretation of state capitalism. More than 25 years after independence, SWAPO has finally recast the uranium infrastructures of colonial, imperial, and neoliberal exploitation into infrastructures of its own vision – even if not its sole ownership.

## Conclusion

Media coverage of contemporary infrastructural investments tends to focus on geopolitical and geoeconomic rivalry between the US and China. This focus is neither historically novel nor unwarranted; infrastructure is indeed a geopolitical realm (Bouzarovski et al, 2015; Bridge et al, 2018; Jia and Bennett, 2018; Barry and Gambino, 2020; Murton and Lord, 2020) and contemporary infrastructural investments are intertwined with broader geopolitical projects, most notably the Belt and Road Initiative (BRI; Liu and Dunford, 2016). An *exclusive* 'chessboard' focus on the agency of great power rivals, however, risks overlooking how infrastructures take material and symbolic shape within, and are in turn shaped *by*, geographic and historical contexts and the agency of non-great power, host country actors.

This chapter analysed how a variety of Namibian actors have leveraged radioactive strategies to pursue their own spatial-political objectives in the context of nuclear geopolitics and uranium mining. Drawing on archival research and interviews spanning the *longue durée* of nuclear geopolitics from the European Scramble for Africa to China's Ecological Civilization, the analysis demonstrated how both state and non-state actors cultivated new infrastructural strategies in response to changing geopolitical and geoeconomic dynamics. These strategies included Herero and Nama resistance to German land expropriation, the apartheid South African state's pursuit of British investment in Namibian uranium to facilitate its nuclear ambitions, SWAPO's simultaneous cultivation of Western activist and communist state allies in the struggle for Namibian independence, and the contemporary Namibian state's leveraging of China's geopolitical rise to pursue its goals for state-centric extraction.

As Harvey and Knox (2012, 524) note, '[i]nfrastructures do not simply reference or represent political ideology but actively participate, in often unexpected ways, in the processes by which political relations are articulated and enacted'. These processes of articulation and enactment are neither static nor the sole purview of contemporary or historical great powers like the US, China, the Soviet Union, or the UK; they are also shaped by non-great power actors, from states and social movements to political parties and commercial enterprises. These actors, in turn, use infrastructures as both material and symbolic tools in the pursuit of their own changing spatial objectives and strategies. By taking agency and context seriously, research over the *longue durée* can shed new light on multiscalar enactments of infrastructural geopolitics – and great power rivalries as well.

## Note

[1] Namibia has had three names during the time period covered in this chapter: German South West Africa (German colonialism, 1884–1915), South West Africa (South African mandate, 1915–66), and Namibia (liberation struggle and independence, 1966–present). For consistency, I use 'Namibia' throughout with the exception of references to formal name changes or specific actors.

## References

Anti-Apartheid Movement (1982) 'Britain: hands off Namibia!/Solidarity with SWAPO!', National Archives of Namibia.

Barry, A. and Gambino, E. (2020) 'Pipeline geopolitics: subaquatic materials and the tactical point', *Geopolitics*, 25(1): 109–42.

Bauer, G. (1998) *Labor and Democracy in Namibia, 1971–1996*, Athens, OH: Ohio University Press.

Bolton, J. (2018) 'A new Africa strategy: expanding economic and security ties on the basis of mutual respect', Lecture no. 1306, Washington, DC: Heritage Foundation.

Bouzarovwki, S., Bradshaw, M., and Wochnik, A. (2015) 'Making territory through infrastructure: the governance of natural gas transit in Europe', *Geoforum,* 64: 217–28.

Braudel, F. (2009) 'History and the social sciences: the *longue durée*', *Review (Fernand Braudel Center)*, 32(2): 171–203.

Bräutigam, D. (2020) 'A critical look at "debt-trap diplomacy": the rise of a meme', *Area Development and Policy*, 5(1): 1–14.

Brenner, N. (2003) 'Glocalization as a state spatial strategy: urban entrepreneurialism and the new politics of uneven development in Western Europe', in J. Peck and H.W. Yeung (eds) *Remaking the Global Economy: Economic-Geographic Perspectives*, London: SAGE, pp 197–215.

Bridge, G., Özkaynak, B., and Turhan, E. (2018) 'Energy infrastructure and the fate of the nation: introduction to special issue', *Energy Research and Social Science*, 41: 1–11.

Burke, A. (2017) *Uranium*, Cambridge: Polity Press.

Bush, G.W. (2003) 'State of the Union Address', Washington, DC: White House Archive.

Carmody, P. (2020) 'Dependence not debt-trap diplomacy', *Area Development and Policy*, 5(1): 23–31.

Conde, M., and Kallis, G. (2012) 'The global uranium rush and its Africa frontier: effects, reactions and social movements in Namibia', *Global Environmental Change*, 22(3): 596–610.

Cooper, A.D. (ed) (1988) *Allies in Apartheid: Western Capitalism in Occupied Namibia*, London: Macmillan Press.

DeBoom, M.J. (2017) 'Nuclear (geo)political ecologies: a hybrid geography of Chinese investment in Namibia's uranium sector', *Journal of Current Chinese Affairs*, 46(3): 53–83.

DeBoom, M.J. (2020) 'Who is afraid of "debt-trap diplomacy": geopolitical narratives, agency, and the multiscalar distribution of risk', *Area Development and Policy*, 5(1): 15–22.

DeBoom, M.J. (2021) 'Climate necropolitics: ecological civilization and the distributive geographies of extractive violence in the Anthropocene', *Annals of the American Association of Geographers*, 111(3): 900–12.

DiCarlo, J. and Schindler, S. (2020) 'Will Biden pass the America LEADS Act and start a new Cold War with China?', *Global Policy*, 10 December. Available from: www.globalpolicyjournal.com/blog/10/12/2020/will-biden-pass-america-leads-act-and-start-new-cold-war-china.

Dierks, K. (2002) *Chronology of Namibian History: From Pre-Historical Times to Independent Namibia*, Windhoek: Namibia Scientific Society.

Dobler, G. (2008) 'Solidarity, xenophobia and the regulation of Chinese businesses in Namibia', in C. Alden, D. Large, and R. Soares de Oliveira (eds) *China Returns to Africa: A Rising Power and a Continent Embrace*, London: Oxford University Press, pp 237–55.

Enns, C., and Bersaglio, B. (2020) 'On the coloniality of "new" mega-infrastructure projects in East Africa', *Antipode*, 52(1): 101–23.

Erichsen, C. and Olusoga, D. (2010) *The Kaiser's Holocaust: Germany's Forgotten Genocide and the Colonial Roots of Nazism*, London: Faber and Faber.

Gewald, J.-B. (1999) *Herero Heroes: A Socio-Political History of the Herero of Namibia, 1890–1923*, Columbus, OH: Ohio State University Press.

Goldschmidt, B. (1989) 'Uranium's scientific history 1789–1939', in *Proceedings of the International Symposium on Uranium and Nuclear Energy*, London: Uranium Institute, pp 6–8.

Hartman, A. (2009) 'Police seize stolen uranium', *The Namibian*, 8 September. Available from: www.namibian.com.na/index.php?id=57445&page=archive-read.

Harvey, D. (2005) *A Brief History of Neoliberalism*, Oxford: Oxford University Press.

Harvey, P. and Knox, H. (2012) 'The enchantments of infrastructure', *Mobilities,* 7(4): 521–36.

Hecht, G. (2009) 'Hopes for the radiated body: uranium miners and transnational technopolitics in Namibia', *Journal of African History,* 51: 213–34.

Hecht, G. (2012) *Being Nuclear: Africans and the Global Uranium Trade*, Cambridge, MA: MIT Press.

Hunt, C.D. (1977) 'Canadian policy and the export of nuclear energy', *The University of Toronto Law Journal*, 27(1): 69–104.

Jia, F. and Bennett, M.M. (2018) 'Chinese infrastructure diplomacy in Russia: the geopolitics of project type, location, and scale', *Eurasian Geography and Economics*, 59(3–4): 340–77.

Joint Committee on Atomic Energy (1970) *Prelicensing Antitrust Review of Nuclear Powerplants, Part 1*, Washington, DC: Joint Committee on Atomic Energy.

Koponen, J. (1993) 'The partition of Africa: a scramble for a mirage?', *Nordic Journal of African Studies*, 2(1): 117–35.

Kristof, N.D. (2006) 'In praise of the maligned sweatshop', *New York Times*, 6 June. Available from: www.nytimes.com/2006/06/06/opinion/06kristof.html.

League of Nations (1922) 'Permanent Mandates Commission: Minutes of the Second Session, 1–11 August 1922', C-548-M-330, Geneva: United Nations Archive.

Lee, C.K. (2017) *The Specter of Global China: Politics, Labor, and Foreign Investment in Africa*, Chicago, IL: University of Chicago Press.

Liu, W. and Dunford, M. (2016) 'Inclusive globalization: unpacking China's Belt and Road Initiative', *Area Development and Policy,* 1(3): 323–40.

Louw, G. (2018) *A Tiger by the Tail: The Story of the Discovery of Rössing Uranium*, self-published.

Melber, H. (2013) 'Africa and China: old stories or new opportunities?', in T. Murithi (ed) *Handbook of Africa's International Relations*, Abingdon: Routledge, pp 333–42.

Melber, H. (2014) *Understanding Namibia: The Trials of Independence*, Oxford: Oxford University Press.

Murton, G. and Lord, A. (2020) 'Trans-Himalayan power corridors: infrastructural politics and China's Belt and Road Initiative in Nepal', *Political Geography*, 77(102100): 1–12.

Reddy, E.S. (ed) (1994) *Anti-Apartheid Movement and the United Nations: Statements and Papers of Abdul S. Minty*, New Delhi: Sanchar Publishing House.

Roberts, A. (1980) *The Rössing File: The Inside Story of Britain's Secret Contract for Namibian Uranium*, London: Namibia Support Committee.

Saunders, C. (2009) 'Namibian solidarity: British support for Namibian independence', *Journal of Southern African Studies*, 35(2): 437–54.

Saunders, C. (2018) 'SWAPO, Namibia's liberation struggle and the Organisation of African Unity's Liberation Committee', *South African Historical Journal*, 70(1): 152–67.

Saunders, C. (2019) 'SWAPO's "Eastern" connections, 1966–1989', in L. Dallywater, C. Saunders, and H.A. Fonseca (eds) *Southern African Liberation Movements and the Global Cold War 'East': Transnational Activism 1960–1990*, Berlin: Walter de Gruyter, pp 57–76.

Schindler, S. and Kanai, J.M. (2019) 'Getting the territory right: infrastructure-led development and the re-emergence of spatial planning strategies', *Regional Studies*, 55(1): 40–51.

Steinmetz, G. (2007) *The Devil's Handwriting: Precoloniality and the German Colonial State in Qingdao, Samoa, and Southwest Africa*, Chicago, IL: University of Chicago Press.

SWAPO (1981) *To Be Born a Nation: The Liberation Struggle for Namibia*, London: Zed Press.

Taylor, I. (1997) 'China and SWAPO: the role of the People's Republic in Namibia's liberation and post-independence relations', *The South African Journal of International Affairs*, 5(1): 110–22.

United Nations (1974) *Report of the United Nations Council for Namibia, Volume 1*, New York: United Nations. Available from: https://undocs.org/pdf?symbol=en/A/35/24%5BVOL.III%5D(SUPP).

United Nations Council on Namibia (UNCN) (1980) *Hearings on Namibian Uranium*, New York: United Nations Digital Library.

Voeltz, R.A. (1988) *German Colonialism and the South West Africa Company, 1884–1914*, Athens, OH: Ohio University Press.

Zimmer, J. and Zeller, J. (2008) *Genocide in German South-West Africa: The Colonial War (1904–1908) in Namibia and Its aftermath*, Monmouth: Merlin Press.

# 10

# Argentina and the Spatial Politics of Extractive Infrastructures under US–China Tensions

*Marcelo I. Saguier and Maximiliano F. Vila Seoane*

Tensions between the US and China over the role that infrastructure plays in their geopolitical imaginaries and prospective scenarios of global governance and (geo)politics are central to ongoing academic and public debates. Disputes revolve around the politics and competition to lead, influence, and control the technical capacities to forge global interconnectivity through infrastructure integration (Brenner, 2004, 2019; Schiller, 2011; Schindler et al, 2021). This raises questions surrounding the relationship between infrastructure integration and shifting configurations of global power, including the implications of US–China competition on industrializing countries in the Global South (Mora, 2018, 2019; Vila Seoane, 2020, 2021; Vila Seoane and Saguier, 2020).

We explore how US–China tensions influence the politics of infrastructure in Argentina. Specifically, we look at the shale gas and oil and lithium mining sectors where the US and Chinese governments have expressed particular interest in promoting infrastructure projects or resisting rival efforts to do so. We ask: how is the manifestation of these infrastructure projects related or attributable to US–China competition? Second and more importantly, what can we learn about the processes around which spatial politics are structured and contested within ongoing trajectories of global infrastructure integration?

Mainstream media and some scholarship have identified an emerging 'new Cold War' between the US and China, but in the case of Argentina, this discourse obscures rather than illuminates the ongoing processes of state restructuring (Cox and Schechter, 2002; Brenner, 2004, 2019). Argentina's state spatial strategies are geared towards the transformation of state–society–nature relations in the context of a global political economy no longer centred in the US with increasing orientation towards China. Drawing on the concept of the infrastructure state, which exhibits agency in pursuit of state spatial strategies, we highlight its multiscalar nature through examples that: (a) are publicly supported by US and/or Chinese authorities; (b) are paradigmatic in their sector; and (c) generate connectivity to global value chains where the US and/or Chinese companies have leading and/or competing roles. We show how spatial politics are (re-)organized as part of the ongoing restructuring of the state in response to US–China competition. In this context, actors situated at multiple scales pursue diverse strategies of territorial integration in attempts to connect with global value chains. The infrastructure state is therefore multiscalar and adept at navigating the complexity of international relations, yet its domestic policy has sought to transform state–society–nature relations, and this has provoked socio-ecological conflicts.

## Argentina in the global politics of infrastructure: in the midst of a 'new Cold War'?

Argentina experienced significant economic challenges during the last decade, such as difficulty in sustaining economic growth and persistent and high levels of poverty (approximately 30 per cent in 2019, which increased up to 40 per cent after COVID-19). Reducing inflation and attracting foreign direct investment (FDI) has been an enormous task. Although the agricultural sector that is focused on soybean production remains Argentina's largest source of export revenue, its current accumulation regime is skewed in favour of a small minority, rather than satisfying the needs of the country's 45 million inhabitants. Consequently, diversifying the country's economy and increasing exports is an overarching objective regardless of the political party in power. Multiple scales of government within Argentina are developing spatial strategies with these ends in mind.

Argentina's relationship with the US was strained during the centre-left administration of Néstor Kirchner (2003–07). Emerging from the crisis that erupted in 2001 after decades of neoliberal policies, Kirchner's government was characterized by a neo-Keynesian and redistributive orientation and advocated South American regionalism as a developmental alternative to the US-sponsored market-led integration such as the Free Trade Area for the Americas (Saguier, 2007; Riggirozzi and Wylde, 2018; Vivares, 2018). During the subsequent administrations of President Cristina Kirchner (2008–15), the US sided with debtors in escalating financial disputes (Míguez, 2016), which locked the country out of international financial markets. As part of a policy of not borrowing from the International Monetary Fund (IMF), the Argentine government turned to other funding sources, mainly

China, which offered the country's central bank several currency swap deals in Renminbi (Brenta and Larralde, 2018). Furthermore, in 2014 Argentina and China signed a comprehensive strategic partnership, which paved the way for increasing Chinese investments in key strategic infrastructure projects such as the upgrading of railways, the construction of hydroelectric dams (Mora, 2018), and the installation of a deep space station in Patagonia (Frenkel and Blinder, 2020).

During the campaign for Argentina's 2015 national elections, the centre-right opposition coalition, led by ex-businessman Mauricio Macri, promised a realignment with the US (Frenkel and Azzi, 2018). His victory heralded a reversal of the country's foreign policy, and his anti-China stance was welcomed by US analysts (Ellis, 2017). The new ruling coalition, Cambiemos, believed that the political realignment with the Obama administration would precipitate much awaited 'investments rain' from Western foreign investors. However, these expectations were largely unmet, while the context changed with the arrival of Donald Trump and the so-called US trade war with China. In a sharp turnaround from his campaign promises, Macri reprioritized relations with China (Busso, 2017). He participated in the first Belt and Road Forum, citing the BRI as an 'an opportunity that we do not want to miss' (Casa Rosada, 2017). During the trip, the administration signed the '2017–2021 China-Argentina comprehensive five-year plan for cooperation in infrastructure' (Ministerio de Finanzas, 2017), spanning infrastructure projects in transport, energy, natural resources, and the information and communications technology sectors.

Macri's policy towards China provoked consternation in Washington, where persuasion and coercion were used in an attempt to push Argentina away from China. For instance, after a bilateral meeting between President Trump and President Macri at the 2018 G20 summit in Buenos Aires, the US spokeswoman claimed that both had agreed to combat China's 'predatory behavior', a statement that Argentina's chancellor rushed to deny (Rodriguez, 2018). Likewise, since 2018 the US has made several pledges to invest in Argentina to compete with China, some of which we detail in the following sections. Perhaps the most significant support to the Cambiemos administration was US pressure on the IMF to approve the largest loan in its history to Argentina. According to the US Executive Director at the IMF, its purpose was to prevent Peronists' return to power (Lejtman, 2020). Despite the IMF loan, however, Peronists won the following election and have been in power since December 2019. Immediately after the election, the US expressed concerns over China to a leading member of the new ruling coalition, Sergio Massa, who answered that Argentina is 'friends of all, but satellite of none' (Lugones, 2019). In 2020 Argentina became a member of the Asian Infrastructure Investment Bank, a process initiated by Cristina Kirchner in 2015.

This summary of a mainstream reading of Argentine foreign policy vis-à-vis China and the US suggests the country has struggled with the strategic decision to either remain unaligned or side with one or the other. Argentina is not the only country to face this difficult balancing act, as tensions between the US and China have escalated in the domains of trade, currency, technological innovation, and military cooperation. Some have interpreted these tensions as evidence of a 'new Cold War', which is accompanied by rhetoric that identifies China as a threat to US vital interests (Mearsheimer, 2006; Pillsbury, 2016; Allison, 2017; Pompeo, 2020). The diffusion of this narrative has direct implications for countries like Argentina that have enhanced their economic and political relations with China in recent decades. However, in many ways it is reminiscent of accounts offered by right-wing leaning publications at the turn of the century, which perceived China as an ideological, strategic, and economic challenge to the US (Broomfield, 2003). Its contemporary resurrection obscures many of the state restructuring processes taking place in Argentina to enable infrastructure projects which can be interpreted as responses to changing configurations of the global political economy that have rendered infrastructure integration a field of geopolitical contestation.

Language and narratives define roles, legitimacy, expectations, and shape reality. They capture the political imagination on the prospects and options that countries like Argentina face in a context of a transforming global political economy no longer underpinned by US leadership. In this case, a new Cold War narrative preconfigures options and forces countries to align with the US or China. In response to this pressure, a group of prominent intellectuals associated with former progressive governments in Latin America warn about the dangers of falling pray of this polarizing discourse, particularly at a time of unprecedented economic crisis amid weak and fragmented South American regional governance (Fortín et al, 2020). They call for an 'active non-alignment' foreign policy that is capable of calibrating a middle path between Washington and Beijing, where states can maximize benefits in their links with the global economy and establish their own development models (Fortín et al, 2020, 122). This view obscures the spatial dimension of the internationalization of the state, where changing relations to global capitalism historically reconfigure relations between politics and space (Brenner, 2004; Sassen, 2006).

The interplay between the scaling of political space and state spatial projects (Brenner, 2004) is increasingly influenced by developmental challenges. As noted earlier, Argentina has experienced a prolonged economic crisis. This has deepened its dependence on extractive sectors to generate export revenues to manage ongoing macroeconomic tensions. Extractivist-led policies have generated increasing strains on the environment leading to a social mobilization and resistance; notably in opposition to mining projects

(Saguier and Peinado, 2016), hydroelectric dams (Saguier, 2017; Mora, 2018) and industrial agriculture of the soja complex. With the consolidation of the extractivist elements of a development strategy, space politics relations are shaped by governments' responses at multiple scales driven by the opportunities to link up in global chains that open in a complex geopolitical context. To explore these dynamics, we analyse two sectors: shale gas and oil in which US firms are the primary foreign investors, and lithium extraction in which Chinese actors are the main investors. We argue that Argentina is an 'infrastructure state' in the sense that it understands its developmental objectives to be dependent on significant infrastructure investments, which in turn (re)produce relations between space and politics at multiple scales (Brenner, 2004; Sassen, 2006; Schindler et al, 2021). We find that state actors strategically employ opportunities offered by the initiatives that the US and China provide amid their rivalry, yet rather than choosing sides, this is part of a dynamic multiscalar spatial strategy designed to foster export-oriented industrialization and economic growth.

## Shale gas and oil: the allure of *Vaca Muerta*

Argentina became a net importer of gas and oil for the first time in 2011, causing even more damage to the precarious financial situation of the country. In this context, former President Cristina Fernández de Kirchner introduced a bill in Congress to renationalize Yacimientos Petrolíferos Fiscales (YPF), the main gas and oil firm. It had been under the control of Repsol, a Spanish firm, since 1999, which had been accused of underinvesting in infrastructure and compromising national energy sovereignty. In 2012, Congress passed Law 26.741, reasserting state sovereignty over hydrocarbons by nationalizing Repsol YPF.

Most consequentially, the law unleashed a state-led spatial strategy to invest in developing the Vaca Muerta reserve, a geological formation that spans 30,000 km$^2$ in the Patagonian province of Neuquén. By some estimates, Vaca Muerta has the second-largest shale gas and the fourth-largest shale oil reserves in the world (US Energy Information Administration, 2015). Unsurprisingly, the development of Vaca Muerta has become the main state spatial strategy for every administration in power, and YPF its main instrument of exploration and extraction. Indeed, the national company has approximately 40 per cent of the basin in concessions. Similarly, the administrations of Neuquén Province, whose GDP depends largely on hydrocarbons, envision Vaca Muerta as a key driver of the provincial economy.

The extraction of shale gas and oil requires the development of infrastructure and significant investments, the cost of which is prohibitive to the national and provincial governments. The main political forces that govern Argentina accept that more foreign investments are needed, and efforts to woo investors have focused on US actors given their managerial and technical experience pioneering the shale gas and oil industry in the Permian Basin in Texas and New Mexico. Partnerships with US firms did not take long to materialize. In 2013, YPF signed its first collaboration agreement to exploit Vaca Muerta with Chevron. The US firm committed about US$1.24 billion to a pilot project to drill 100 wells and expand activity in subsequent years. The text of the agreement was confidential, and anti-fracking activists protesting its ratification by the provincial legislature were violently repressed, exposing a nasty face of the state in its pursuit of energy self-sufficiency. It is important to stress that the agreement was signed during a period in which, from a foreign policy perspective, bilateral relations between Argentina and the US were strained, with the former more aligned with China.

Since then, international gas and oil firms from other nations – such as Shell, Total, and Petronas – have invested in Vaca Muerta. Even Chinese and Russian state-firms, like Sinopec and Gazprom, have attempted to partner with YPF, though unsuccessfully thus far. Nonetheless, the state has prioritized partnerships with US firms. For instance, in 2017, Exxon Mobile confirmed a US$200 million investment in Vaca Muerta in partnership with a provincial state firm. Furthermore, Argentina and the US signed an Energy Cooperation Framework Agreement in 2018 that contains a core objective of exploiting shale natural gas, primarily from Vaca Muerta. According to the US State Department (2020), this agreement is part of the Growth in the Americas Initiative, which was designed to compete with China's BRI in the region. It was introduced as a whole-of-government approach to 'support economic development by catalyzing private sector investment in energy and other infrastructure projects across Latin America and the Caribbean' (US State Department, 2019), supposedly in a sustainable manner, something that the US accuses the BRI of lacking. Similarly, in 2019, the former Overseas Private Investment Corporation (OPIC) approved a US$450 million loan for two Argentine shale gas and oil companies operating in Vaca Muerta (US Embassy in Argentina, 2019).

The state spatial strategy of developing Vaca Muerta has faced staunch resistance due to its socio-environmental impacts (Riffo, 2017; Delgado, 2018). Indeed, the main technology of extraction, hydraulic fracturing (fracking), causes serious environmental harms, such as polluting water and air, or even inducing earthquakes, with grave consequences for nearby inhabitants. In response to these potential effects, neighbouring indigenous populations, such as the Mapuche, have mobilized in protest and been at the forefront of resisting the state fracking strategy (Riffo, 2017). At the same time, civil society organizations from the US organized opposition to the project on the grounds that shale gas and oil extraction is unsustainable (FARN, 2019), yet these efforts were unsuccessful, perhaps because of the project's geopolitical implications.

Analysts from the Center for Strategic and International Studies have argued that the US should provide funding to develop Vaca Muerta, and prevent other nations such as China or Russia from operating there (Runde and Schreyer, 2019). Similarly, in 2018, the US Embassy in Argentina announced that the US Southern Command would build a US$1.3 million Emergency Operations Center to assist the province of Neuquén in disaster preparedness and response (US Embassy in Argentina, 2018). This news unleashed a flurry of criticism, protests, and demands from members of parliament, who denounced the project as a mere façade designed to protect American investments in Vaca Muerta and to spy on a nearby Chinese satellite deep space station. As a result, the project stalled due to political pressure and resistance, together with the change in national administration in Argentina, yet US concerns over Chinese investments in the country remain.

Irrespective of political leanings, the urge to develop Vaca Muerta to secure energy self-sufficiency has tightened the links between provincial and state actors with firms and investors from the US. Argentina's ambassador to the US under President Fernández has already pledged to seek new investments in Vaca Muerta. However, it remains to be seen whether Biden's policies to address climate change will influence US investments in fracking. In this context, Chinese interest in funding infrastructure projects for Vaca Muerta have resurfaced (Garrison, 2020), such as in a rail project to export gas and oil to Brazil. This shows how the need for funds for infrastructure is a better guide to understand Argentina's geopolitical strategy vis-à-vis the US and China, rather than a new Cold War frame. Indeed, as the following case shows, the malleable way of linking territories makes it hard for Argentine authorities to choose between one side or the other.

## Lithium: white gold and electromobility

When Xi Jinping arrived in Buenos Aires for the 2018 G20 summit, he was received at the airport by Gerardo Morales, the Governor of Jujuy, one of Argentina's smallest and poorest provinces in the remote northeast. Since assuming office in December 2015, Morales has made continuous efforts to cultivate links with China. His efforts paid off in the form of a US$331.5 million loan by the Export–Import Bank of China for the construction of three solar power plants to be built by Power China and Shanghai Electric Power Construction (*El Tribuno*, 2017). Likewise, the Chinese firm ZTE won a US$24 million tender to build communications and video surveillance systems for the province (Gobierno de Jujuy, 2019b), a deal that international media quickly portrayed as an extension of China's surveillance state. One of the most significant links between Jujuy and Chinese capital is related to the extraction of lithium.

Known as 'white gold', this alkali metal is the key component of lithium-ion batteries, which are central to current and future consumer electronics and electric cars. In light of the shift in the transport industry towards electromobility, the control of lithium extraction and supply chains has become a much-coveted target for mining companies and states. However, lithium is very unevenly distributed. Almost 60 per cent of global lithium reserves are located in South America (Martin et al, 2017), mainly in the so-called Lithium Triangle that spans territories across Argentina, Bolivia, and Chile. The province of Jujuy contains 37 per cent of Argentina's known lithium reserves (López et al, 2019).

The Jujuy government's development strategy increasingly centres on the extraction of lithium. International mining firms from Australia, Canada, China, Japan, Italy, among others already operate in the province or have expressed interest in doing so. For instance, the Canadian firm Lithium Americas Corporation has been working in Jujuy since 2007. The provincial government also formed a joint venture with an Italian firm, SERI Group, to produce batteries in the province and add value to its exports (Gobierno de Jujuy, 2017). Despite these partnerships and further interest from Western enterprises, the provincial government identified China as the central partner to develop its lithium ecosystem (Gobierno de Jujuy, 2018). There are two reasons for this decision. First, the Jujuy government anticipates that China will not only invest heavily in the lithium sector, but that it will also support other state spatial projects in the province. Second, the preference for China is based on a forecast of future demand, which will be driven by its potential electric car market.

Jujuy's public officials have made frequent business and political trips to China in attempts to attract investments. Upon their return they stress the unbounded opportunities such relations offer the province's development, in particular in lithium mining. These efforts have borne fruit, and in 2018 the Chinese firm Ganfeng Lithium entered into partnership with Lithium Americas Corporation. Together they invested US$160 million in the Cauchari–Olaroz project to increase lithium production from 25,000 to 40,000 tons per year (Gobierno de Jujuy, 2019a). Ganfeng Lithium subsequently acquired a majority stake over its joint venture with Lithium Americas in 2020, exemplifying increasing Chinese interest in Argentina. Other firms from China's Guizhou province, with which Jujuy entered into a twinning agreement, also expressed interest in investing in the province's lithium mines (Gobierno de Jujuy, 2019c).

Argentina's national government is proactively supporting Jujuy's state spatial strategy, and it considers lithium a potential driver of export-oriented growth. Indeed, in 2021, the Fernández administration signed a memorandum of understanding with Guoxuan Hi-Tech, a Chinese company specializing in lithium-ion power batteries, to manufacture batteries and electric cars in the country. Thus, lithium extraction and joint ventures with Chinese firms have becomes central components of a broader industrial strategy (Télam, 2021). Moreover, the national government plans to introduce an electromobility law designed to attract investment and foster industrial growth, thereby making Argentina a regional leader in electric vehicle value chains.

Similar to the previous case, the exploitation of white gold has considerable environmental costs. The lithium extraction process uses evaporation ponds that require huge amounts of water. Although there is no consensus on the sector's long-term environmental impacts on the region, critics contend that mining companies will jeopardize access to freshwater in what is already an arid area (Wanger, 2011; Flexer et al, 2018). Furthermore, the extraction process employs chemicals that may cause irreversible damage to groundwater, biodiversity, and human health if not properly managed (Wanger, 2011), which explains why indigenous communities living nearby oppose lithium extraction (Gullo and Fernández Bravo, 2020). Yet, the government has so far subordinated these concerns in the name of development and progress (Fornillo, 2018).

Although the US–China rivalry has thus far not influenced Argentina's state spatial strategy with regard to lithium extraction and processing, the geopolitics of lithium extraction influenced events in neighbouring Bolivia. It has the largest known reserves of lithium in the world and according to former President Evo Morales, US support of Bolivia's 2019 *coup d'état* was a form of retaliation for Bolivia's alignment with China and Russia in the lithium sector (AFP, 2019). Even Tesla CEO Elon Musk infamously tweeted support for the US intervention in Bolivia. Although this allegation was dismissed as mere conspiracy by American analysts and other critics of Morales, it is clear that China's clout over global lithium supply chains troubles US policy makers who fear the military applications of lithium-ion batteries and what an increasing dependence on China for acquiring such a strategic resource might mean (US–China Economic and Security Review Commission, 2019).

Overall, attempts by Jujuy's government to foster close links with Chinese actors and firms to develop the lithium industry exemplify the rescaling of Argentina's infrastructure state and the increasing importance of subnational diplomacy with China and Chinese provinces.

## Conclusion

This chapter has explored how Argentina is positioned vis-à-vis the rivalry between China and the US. According to many commentators this rivalry is structuring an emergent bipolar international order in which states must side with the US or China, or stake out a non-aligned position. This model centres the analysis on states and neglects other actors shaping inter- and transnational relations such as firms and civil society organizations. Furthermore, it assumes that the state is a unitary actor whose foreign policy is driven by external pressures, neglecting the spatiality of the state and its spatial strategies to transform how territories link with global value chains.

Argentina is geographically distant both from China and the US, and its relations with each are neither adversarial nor cooperative. Government at multiple levels has sought to expand Argentina's connectivity to global value chains, and attract investment from both countries. In the case of shale gas and oil in Neuquén Province, partnerships with American actors are prioritized, while Chinese actors have been identified as the most suitable partners in the lithium sector in Jujuy. Therefore, the 'active non-alignment' proposal of progressive thinkers of the region is not an outcome of the 'new Cold War' reality, but rather it reflects the heterogeneous ways of linking with China and the US. Rather than a security state at the centre of geopolitical struggles as some propose, Argentina is emerging as a multiscalar infrastructure state in pursuit of spatial strategies whose overriding objective is to stabilize and diversify the economy. Rather than geopolitics, this strategy is animated by contestations surrounding state–society–nature relations that are configured to accommodate resource extraction. Indeed, Argentina's infrastructure state overwhelmingly focuses on the promise of economic growth through the exploitation of nature, neglecting the serious environmental and social consequences of extractivist development.

## References

AFP (2019) 'Morales claims US orchestrated "coup" to tap Bolivia's lithium', 24 December. Available from: https://news.yahoo.com/morales-claims-us-orchestrated-coup-tap-bolivias-lithium-211124837.html

Allison, G. (2017) *Destined for War: Can America and China Escape Thucydides's Trap?* New York: Houghton Mifflin Harcourt.

Brenner, N. (2004) *New State Spaces: Urban Governance and the Rescaling of Statehood*, Oxford: Oxford University Press.

Brenner, N. (2019) *New Urban Spaces: Urban Theory and the Scale Question*, New York: Oxford University Press.

Brenta, N. and Larralde, J. (2018) 'La internacionalización del renminbi y los acuerdos de intercambio de monedas entre Argentina y China, 2009–2018', *Ciclos*, 25(51): 55–84.

Broomfield, E.V. (2003) 'Perceptions of danger: the China threat theory', *Journal of Contemporary China*, 12(35): 265–84.

Busso, A. (2017) 'El rol de los Estados Unidos en el diseño de política exterior del gobierno de Mauricio Macri. Conceptos básicos para su análisis', *Anuario de Relaciones Internacionales*, 23, Universidad Nacional de La Plata.

Casa Rosada (2017) 'El Presidente expuso en el foro 'Una Franja y una Ruta para la Cooperación Internacional'. Available from: www.casarosada.gob. ar/slider-principal/39552-el-presidente-expuso-en-el-foro-una-franja-y-una-ruta-para-la-cooperacion-internacional.

Cox, R. and Schechter, M.G. (2002) *The Political Economy of a Plural World: Critical Reflections on Power, Morals and Civilization*, London: Routledge.

Delgado, E. (2018) 'Fracking Vaca Muerta: socioeconomic implications of shale gas extraction in northern Patagonia, Argentina', *Journal of Latin American Geography*, 17(3): 102–31.

El *Tribuno* (2017) 'Se firmó en China el convenio para la construcción del parque de energía solar en Cauchari', 23 November. Available from: https://www.eltribuno.com/jujuy/nota/2017-11-23-16-13-2-se-firmo-en-china-el-convenio-para-la-construccion-del-parque-de-energia-solar-en-cauchari.

Ellis, R.E. (2017) 'Argentina at the crossroads again: implications for the United States and for the region', *Military Review*, 97(2): 27–37.

FARN (2019) 'Members of Congress, Argentinian and US groups call on OPIC to reject financing for fracking in Argentina', 10 September. Available from: https://farn.org.ar/piden-a-opic-que-rechace-el-financiamiento-para-el-fracking-en-argentina/.

Flexer, V., Baspineiro, C.F., and Galli, C.I. (2018) 'Lithium recovery from brines: a vital raw material for green energies with a potential environmental impact in its mining and processing', *Science of the Total Environment*, 639(15): 1188–204.

Fornillo, B. (2018) 'La energía del litio en Argentina y Bolivia: comunidad, extractivismo y posdesarrollo', *Colombia Internacional*, 93: 179–201.

Fortín, C., Heine, J., and Ominami, C. (2020) 'Latinoamérica: no alineamiento y la segunda', *Foreign Affairs Latinoamérica*, 20(3): 107–15.

Frenkel, A. and Azzi D. (2018) 'Cambio y ajuste: la política exterior de Argentina y Brasil en un mundo en transición (2015–2017)', *Colombia Internacional*, 96: 177–207.

Frenkel, A. and Blinder, D. (2020) 'Geopolítica y cooperación espacial: China y América del Sur', *Desafíos*, 32(1): 1–30.

Garrison, C. (2020) 'PowerChina eyes rail project to transport oil and gas from Argentina's Vaca Muerta', Reuters, 17 September. Available from: www.reuters.com/article/argentina-shale-rail-idUKL1N2GE2CN

Gobierno de Jujuy (2017) 'Morales concreta la instalación de la fábrica de baterías de Litio', 9 June. Available from: https://prensa.jujuy.gob.ar/2017/06/09/morales-concreta-la-instalacion-de-la-fabrica-de-baterias-de-litio/.

Gobierno de Jujuy (2018) 'Jujuy apuesta a un ecosistema de litio y más inversiones desde China', 29 May. Available from: https://prensa.jujuy.gob.ar/2018/05/29/jujuy-apuesta-a-un-ecosistema-del-litio-y-mas-inversiones-desde-china/.

Gobierno de Jujuy (2019a) 'Afianzan la producción de Litio en la Puna Jujeña', 10 September. Available from: https://prensa.jujuy.gob.ar/2019/09/10/afianzan-la-produccion-de-litio-en-la-puna-jujena/.

Gobierno de Jujuy (2019b) 'Avanza el Plan Jujuy seguro e interconectado'. Available from: https://prensa.jujuy.gob.ar/ekel-meyer/avanza-el-plan-jujuy-seguro-e-interconectado-n64377.

Gobierno de Jujuy (2019c) 'Jujuy y China tienden lazos en el campo minero', 10 July. Available from: https://prensa.jujuy.gob.ar/2019/07/10/la-mineria-acerca-lazos-comerciales-entre-jujuy-y-china/

Gullo, E. and Fernández Bravo, E. (2020) 'White gold: the violent water dispute in Argentina', *Diálogo Chino*, 20 May. Available from: https://dialogochino.net/en/extractive-industries/35354-white-gold-the-violent-water-dispute-in-argentina/

Lejtman, R. (2020) 'Un asesor de Trump reveló por qué ayudaron al gobierno de Macri a acceder a un rescate del Fondo Monetario Internacional', *infobae*, 28 July. Available from: www.infobae.com/politica/2020/07/28/un-asesor-de-trump-revelo-por-que-ayudaron-al-gobierno-de-macri-a-acceder-a-un-rescate-del-fondo-monetario-internacional/

López, A., Obaya, M., Pascuini, P., and Ramos A. (2019) *Litio en la Argentina. Oportunidades y desafíos para el desarrollo de la cadena de valor*, Secretaría de Ciencia, Tecnología e Innovación Productiva.

Lugones, P. (2019) 'Reunión reservada en Washington Estados Unidos le planteó a Massa su preocupación por cómo sería un posible gobierno de Alberto Fernández', *Clarín*, 8 October. Available from: www.clarin.com/politica/unidos-planteo-massa-preocupacion-posible-gobierno-alberto-fernandez_0_2F2d5qub.html

Martin, G., Rentsch, L., Höck, M., and Bertau, M. (2017) 'Lithium market research: global supply, future demand and price development', *Energy Storage Materials*, 6: 171–9.

Mearsheimer, J.J. (2006) 'China's unpeaceful rise', *Current History*, 105(690): 160–62.

Míguez, M.C. (2016) 'La política exterior Argentina y su vinculación con los condicionamientos internos del Siglo XXI', *Relaciones Internacionales*, 89(2): 125–42.

Ministerio de Finanzas (2017) 'Plan Quinquenal Integrado China-Argentina para la Cooperación en Infraestructura (2017–2021)'. Available from: https://servicios.infoleg.gob.ar/infolegInternet/anexos/270000-274999/274612/norma.htm

Mora, S. (2018) 'Resistencias sociales a la cooperación de China en infraestructura: Las represas Kirchner-Cepernic en Argentina', *Colombia Internacional*, 94: 53–81.

Mora, S. (2019) 'El *Going Out* agrícola de China. Un análisis de su desarrollo en Argentina', *Si Somos Americanos*, 19(2): 89–103.

Pillsbury, M. (2016) *The Hundred-Year Marathon: China's Secret Strategy to Replace America as the Global Superpower*, New York: St. Martin's Griffin.

Pompeo, M.R. (2020) 'Communist China and the free world's future', US Department of State. Available from: https://2017-2021.state.gov/communist-china-and-the-free-worlds-future-2/index.html

Riffo, L.N. (2017) 'Fracking and resistance in the Land of Fire', *NACLA Report on the Americas*, 49(4): 470–75.

Riggirozzi, P. and Wylde, C.(eds) (2018) *Handbook of South American Governance*, London: Routledge.

Rodriguez, J. (2018) 'White House statement drags Argentina into US–China brawl', *Politico*, 30 November. Available from: www.politico.com/story/2018/11/30/us-china-argentina-g20-1036939.

Runde, D.F. and Schreyer, W.A. (2019) 'OPIC should strengthen US–Argentine partnership with Vaca Muerta deal', *The Hill*, 10 September. Available from: https://thehill.com/opinion/finance/460621-opic-sho uld-strengthen-us-argentine-partnership-with-vaca-muerta-deal.

Saguier, M.I. (2007) 'The Hemispheric Social Alliance and the Free Trade Area of the Americas process: the challenges and opportunities of transnational coalitions against neo-liberalism', *Globalizations*, 2(4): 251–65.

Saguier, M.I. (2017) 'Transboundary water governance in South America', in P. Riggirozzi and C. Wylde (eds) *Handbook of South American Governance*, London: Routledge, pp 373–84.

Saguier, M.I. and Peinado, G. (2016) 'Canadian mining investments in Argentina and the construction of a mining-development nexus', *Latin American Policy*, 7(2): 267–87.

Sassen, S. (2006) *Territory, Authority, Rights: From Medieval to Global Assemblages*, Princeton, NJ: Princeton University Press.

Schiller, D. (2011) 'Geopolitical-economic conflict and network infrastructures', *Chinese Journal of Communication*, 4(1): 90–107.

Schindler, S., DiCarlo, J. and Paudel, D. (2021) 'The new cold war and the rise of the 21st century infrastructure state', *Transactions of the Institute of British Geographers*, 47(2): 331–46.

Télam (2021) 'Argentina y una empresa china firman acuerdo para producir vehículos eléctricos en el país', 3 February. Available from: www.telam.com.ar/notas/202102/543541-argentina-y-empresa-china-firman-acuerdo-para-producir-vehiculos-electricos-en-el-pais.html.

US–China Economic and Security Review Commission (2019) 'Report to Congress'. Available from: www.uscc.gov/annual-report/2019-annual-report-congress

US Embassy in Argentina (2018) 'United States funds the construction of an Emergency Operations Center for Neuquén Province', 1 June. Available from: https://ar.usembassy.gov/united-states-funds-the-construction-of-an-emergency-operations-center-for-neuquen-province/.

US Embassy in Argentina (2019) 'OPIC approves financing for energy projects in Argentina', 11 September. Available from: https://ar.usemba ssy.gov/opic-approves-financing-for-energy-projects-in-argentina/.

US Energy Information Administration (2015) 'World Shale Resource Assessments', 24 September. Available from: www.eia.gov/analysis/studies/ worldshalegas/

US State Department (2019) 'Growth in the Americas'. Available from: www. state.gov/wp-content/uploads/2019/11/America-Crece-One-Pager-003-508.pdf.

US State Department (2020) 'US Relations with Argentina, Bilateral Relations Fact Sheet', Bureau of Western Hemisphere Affairs, 10 August.

Vila Seoane, M.F. (2020) 'Alibaba's discourse for the digital Silk Road: the electronic World Trade Platform and "inclusive globalization"', *Chinese Journal of Communication*, 13(1): 68–83.

Vila Seoane, M.F. (2021) 'Chinese and US AI and cloud multinational corporations in Latin America', in T. Keskin and R.D. Kiggins (eds) *Towards an International Political Economy of Artificial Intelligence*, London: Palgrave Macmillan, pp 85–111.

Vila Seoane, M.F. and Saguier, M.I. (2020) 'Cyberpolitics and IPE: towards a research agenda in the Global South', in E. Vivare (ed) *Routledge Handbook of Global Political Economy*, London: Routledge, pp 702–18.

Vivares, E. (ed) (2018) *Regionalism, Development and the Post-Commodities Boom in South America*, London: Palgrave Macmillan.

Wanger, T.C. (2011) 'The lithium future: resources, recycling, and the environment', *Conservation Letters*, 4(3): 202–6.

# 11

# Turkey between Two Worlds: EU Accession and the Middle Corridor to Central Asia

*Mustafa Kemal Bayırbağ and Seth Schindler*

The Chinese and Turkish people, both representatives of ancient civilizations on the opposite ends of the Asian continent, have served humanity greatly by watching over the Silk Road and promoting commercial and cultural interaction. ... As late modernizers, Turkey and China are among countries seeking to bridge their development gap with Western nations in the 21st century. In other words, the 'Chinese dream' is to see China where it deserves to be on the world stage, just as the 'Turkish dream' is to witness our nation secure the place it deserves in the international arena.

President of Turkey Recep Tayyip Erdoğan, 2019

Turkey's contemporary spatial strategy has evolved out of tumultuous domestic politics and global events of the early 1990s. Turkey's 1994 municipal elections were a political earthquake. For the first time in the history of the Republic, an Islamist party (*Refah Partisi*) won municipal elections in İstanbul and Ankara. It was the culmination of a long transformation of political Islam in Turkey from a rural to an urban-based movement. It is noteworthy that political Islam was ultimately most successful in Turkey's cosmopolitan metropolis. Under the leadership of Recep Tayyip Erdoğan, the mayor of İstanbul from the Refah Partisi, the Justice and Development Party (the AKP) was established in 2001, and won national elections in 2002. It remains in office to this day. Its platform enjoyed support among secular voters because of its pro-Western agenda, which included seeking membership of the EU. A cornerstone of Turkey's lengthy campaign for EU membership (since 1963) was to demonstrate to Brussels that it was a reliable partner.[1] One of the ways Ankara sought to do this was to advance the spread of liberal democracy and free market liberalism by acting as a conduit between Brussels and post-Soviet Central Asia. The institutional mechanism through which Turkey would play this role was 'Transport Corridor Europe-Caucasus-Asia' (TRACECA), an EU-financed regional integration initiative launched in 1993. Its primary aim was to support infrastructure projects designed to enhance connectivity between the EU and post-Soviet countries (both the Caucasus and Central Asia). This spatial integration unfolded alongside economic, diplomatic, and political cooperation between Turkey and the EU.

As noted, the AKP drew considerable support from Turkey's secular elite in part because of its commitment to continue campaigning for Turkey's accession to the EU. By the mid-2000s it was clear that Turkey was unlikely to join the EU in the near future, and this contributed to the erosion of the domestic consensus that the AKP had cultivated. Since about 2010 it has faced significant domestic opposition for a number of reasons.[2] Massive demonstrations rocked Turkish cities and erstwhile allies called for Erdoğan's resignation. Turkey's embattled president turned to a powerful section of the bourgeoisie associated with the domestic construction sector for support. This class had become affluent in the previous decade and it owed its meteoric rise to the AKP's nationwide urban regeneration schemes and infrastructure projects. In 2011 Erdoğan campaigned to undertake 'crazy' infrastructure projects, which signalled that public investment in infrastructure would remain the primary mechanism for the government to foster demand and manage the economy.

Meanwhile, the EU's rebuff of Turkish entreaties fuelled desire among Turkish nationalists to pursue an eastward-oriented 'Eurasianist' foreign policy (Çolakoğlu, 2019; also see Oğuzlu, 2008), and (re-)establish a sphere of influence in areas that they consider pan-Turkic. To this end, they supported the Turkish government's regional integration initiative called the Middle Corridor to the Caucasus and Central Asia (hereafter 'Middle Corridor'). Published in 2011, the Middle Corridor initiative extends Turkey's domestic infrastructure abroad and 'passes by rail and road respectively through Georgia, Azerbaijan and Caspian Sea (crossing the Caspian transit corridor) and reaches China by following Turkmenistan–Uzbekistan–Kyrgyzstan or Kazakhstan route'.[3] As a spatial project, the Middle Corridor represents an evolution of TRACECA, but it is animated by a very different rationale. While the underlying objective of TRACECA was to integrate post-Soviet space with the EU via Turkey, the Middle Corridor is oriented eastward and emphasizes connecting İstanbul to China via Central Asia. It is in this context that Turkey signed a memorandum of understanding (MoU) with

China in 2015 entitled 'Jointly Promoting the construction of the Belt and Road Initiative' (taking effect in Turkey in 2018). The MoU purportedly streamlines the BRI with the Middle Corridor, and subsequent agreements deepened the Sino-Turkish partnership.

Our primary argument in this chapter is that the AKP – which at the time of writing controls the Turkish government – is pursuing state spatial objectives that are designed to entrench it within the state apparatus. The alignment of the Middle Corridor with the BRI has enabled the AKP to pursue its domestic political objectives, which include (1) the cultivation of a supportive bourgeoisie whose fortunes are connected to the construction sector, and (2) institutional reforms that enhance the executive branch's regulatory powers over economy and society. We begin by recounting how the TRACECA evolved into the Middle Corridor and became aligned with the BRI. We then demonstrate that Turkey has simultaneously maintained – and in some ways deepened – its institutional alignment with the EU, while it has implemented a series of reforms and initiatives aimed at accessing capital from China for infrastructure projects. We conclude that the AKP pursues international alliances that support particular domestic constituencies and allow it to justify institutional reform.

## Turkey's state spatial objectives: from TRACECA to the Middle Corridor

Turkey's repertoire of state spatial strategies[4] is shaped by a number of deeply rooted imperatives established throughout its imperial and republican histories. First, there is an innate awareness among successive Turkish governments that the society is at the crossroads of a number of powerful civilizations (see Fisher-Onar 2013). During the Ottoman era, this posed challenges for the Sublime Porte[5] whose leaders struggled to balance between competing neighbours. These great power politics played out in borderlands where the Sublime Porte sought to limit foreign influence, such as the Balkans, Caucasus, and the Levant (see Rieber, 2014). With the collapse of the Ottoman Empire and the establishment of the Republic in 1923, Ankara's focus shifted inward as it sought to fundamentally transform and modernize economy and society (Zürcher, 1993). The state aimed to integrate, act upon, and transform historically isolated and less 'modern' areas, particularly in rural Anatolia (Ertürk-Keskin, 2009; Bayırbağ, 2013a; Türel and Altun, 2013). Thus, Turkey's state spatial objectives have historically blended these components – bridging civilizations, balancing great powers, managing borderlands, and integrating national territory – in various combinations.

Turkey enhanced its standing in Brussels at the end of the Cold War by burnishing its credentials as a responsible partner that could bridge the EU and Post-Soviet Central Asia, as well as the Middle East. This appealed to

Turkish policy makers eager to expand Ankara's influence over erstwhile borderlands, and it led to optimism in Ankara that accession to the EU was imminent. Thus, for a brief period Turkish policy makers believed that they were in an interregnum between two tectonic shifts of world historical significance in which they would play an influential role – the break-up of the Soviet Union in the recent past and enlargement of the EU in the near future. Turkey's state spatial objectives were shaped with these events in mind, and geared towards (1) asserting control over areas that were supposedly part of a 'natural' Pan-Turkic sphere of influence and (2) hastening domestic development through economic integration with the EU. It was in this context that TRACECA was launched.

TRACECA was the EU's spatial response to the dissolution of the Iron Curtain on its eastern frontier. It was one of a number of regional integration initiatives launched in Central Asia after the Cold War (Kaw, 2019), and its objective was to foster cooperation by connecting existing infrastructure networks (see Üzümcü and Akdeniz, 2014; Bahçecik, 2017).[6] Its membership initially comprised Kazakhstan, Kyrgyzstan, Tajikistan, Turkmenistan, Uzbekistan, Armenia, Azerbaijan and Georgia, and it was ultimately expanded to include Ukraine, Moldova, Bulgaria, Romania, Iran and Turkey. According to one analysis, TRACECA was the EU's 'flagship connectivity initiative', and the Central Asia–Caucasus transport axis was designed to parallel, and provide an alternative to, the northern Eurasian route that passed through Russia (Kaw, 2019, 419).

Turkey joined TRACECA in 2002, and from the beginning its participation was conditioned by Ankara's spatial objectives. A number of large-scale infrastructure projects that remain central to Turkey's national spatial plans were mobilized by TRACECA – one example is the Kars-Tbilisi-Baku railway project designed to link Turkey, Georgia, and Azerbaijan while skirting Armenia (see Schindler et al, 2021). TRACECA also served as a policy jump-starter for Ankara to develop a coherent national spatial strategy. Launched in 2011 the Middle Corridor represented the evolution and consolidation of a bundle of TRACECA projects. From a spatial perspective the Middle Corridor was aligned with TRACECA, yet it heralded a shift of emphasis in Ankara's international spatial strategy from bridging the EU and post-Soviet countries to the management of its borderlands.

The Middle Corridor's inception in 2011 is significant because it coincided with a marked drift towards authoritarianism in Turkey (see Bayırbağ and Penpecioğlu, 2017). The 2008 financial crisis sent shockwaves through Turkey, and a series of institutional and constitutional reforms empowered Ankara to simultaneously quell dissent and shore up its accumulation strategy. Heavy-handed responses to demonstrators in İstanbul's Gezi Park and other cities across Turkey received significant attention from the international media, but less visible was the AKP's unmistakable attempt to entrench itself within the state apparatus. To this end it sought to strengthen the domestic regime of accumulation whose motive force was the transformation of the built environment. Since coming to power at the national scale in 2002, the AKP had centred itself within a network of stakeholders committed to territorial transformation. These groups included regulatory agencies, financiers, construction firms, landowners, and community groups, and the AKP's role was to coordinate the complex negotiations surrounding land acquisition and finance. It gained immense power from its network centrality because it was positioned to manage the surplus generated by the proliferation of construction projects. It entrenched its political position by channelling this surplus to two constituencies. First, many poor urban residents who had lived in informal settlements known as *gecekondu* were given the opportunity to own flats in newly built apartment buildings. This constituency historically tended to vote for left and centre-left parties, but the AKP's housing programme – coupled with its vision of a modern-conservative Islamic society – proved popular. In addition to this section of the electorate, the construction sector benefitted from a series of laws that streamlined land acquisition and urban redevelopment. In the early stages of Turkey's nationwide push to redevelop cities, this was considered a 'win-win-win' situation – that is the AKP managed urban transformation and established a loyal constituency, developers earned a handsome profit and residents gained ownership over assets (Bayırbağ et al 2021; also see Doğru, 2021; Gülhan, 2021). The AKP's political fortunes depended on the loyalty of these two constituencies, and their support was contingent on the continuation of a regime of accumulation whose core component was the transformation of the built environment. This distinguishes the bourgeoisie connected to the construction sector from other fractions of Turkey's bourgeoisie whose fortunes are tied to production and trade (particularly with the EU).

The limits to the AKP's ability to maintain a construction-fuelled regime of accumulation became increasingly apparent in the late 2010s (Bayırbağ, 2013b; Penpecioğlu and Bayırbağ, 2015). The public and private sectors in Turkey had accumulated debts that attracted significant attention from global investors, and Erdoğan was under intense pressure to rein in spending and tighten monetary policy (Akçay and Güngen, 2014). This precipitated a shift in Ankara's spatial strategy, which became increasingly focused on a smaller number of large-scale megaprojects at the expense of urban regeneration schemes. In other words, the regime was initially touted as a 'triple win' because its surplus was shared among the state, construction firms and low-income residents, but as the focus shifted to large-scale infrastructure it became a bilateral affair between the AKP and construction firms. It is in this context that the Middle Corridor began to take shape.

İstanbul has emerged as the heart of the Middle Corridor, and thus the territorial focus of the AKP's desire to establish 'Turkey as a Eurasian

Transport Hub' (see Atlı, 2018). The infrastructure projects that are under construction or planned are slowly weaving the whole Marmara Region surrounding İstanbul into a hub of global connectivity. These include a railway underpass, highway tunnel and a new bridge over the Bosphorus, the Osmangazi Bridge over the İzmit Bay, and the 1915 Bridge over the Dardanelles, a new international airport, and the North Aegean Port. There is inter-regional infrastructure as well, such as the Gebze–Orhangazi–İzmir highway, and east–west high-speed railways for passengers (to Ankara and Konya, for instance) and transnational freight rail lines to Kars that continue to the Caucasus. Finally, the Middle Corridor includes 21 new logistics hubs (Turhan, 2019). In sum, the Middle Corridor enhances intra-regional connectivity within the Marmara Region as well as national integration along an east–west axis, and transnational integration with Central Asia, the Caucasus and ultimately China. This eastward spatial reorientation allows the AKP shore up an important ally, the construction sector, and as we show in the following section, it justifies institutional reforms that ultimately entrench the AKP within the state apparatus.

## State restructuring in pursuit of the Middle Corridor

Turkish politics in the 21st century have been animated by a series of reforms designed to centralize power in the executive branch. These rounds of state restructuring have enabled Ankara's pursuit of spatial objectives, which, as discussed, have changed over time in accordance with geopolitical events. Efforts to accelerate its EU accession campaign from the early 1990s onward informed Turkey's spatial strategy to bridge the EU and post-Soviet republics. When the AKP came to power in 2002 it was faced with a recalcitrant opposition deeply embedded within the state bureaucracy. Its response was twofold. First, as mentioned, the AKP sought to cultivate a loyal constituency among sections of the urban poor and the bourgeoisie connected to the construction sector. Second, they introduced a series of reforms whose purpose was to situate power within the executive organs of the state, which precipitated a prolonged struggle for control of the state apparatus. The AKP proposed two major constitutional referenda of 2007 and 2010, which were presented as a popular fight against the so-called Secular-Kemalist establishment. Both referenda were passed, and feeling emboldened, the AKP secured approval from parliament (via Law numbered 6223, 06 April 2011) for the council of ministers to pass 'decrees with the power of law' in 2011. This was a supra-reform that allowed the executive branch to unilaterally issue 'decree laws' and re-organize ministries and other state institutions. A series of institutional reforms were introduced the following year with the intention of further centralizing power within the executive branch. Many of these reforms were in discord with one another or other regulations, and behind a veneer of institutional thickness was a chaotic governance regime in which powerful individuals ruled through ad hockery (Aksu, 2018).

National elections were scheduled for 2011, and the AKP positioned itself as a party of stability in the context of domestic political opposition and the global financial crisis. In this context it touted its infrastructural achievements and announced the Middle Corridor (Schindler et al, 2021). In order to deliver political and economic stability, the AKP argued that it was necessary to concentrate executive power into the hands of the prime minister, while freeing that post from rigid parliamentary oversight for the sake of country's development. The AKP was returned to power in a landslide victory, but despite – or because of – its ongoing effort to centralize power and restructure the state, stability proved elusive. A bottom-up groundswell of discontent exploded in 2013 over the proposed redevelopment of İstanbul's Gezi Park. This was followed in 2016 by a coup attempt orchestrated by followers of exiled cleric Fethullah Gülen. In response, the AKP launched a renewed attempt to further centralize power, which culminated in yet another constitutional referendum in 2017 that transformed Turkey's parliamentary political system into a presidential one in the most significant overhaul in the history of the Republic.

This cycle of crisis and state restructuring unfolded against backdrop of dynamic geopolitics. Brussels was weary of the AKP's centralization of power, yet it sought Ankara's help in dealing with the humanitarian crisis precipitated by the Syrian civil war. Rather than a bridge, Brussels hoped Turkey could act as a buffer and inhibit the transit of refugees. What was not on the table was EU accession, and Ankara's hope of joining the bloc had been replaced by a desire to renew its status as a powerful regional actor. This neo-Ottomanism animated an active and east-oriented foreign policy geared towards the re-establishment of a sphere of influence (see Fisher-Onar, 2013). The Middle Corridor is a manifestation of this eastward turn, and according to Erdoğan it is a prerequisite for the achievement of the 'Turkish Dream'. The achievement of this vision, however, requires the continuation of Turkey's construction-centric regime of accumulation, the expansion of its influence over its northeastern borderlands, and the enhancement of Eurasian connectivity. This explains why the Middle Corridor has been aligned with the Belt and Road Initiative.

China and Turkey signed an MoU to integrate the BRI and the Middle Corridor in 2015. In some ways it institutionalized trends that were well under way. Chinese companies contributed to the construction of the high-speed rail linking İstanbul and Ankara. The project was initially funded by the European Investment Bank, but cost overruns required Ankara to secure further external funding. In 2007, China EXIM Bank provided a loan for a section of the project to be undertaken by a joint venture that included two

Turkish firms (CENGİZ, holding 30 per cent; IC İçtaş İnşaat, holding 30 per cent) and China's Railway Construction Corporation Limited (CRCC) (32 per cent) and China National Machinery Import & Export Corporation (8 per cent). The project was ultimately completed, but not without significant challenges (see Schindler et al, 2021). Contrary to expectations at the time, the project did not precipitate significant involvement of Chinese construction firms in Turkey's territorial transformation. This was due to the fact that most of Turkey's infrastructure is built in accordance with EU standards, which Chinese firms found difficult to meet. Furthermore, the AKP sought to protect the domestic construction sector from Chinese competition, and Chinese firms operating in Turkey complain that the playing field is unequal (Schindler et al, 2021). Thus, the AKP seeks to attract Chinese funding for its grand spatial objectives by aligning the Middle Corridor with the BRI, but to avoid the involvement of Chinese firms in domestic spatial projects.

A close examination of the MoU[7] reveals the emergence of an expansive field of bilateral Turkey–China relations and a stated commitment to pragmatism (Chaziza, 2021). The shared 'cooperation objectives' are broadly defined, and its wide-ranging 'contents of cooperation' include policy coordination, facilities connectivity, unimpeded trade, financial integration and people-to-people bonds. The alignment of spatial objectives is covered under the banner of 'facilities connectivity,' in which the two countries agree to:

> [f]ormulate plans on cooperation in bilateral infrastructure projects in Turkey, China and third countries, including highway, railroad, civil aviation, port as well as oil and gas pipelines, power grid and telecommunication network; developing and implementing plans to strengthen cooperation among seaports for cargo transportation ... strengthen mutual recognition of standards and information sharing.

Thus, it is not an exaggeration to interpret the MoU as a meta-framework for bilateral relations, and importantly, the impetus drawing Turkey and China together is a set of complementary state spatial objectives. In contrast to its relations with the EU, the MoU's sections on institutional and policy harmonization are deliberately open-ended. Emphasis is instead placed on 'pragmatic cooperation':

> (i). ... *the two parties shall respect each other's core interests and major concerns* ...; (ii) To inject new vigor into the economic and social development of the two countries *in accordance with each other's applicable laws and regulations and their international obligations and pledges*; (iii) ... the two parties shall strengthen connectivity and mutual support, draw on

each other's strength and bring out the best in each other *by taking full advantage of the existing bilateral cooperation mechanisms and multilateral mechanisms participated by the two countries as well as effective regional cooperation platforms.* (Emphases added)

The MoU's focus on pragmatism allows the AKP to pursue its domestic political agenda unimpeded, as the party seeks to maintain a regime of accumulation geared towards territorial transformation and to concentrate power within the executive branch. The Türkiye Wealth Fund (TWF) was created in 2016 to facilitate these two objectives. It is a sovereign wealth fund with a mandate to play a significant role in the economy and it 'focuses on the growth targets of its portfolio companies through value creation programmes, investments in key sectors and visionary projects'.[8] Rather than oversight from an independent public authority, its activity is guided by key principles: (1) serving the nation; (2) accountability and transparency; (3) discipline, professionalism and teamwork; and (4) delivery of results. The TWF's board of directors includes Erdoğan and AKP allies.

The TWF has served as a vehicle through which the Turkish state has gained control over domestic capital, financial institutions and also deepened relations with China. For instance, 35 per cent of TWF's investments are concentrated in the domestic financial sector – it owns Ziraat Bankası outright, and it has significant stakes in Halkbank (75 per cent), VakıfBank (36 per cent), Borsa İstanbul (81 per cent), and Türkiye Hayat Emeklilik (100 per cent).[9] These are some of Turkey's largest financial institutions. They have played a fundamental role in support of the state's large-scale spatial projects, and they have also forged linkages with Chinese financial institutions. For example, Ziraat Bankası recently received US$400 million credit from the Export–Import Bank of China, and a group of Chinese banks agreed to provide US$1.6 billion to refinance the Third Bridge over Bosporus and the Northern Marmara Highway (Bloomberg, 2021; Hürriyet, 2021).[10]

In summary, Turkey's MoU with China enabled the AKP to continue to pursue its domestic political objectives of centralizing power and bolstering a regime of accumulation centred on the transformation of territory. As we have noted, the AKP's power previously stemmed from its network centrality among key domestic stakeholders whose negotiations it coordinated, but a shrinking economic surplus and the manifestation of political opposition has forced it to bolster its position with Chinese capital.[11] This explains why Turkish financial institutions were brought under control of the executive branch, and the MoU with China allowed them to gain access Chinese capital in ways that avoid significant public oversight or taking on sovereign debt (instead, debts appear on the balance sheets of financial institutions controlled by the TWF). The notion of pragmatic cooperation between two party states with complementary spatial objectives seems to have reached a new stage via yet another MoU. The TWF and Sinosure signed an MoU in 2020 (TWF, 2020) to support TWFs 'investment plans in these sectors as well as its contribution to the logistics infrastructure investments of Turkey'. It states that: 'Sinosure will recommend relevant Chinese enterprises to TWF as the investors, contractors and financial institutions concerning projects especially in energy, petrochemicals and mining sectors addressing the current account deficit of Turkey. Sinosure will also *consider* providing insurance support up to USD5 billion with respect to the financing activities' (TWF, 2020, emphasis added).

While the precise terms have not been made public, it appears that if Ankara wants to maintain access to Chinese capital, it must reduce barriers that have inhibited Chinese firms from operating within Turkey. The CEO of Sinosure stated:

> With China 'Belt and Road' Initiative and Turkey 'Middle Corridor' Initiative go forward and go deep, the economic, trade and investment cooperation between China and Turkey become more and more frequent. The MoU is an important milestone for deepening cooperation between TWF and Sinosure. Based on the MoU, hope we can continue to strengthen communication and coordination in practical projects. (TWF, 2020)

In conclusion, bilateral cooperation between China and Turkey does not appear to be comprehensive or 'deep', but it is more than mere pragmatic exchanges animated by realpolitik. Certain segments of the Turkish state apparatus under the direct control of the AKP are deeply connected with Chinese institutions with which it shares spatial objectives. Despite its cooperation with the AKP, the Chinese establishment understands full well that Turkey will likely remain deeply integrated with the EU. The Xinhua News Agency (2019) noted that Turkey is not about to 'turn its back on Europe'.[12] Turkey's international orientation is so inseparable from its domestic politics that the nature of its foreign entanglements will surely be a political issue in the next election. The election is more than a referendum on the AKP and the 'Turkish Dream', however, and the choices made by Turkish voters will be shaped by the entreaties from Beijing and Brussels.

## Notes

[1] For a detailed account of the ups and downs of this campaign, see Alpan (2021).
[2] For an analysis see Bayırbağ (2013b), Şengül (2015) and Schindler et al (2020, 292–94).
[3] www.mfa.gov.tr/turkey_s-multilateral-transportation-policy.en.mfa.
[4] See Brenner (2004) on state spatial strategies, projects and objectives, and Brenner (2009) on 'state restructuring'.
[5] The 'Sublime Porte' refers to the Ottoman Empire's state apparatus based in Istanbul, somewhat like saying 'Washington' when referring to the US government.

[6] The project 'included funding from the European Commission for the rehabilitation of transport infrastructure in disrepair and technical assistance. It also sought to harmonize technical standards and legal frameworks' (Schindler et al, 2021).

[7] The MoU was signed in 2015 and gazetted in 2017: www.resmigazete.gov.tr/eskiler/2017/06/20170607-1.pdf.

[8] www.tvf.com.tr/en/who-we-are/about-us/general-info.

[9] www.tvf.com.tr/en/our-portfolio/financial-services.

[10] For instance, some six months after TURKCELL (one of Turkey's largest communication operators) received EUR 500 million credit from the Development Bank of China (see www.aa.com.tr/tr/sirkethaberleri/bilisim/turkcell-cin-kalkinma-bankasi-ile-500-milyon-avro-tutarinda-kredi-anlasmasi-imzaladi/658759).

[11] The AKP's quixotic interest rate policy and overall indebtedness made it increasingly difficult to access Western Capital. As Orhangazi and Yeldan note, towards the end of the 2010s, 'Turkey was faced with a deteriorating external balance due to increased capital inflows leading to wide current account deficits and a rapid accumulation of external debt. This rendered the economy extremely vulnerable to a slowdown or even a reversal of capital inflows' (2021: 473).

[12] Also see Atlı (2020) and Chaziza (2021).

## References

Akçay, Ü. and Güngen, A.R. (2014) *Finansallaşma, Borç Krizi ve Çöküş: Küresel Kapitalizmin Geleceği*, Ankara: NotaBene.

Aksu, S. (2018) '2011 yılında çıkarılan Kanun Hükmünde Kararnameler ile yapılan düzenlemelere ilişkin genel bir değerlendirme', *Yasama Dergisi*, 35: 26–44.

Alpan, B. (2021) 'Europeanization and EU–Turkey relations: three domains, four periods', in W. Reiners and E. Turhan (eds) *EU–Turkey Relations: Theories, Institutions, and Policies*, Basingstoke: Palgrave Macmillan, pp 107–37.

Atlı, A. (2018) 'Turkey as a Eurasian transport hub: prospects for inter-regional partnership', *Perceptions*, 23(2): 117–34.

Atlı, A. (2020) 'The political economy of Turkey's relations with the Asia-Pacific', in E. Parlar-Dal (ed) *Turkey's Political Economy in the 21st Century*, Basingstoke: Palgrave Macmillan, pp 271–95.

Bahçecik, O. (2017) 'OBOR and Turkey's turn to the East', in A. Ekman, F. Nicolas, J. Seaman, G. Desarnaud, T. Kastoueva-Jean, S.O. Bahcecik and C. Nallet (eds) *Three Years of China's New Silk Roads: From Words to (Re)action?*, Paris: Institut Français des Relations Internationales, pp 51–60.

Bayırbağ, M.K. (2013a) 'Continuity and change in public policy: exclusion, redistribution and state rescaling in Turkey', *International Journal of Urban and Regional Research*, 37(4): 1123–46.

Bayırbağ, M.K. (2013b) 'Homo sum, humani nihil a me alienum puto: Kurtuluş yok tek başına!', *Praksis*, August: 7–16.

Bayırbağ, M.K. and Penpecioğlu, M. (2017) 'Urban crisis: limits to governance of alienation', *Urban Studies*, 54(9): 2056–71.

Bayırbağ, M.K., Schindler, S., and Penpecioğlu, M. (2021) 'The limits to urban revolution: the transformation of Ankara, Turkey, under the Justice and Development Party', Unpublished manuscript.

Bloomberg (2021) 'Köprüde hisse devrini Çinli bankalar destekleyecek', Bloomberg, 26 March. Available from: www.bloomberght.com/istanbul-3-kopru-hisse-devrini-cinli-bankalar-destekleyecek-2277432.

Brenner, N. (2004) *New State Spaces: Urban Governance and the Rescaling of Statehood*, Oxford: Oxford University Press.

Brenner, N. (2009) 'Open questions on state rescaling', *Cambridge Journal of Regions, Economy and Society*, 2(1): 123–39.

Chaziza, M. (2021) 'China's new Silk Road strategy and the Turkish Middle Corridor vision', *Asian Journal of Middle Eastern and Islamic Studies*, 15(1): 34–50.

Çolakoğlu, S. (2019) 'The rise of Eurasianism in Turkish foreign policy: can Turkey change its pro-Western orientation?', *The Middle East Institute*, 16 April. Available from: www.mei.edu/publications/rise-eurasianism-turkish-foreign-policy-can-turkey-change-its-pro-western-orientation.

Doğru, H.E. (2021) *Çılgın Projelerin Ötesinde: TOKİ, Devlet ve Sermaye*, İstanbul: İletişim.

Ertürk-Keskin, N. (2009) *Türkiye'de Devletin Toprak Üzerinde Örgütlenmesi*, Ankara: Tan.

Fisher-Onar, N. (2013) 'Historical legacies in rising powers', *Critical Asian Studies*, 45(3): 411–30.

Gülhan, S.T. (2021) 'Neoliberalism and neo-dirigisme in action: the state–corporate alliance and the great housing rush of the 2000s in Istanbul, Turkey', *Urban Studies*, 59(7): 1443–58.

*Hürriyet* (2021) 'Ziraat Bankası, Çin Exim Bank'tan 400 milyon dolarlık kredi temin etti', 31 March. Available from: www.hurriyet.com.tr/ekonomi/ziraat-bankasi-cin-exim-banktan-400-milyon-dolarlik-kredi-temin-etti-41776386.

Kaw, M.A. (2019) 'Theorizing EU-TRACECA relationship in eurasian context', *Strategic Analysis*, 43(5): 418–34.

Oğuzlu, T. (2008) 'Middle Easternization of Turkey's foreign policy: does Turkey dissociate from the West?', *Turkish Studies*, 9(1): 3–20.

Orhangazi, Ö. and Yeldan, A.E. (2021) 'The re-making of the Turkish crisis', *Development and Change*, 52(3): 460–503.

Penpecioğlu, M. and Bayırbağ, M.K. (2015) 'Büyük ölçekli projeler ve neoliberal şehirciligin karsılastırmalı siyaseti: Türkiye metropoliten kentlerinden örnekler', in İ. Coşkun and B. Duman (eds) *Neden, Nasıl, Kimin için Kentsel Dönüşüm*, Istanbul: Litera, pp 333–55.

Rieber, A.J. (2014) *The Struggle for the Eurasian Borderlands: From the Rise of Early Modern Empires to the End of the First World War*, Cambridge: Cambridge University Press.

Schindler, S., Gillespie, T., Banks, N., Bayırbağ, M.K, Burte, H., Kanai, J.M., and Sami, N. (2020) 'Deindustrialization in cities of the Global South', *Area Development and Policy*, 5(3): 283–304.

Schindler, S., Bayırbağ, M.K. and Gao, B. (2021) 'Incorporating the İstanbul-Ankara high-speed railway into the Belt and Road Initiative: negotiation, institutional alignment and regional development', *Journal of Geographical Sciences*, 31(5): 748–64.

Şengül, H.T. (2015) 'Gezi başkaldırısı ertesinde kent mekanı ve siyasal alanın yeni dinamikleri', *METU JFA*, 32(1): 1–20.

Turhan, M.C. (2019) 'Bir Kuşak Bir Yol, coğrafyamızın geleceğini yeniden şekillendirecektir', *Milliyet Newspaper*, 24 April.

Türel, O. and Altun, S. (2013) 'The articulation of the Middle East with the World System in the XXI century: afterthoughts on the "Arab Spring"', Unpublished manuscript.

TWF (Türkiye Wealth Fund) (2020) 'TWF and Sinosure signed MoU for cooperation'. Available from: www.tvf.com.tr/en/contact/disclosures/2020/twf-and-sinosure-signed-mou-for-cooperation

Üzümcü, A. and Akdeniz, S. (2014) 'Yeni İpek Yolu: TRACECA ve Bakü-Tiflis-Kars Demiryolu Projesi', *Avrasya Etüdleri,* 45: 171–99.

Xinhua News Agency (2019) 'Spotlight: Turkey sets sights on Asian markets with Asia Anew initiative', 17 August. Available from: www.xinhuanet.com/english/2019-08/17/c_138316832.htm.

Zürcher, E.J. (1993) *Turkey: A Modern History*, London: I.B. Taurus.

# 12

# Multipolar Infrastructures and Mosaic Geopolitics in Laos

*Jessica DiCarlo and Micah Ingalls*

Three decades since the fall of the Soviet Union, the media is again rife with Cold War rhetoric as China takes a more assertive stance globally. China's expansion in, for example, lending, infrastructure, and diplomacy, has prompted debates on whether China's rise and US reactions to it will create new Cold War dynamics. Although the tensions between China and competing powers – principally the US – have global dimensions, they arguably play out more concretely in third countries. This has become particularly apparent in Southeast Asia and in the infrastructure sector. Indeed, a 2021 article in *The Economist* contends that 'the rivalry between America and China will hinge on southeast Asia'. According to media and some recent academic scholarship, Southeast Asia has become a linchpin in China's strategic and infrastructural engagement, particularly through the Belt and Road Initiative (BRI) (Hiebert, 2020; Lampton et al, 2020; Strangio, 2020). Shambaugh (2020) adds to these perspectives the need to consider US competition in the region alongside China.

Laos – low-income, landlocked, and historically considered isolated and peripheral – is centrally located on a primary BRI corridor to Southeast Asia, ostensibly situating the country at the heart of US–China competition. However, in Laos, like many other countries in Southeast Asia, there is little desire within the government to choose between powers – Washington, Beijing, or otherwise. However, at the moment in the infrastructure sector, Laos does not need to choose between the US and China because it is already dominated by Chinese actors, the US is virtually non-existent, and the Lao government instead draws on other regional actors to balance China. These other regional and multilateral actors thus end up playing a far more significant role in the Lao government's spatial projects than the US.

China is by far the largest and most significant investor in infrastructure, the largest provider of aid, largest holder of Lao sovereign debt, and second-largest trading partner. While Laos is typically viewed as a weak state susceptible to Chinese control, it has historically used infrastructure, investment, and aid to balance geopolitical relations and continues to do so. Yeh (2016, 227) points out that 'the crux of the issue is less about "weak" and "strong" states than it is about how Chinese investment may enable a reconfiguration or reinforcement of state power'. Laos shows that host countries can capitalize on geopolitical competition and regional politics to pursue their own spatial projects. Here, the BRI's emphasis on infrastructure aligns with Laos's primary spatial and development goals: to extend its reach to rural areas and their extractable resources, and to connect to its regional neighbours by becoming 'landlinked' and the 'battery of Southeast Asia'. These strategies give rise to overlapping, often competing, 'jurisdictions', and effectively shape new forms of governance. As Laos leverages international commitments to infrastructure as a means of geopolitical balancing and (re)orients toward Sino-centric value chains to pursue domestic development agendas, the scale of the projects involved require massive oversight and expertise that the current system lacks. The strengthening of institutional capacity within specific bureaucratic entities that can manage these spatial projects thus becomes the objective of state restructuring (Brenner, 2004). Institutional reforms allow the Lao state to pursue new spatial strategies and expand infrastructure networks at an accelerated pace and scale, financed and supported by the BRI.

This chapter offers two contributions. First, it problematizes grandiose new Cold War narratives by contextualizing the processes that are discussed in faraway offices in Washington and Beijing. We argue that bipolar new Cold War rhetoric does not aptly capture the Lao infrastructural experience, which is instead characterized by an array of actors and interests that the Lao government has negotiated since well before the BRI. The Lao government has demonstrated an aptitude for successfully manoeuvring between various countries, donors, and international organizations who rush to offer aid or development finance and secure influence. Within Laos's infrastructural balancing act, the US is ostensibly absent. Instead, 'Japan, Thailand, and Vietnam [are vying] with China for influene' (Jennings, 2021).

Second, we propose reframing debates on infrastructural competition to focus on the ways in which BRI investments align with host state interests and prompt internal restructuring. When a host state like Laos deems a project desirable, it can often quickly adopt new ideas, policies, hopes and expectations as a means of mobilizing domestic support. Indeed, infrastructure is often desired by host countries – here this could mean local elites, specific ministries, or central governing bodies – who identified infrastructures gaps and may benefit from certain projects. In

Laos, Chinese investment has previously been shown to reinforce Lao state control in peripheries (Tan, 2012). Today, with the BRI, national development goals are not giving in to Chinese demands, but rather incorporate problems and agendas that have been on the table much longer than the BRI. The key question that arises across Lao government ministries is how to juggle and benefit from a deluge of projects and investment capital.

The next section traces some of the historical legacies that inform infrastructural geopolitics today, drawing out the heterogeneity of actors with long relations in the country. This is followed by a section that explores how state restructuring takes place in connection to Laos's main development objectives concerned with connectivity and energy. These examples show how the Lao state engages infrastructure development on its own accord, as well as in response to the array of interests from other countries such as Vietnam, China, and Japan that observers have suggested are 'pulling Laos apart'. We conclude that Laos draws on longstanding regional relations for investment capital and political alliance in ways that allow for internal institutional reforms not bound by the kinds of conditionalities that often come with Western development aid and investments.

## Historical legacies of contemporary infrastructural geopolitics

Multiple influences, including elite politics, infrastructural legacies, and competition among regional powers and organizations, have led to a diverse landscape of variously complementary, conflicting, and overlapping personal and institutional relations that we refer to as *mosaic geopolitics*. First, contemporary geopolitical relations between Laos, its neighbours, and the world have been shaped and informed by elite political relations that are rooted in the past. As Laos emerged from its colonial relations with Europe – significantly aided by Vietnam, and to a lesser degree China – postcolonial restructuring also entailed close relations with the Soviet Union. With the end of the Cold War, new patterns of aid and alliance emerged as Western governments sought to shore up the Lao state and secure access and influence. As a direct result of these political and aid-dependent relations and Cold War era restructuring, the contemporary Lao elite have been educated across the world – in Russia and the former Soviet republics, in China, Vietnam, Japan, the US, and Australia, to name a few. Virtually every senior government person was educated abroad – most in the former Soviet Union, China, or Vietnam. For example, the Minister of Foreign Affairs Saleumxay Kommasith studied at Moscow State University; President Sisoulith studied at the Leningrad State Pedagogical Institute and later was the head of the National University of Laos's Russian language programme; while the children of Quinim Pholsena grew up in China and now hold positions as, variously, the Vice President of the National Assembly, Minister of Industry and Commerce, Head of the President's Office, and Member of the Central Committee of the Lao People's Revolutionary Party (LPRP) and Attaché of the Lao Military to Beijing. These relations affect the channels of access to decision makers.

In addition, Laos's contemporary geopolitics are informed by infrastructural legacies – from the colonial period through the Indochinese conflicts to Cold War era restructuring. Discursively framing Laos as disconnected and backward is not new. As far back as the 17th century, the kingdom of Lan Xang (part of which now constitutes Laos) was referred to as 'the land in between' (Evans, 2002). Considered a 'backwater' (Gunn, 1990) under French colonialism, Laos was not recognized as a state, but rather perceived as a part of Indochina, as a Vietnamese hinterland, and as a route to China. During this time, the French aimed to achieve the *mise en valeur* (value creation) of Lao territory primarily through the construction of transportation infrastructures and trading posts (Stuart-Fox, 1995). However, when rivers failed as navigable routes to southern China, the Colonial Council of Cochinchina announced a plan to build a railway to Yunnan. This, too, did not materialize due to insurmountable costs, terrain, and distance (DiCarlo, 2021). As a result of the mountainous terrain in the north and east as well as challenges of river travel on the Mekong, the goal of *débloquement*, or unblocking Laos, turned to road development in order to access natural resources (Dwyer, 2016). This expansion depended on both the dream of accessing Lao territory and the physical infrastructures themselves. While *mise en valeur* was not realized by the French, their 1930s road construction programme – which aimed to delink Laos from Siam in favour of roads to Indochina – was extolled as a success for making Laos part of Indochina (Ivarsson, 2008). Visions of infrastructural 'success' coupled with their failures foreshadow a recurrent vision of Lao territory as a bridge and pivot point in the region.

The proclamation of the Lao People's Democratic Republic on 2 December 1975 by the LPRP placed the country firmly in the socialist sphere, solidifying 'fraternal' relations with communist neighbours China and Vietnam, as well as the Soviet Union. These fraternal relations were by no means smooth. In particular, tensions between Vietnam and China were rife during Laos's early postcolonial years. Vietnam took on a more directive role in supporting and building up the bureaucratic and military capacities of the nascent state. China, meanwhile, was actively involved in fomenting anti-LPRP resistance in Laos until the mid-1980s, including the promotion of an alternative, pro-China communist party, and the arming of Hmong insurgency (Dwyer et al, 2016). While Sino-Viet tensions cooled through the 1990s and early 2000s, their historical antagonism nevertheless shaped their differential relations with Laos.

Laos's 'special relationship' with Vietnam has long been a dominant feature in Laos's political life. While this relationship is at its root largely political, it has found material expression in preferential trade deals and the favourable treatment of Vietnamese investment companies. With China's growing interest in Laos as both an important geopolitical partner and a frontier of economic opportunity, China has increased not only its capital investments in infrastructure and other sectors but also its investment in the political organs of the Lao state. In a significant show of solidarity, for instance, the Chinese Communist Party (CCP) donated millions of dollars to support the LPRP's 10th Party Congress in 2016 (Sayalath and Creak, 2017), assuming the role of political benefactor alongside Vietnam. The degree to which overtures such as this are sufficient to destabilize Vietnam's special place in Laos's political calculus remains to be seen, but the winds of change, it would appear, are certainly blowing.

Beyond fraternal relations, the Lao government has pursued integration into the global economy through regional organizations and trade agreements. Within regional frameworks, Laos plays a key role in transportation and energy strategies. The two most infrastructurally significant regional initiatives to which Laos is a party, the Association of Southeast Asian Nations (ASEAN) and the Asian Development Bank's (ADB) Greater Mekong Subregion Initiative (GMS), focus on trade, logistics and enhanced connectivity through roads construction. Laos joined ASEAN in March 1996, and infrastructure is one of the four pillars of the ASEAN Integration Framework (the others are human resource development, information and communication technology, and regional economic integration). This assumes that investment in roads, energy systems, and public projects will develop the supply side of the Lao economy, reduce its reliance on the mineral sector, and foster pro-poor growth.

In earlier years, ASEAN was commonly seen as a regional counterbalance to China. However, this appears to be changing. January 2010 saw the launch of the China–ASEAN Free Trade Agreement (Creak, 2011), and in 2015 ASEAN and China agreed to higher levels of economic cooperation. In line with South–South development, some ASEAN members position China as an ally against Western powers, whereby: 'China serves as a kind of screen from Western countries and non-governmental organizations which seek progress in the areas of human rights and the democratization of their political system' (Pholsena and Banomyong, 2006, 44). Vietnam, Laos, and Cambodia tend to fall in this camp as they demonstrate support for China through, for example, the 2019 letter supporting China's Xinjiang policies (Yellinek and Chen, 2019) or defence of the 'one China policy'. At the same time, Laos's political proximity to Vietnam leads the government to take the middle ground on some sensitive regional disputes; Laos and Cambodia have demonstrated support for China through, for example, actions to block anti-Chinese positions in ASEAN, such as those relating to the South China Sea (Wong, 2017). Chinese officials continue to emphasize their fraternal relations within Laos and the pivotal role it plays in BRI in Southeast Asia. During a state visit to Laos in 2016, Xi Jinping referred to Laos and China as an 'unshakable community of common destiny' (FMPRC, 2016). Later in November 2017, Xi met with Lao President Vorachit and signed multiple cooperation pacts on infrastructure, economic zones, trade, and energy, as well as military contracts. During his visit, Xi referred to Laos as China's 'iron brother', the same language used (only) of Pakistan, also in reference to the country's crucial role in geopolitics, in that case on the western front.

China's strategy appears to focus on finding areas of common interest surrounding trade and investment, bringing countries more squarely into its economic orbit through networks of investment, production, and trade wherein China functions as the hub. Ultimately, this creates relations of dependency that forestall efforts to articulate independent policies (Goh, 2014). This is clear from Laos's enthusiastic embrace of the BRI (Lampton et al, 2020; Kuik, 2021). Lampton et al (2020, 84) characterize the Lao state response to 'big economic power inducement' as that of 'receptive embrace'. While this may be the case, large-scale infrastructure projects in Laos are outward-oriented in multiple directions, and also driven by domestic developmental objectives. The BRI, for example, is doubly appealing because infrastructure has been at the centre of Lao national development strategy since the first National Socio-Economic Development Plan (1981–85). Prior to the BRI and US–China rivalry, Stuart-Fox noted in 2009 that China was a rising force in the Lao economy, offering development projects from agribusiness to hydropower and urban infrastructures that claim to lift the country out of poverty. These projects were not as transnationally oriented as the BRI, so today Laos explicitly engages infrastructure-led development to 'get the territory right' and enhance its regional integration (Schindler and Kanai, 2021), looking to China as a role model.

While China is the clear leader, the broader picture of infrastructure construction and finance is a mosaic of bilateral and multilateral relations. Competition in the field of infrastructure is backed by two sources of capital: foreign direct investment (FDI) and development aid and finance. China is the largest provider of FDI, followed by Vietnam and Thailand. Chinese FDI in Laos has grown as Chinese firms seek alternative manufacturing sites to avoid US tariffs that have resulted from the US–China trade dispute (Yuvejwattana, 2019). In terms of development aid and finance, the Japan-led Asian Development Bank (ADB) is the largest donor, followed by Japan (the largest bilateral donor), followed by South Korea. In the field of development finance, competition between Japan and China is evident. However, it follows the familiar pattern of competition–collaboration even in Laos, such as in the upgrading of the nation's key international airport in

Vientiane that involved both Japanese and Chinese finance. Meanwhile the mosaic pattern of investors can be traced throughout the capital, where for example South Korea improved tourist infrastructure on the Mekong, and Chinese capital built international conference halls and is currently upgrading Vientiane's Mahosot Hospital. Between FDI and aid, investment is spatially distributed, with much Chinese investment clustered in the north near the border, and Vietnamese investment in the south. Some observers suggest that this shows that Laos is being 'carved up' by the rush of international investors and actors. However, this does not consider how the Lao government may or may not be engaging in how such projects take shape, nor project proximity to border crossings from Vietnam and China that eases capital flows.

Although the US has yet to directly compete in infrastructure construction or finance in Laos, there are competing visions for the integration of Southeast Asia. The US supports the Asian Development Bank's (ADB) regional spatial plan under the Greater Mekong Subregion (GMS). Although Japan has a lead operational role in the ADB and is the largest donor, the US and Japan are the largest shareholders in the ADB, each with a 15.6% stake (Runde and McKeown, 2019). The GMS largely began in 1992 with dominant support from Japan, an important piece in this geopolitical infrastructure puzzle, as a strategic counterbalance to China (Soong, 2016). Glassman (2010, 32) notes it is 'driven by more global, but highly uneven, capitalist investment, production, and trade, leading less to the integration of the GMS per se than to the integration of GMS countries into a much larger East Asian regional system'. Meanwhile, China's BRI as well as the Lancang-Mekong Cooperation links Chongqing with Singapore and would integrate the regional system along a north–south Sino-centric axis.

Western actors have begun to explicitly try to counter Chinese largesse. In Laos, US launched initiatives – such as Asia EDGE, Infrastructure Transaction and Assistance Network, and Clear Choice – constitute a BRI counter strategy. According to a USAID officer in Laos, "[n]o country can compete with China in terms of money for infrastructure", hence US programmes aim to "build [Lao] government capacity to make better decisions and negotiate better terms [with Chinese actors]" (Interview, March 2019). In the words of USAID Administrator Mark Green:

> Whenever you're dealing with China there's the great power competitions ... what we we're trying to do is help countries understand, from a development perspective, if they do choose the authoritarian model ... what the fine print is ... it's unsustainable debt very often. It's tying up strategic assets. In some cases, it's robbing particularly young citizens of their birth right ... access to natural resources. (USAID, 2018)

The Australian government also recognizes the BRI as an 'important catalyst' in regional infrastructure development. In 2019, it proposed the Southeast Asian Economic Governance and Infrastructure Facility, 'to help improve infrastructure decision-making ... by building and enhancing government-to-government partnerships' and focusing on planning, procurement, implementation support, policy, and regulations (Australian Government, 2019). The most recent G7 Build Back Better World (B3W) may reorient geopolitical competition in infrastructure construction or alternatively, as the Lao experiences suggests, with so many other actors and already cheap finance available the B3W might become one more 'ball to juggle'.

A common perspective among Lao officials is that as China finances large infrastructures, then Laos needs complementary projects to be led by Japan and other actors. However, despite Laos's complex geopolitics, China's has come to exert a crucial influence on the apparatus of the Lao state, particularly regarding infrastructure investment and construction. At the same time, Laos pursues spatial objectives that enhance transnational integration through the mobilization of foreign loans and grants for infrastructure projects. The government has staked its legitimacy in part on its ability to realize developmental dividends through spatial projects while simultaneously navigating fraught politics and geopolitical tensions (Schindler et al, 2020). Thus, rather than a US-China new Cold War, on the ground infrastructural competition is better categorized through a lens of mosaic geopolitics. To further nuance the contours of powers at play the next section turns to how state restructuring has taken place to specifically accommodate Chinese projects.

## State restructuring with an eye to China

State restructuring has been undertaken to pursue Laos's primary spatial and developmental strategies: to transition from landlocked to 'landlinked' and achieve aspirations to become the 'battery of Asia'. While the BRI is consistent with Laos's domestic development agenda, this does not mean that alignment is a straightforward affair. Even as the BRI precipitated a proliferation of projects that the state was under-prepared to manage, Laos proved to be not simply a playground for geopolitical heavyweights. Rather, Laos undertook a substantial restructuring of key agencies to manage the influx of investments. Departments, advisory committees, and research groups have been formed or re-tasked with BRI-related mandates. Prior to the BRI, the Ministry of Planning and Investment (MPI) established the Laos–China Department to oversee a large portfolio of projects, including mining, hydropower, and plantations. There are three foreign investment departments in MPI: Laos–China Cooperation Department, Laos–Vietnam Cooperation Department, and International Cooperation Department.

An MPI official explained, "there is just more certainty that we will work closely on big projects with China and Vietnam, so these offices [allow us to act] bilaterally and discuss issues directly if we have challenges" (Interview, June 2019). Other development actors, such as multilateral development banks, have expressed frustration that Chinese and Vietnamese actors do not attend the Annual Development Roundtable Meetings, preferring rather to maintain direct lines with Lao institutions, particularly those managing BRI projects.

An official in the National Institute for Economic Research (NIER) – which was created at the behest of former Prime Minister Bousaone Bouphavanh wherein he served as president – described a central purpose of the Institute as "supporting and managing Chinese investment in Laos. We are looking now to China for better investments. Vietnamese investments have not been so successful by comparison, and we don't see much of a future in these." (Interview, 2019) While such a mandate cannot be found in the formal charter of NIER, this informal characterization is meaningful, and largely supported by Bouphavanh's first actions as its president. In 2016, just weeks before Xi Jinping's first state visit to Laos, Bouphavanh visited China during which time a major investment deal on the Bolaven Plateau was agreed. While the outcome of this agreement remains to be seen, its scale and location – in Vietnam's backyard – is symbolically significant.

State restructuring in Laos has also been undertaken for specific projects, for example the Laos–China Railway (LCR) and Laos–China Economic Corridor (LCEC) – both flagship BRI initiatives to link Laos with China and Southeast Asia. Rail is a well-recognized sector in which China and Japan compete, particularly in Southeast Asia (as other chapters in this volume explore). With the LCR, the Department of Railway under the Ministry of Public Works and Transport of Laos was rebranded the Laos–China Railway Department, and tasked to oversee the LCR. However, most of the planning, construction, and management fell to the Laos-China Railway Company, a Lao–Chinese joint venture established to develop, manage, and operate the project. In addition, 'six special policies' were instituted that allow state land to be used at no additional cost, tax exemptions are provided, domestic resource charges and import duties are waived, foreign residence fees for workers are reduced, related investments are encouraged.

Lao bureaucracies have been created and redirected to negotiate and manage projects with Chinese government counterparts and to coordinate different levels of the Lao government. Ministries focus on how to benefit from the deluge of projects and capital. Research desks have been formed to answer key questions related to the BRI. The Laos–China Committee, for example, was created to analyse and negotiate cooperation frameworks and in response to the LCR and the LCEC that was proposed to follow it. Its main task is macroeconomic analysis of agriculture, processing, services and tourism, finance, and customs in northern Laos. The prime minister's office established a BRI committee for decisions regarding costs, benefits, how to move forward with the BRI, and which sectors to prioritize. The BRI is the starting point of the most recent five-year plan (2021–25), and national research institutes within the government were directed to conduct BRI-focused research to inform this plan.

Turning to the Mekong River offers an example of more active US involvement. The Mekong, much like the South China Sea, has become a geopolitical issue grounded in infrastructure construction. While the US does not compete in terms of infrastructure construction and finance, the Mekong is a central forum of US interest in Laos. The 1957 founding of the Coordination of Investigations of the Lower Mekong Basin (later known as the 'Mekong Committee') was a precursor of the Mekong River Commission, which the US saw as an integral component of its military strategy in Indochina – the carrot to the stick (Menon, 1971). Building on the Mekong Committee, the Lower Mekong Initiative was created in 2009 under Secretary of State Hillary Clinton and Foreign Ministers of Lower Mekong countries, offering a means of expanding US government presence in the region. In September 2020, it was updated to the Mekong–US Partnership, Washington's aid plan and counterweight to China in continental Southeast Asia, including Laos. According to the US Department of Defense (2019, 40), 'China is increasingly focused on Laos, and Beijing continues efforts to expand its strategic footprint through large debt-fuelled investments, especially in infrastructure and energy'. However, the US Department of Defense has cited Laos as cautious, maintaining strong ties with Vietnam and Russia as defence partners.

Meanwhile, as the US attempts to increase engagement, the Lao government undertook perhaps the most significant restructuring efforts connected to China around Électricité du Laos (EdL), the state energy corporation. While EdL's finances are largely a black box, its debt levels were so high (largely due to sunk investments in hydropower projects) that the Lao state restructured its ownership stakes in EdL to keep it afloat. The government also created a new entity, the Électricité du Laos Transmission Company Limited (EdL-T). Assets were privatized and debts were renegotiated with China as the government attempted to secure new sources of short-term credit (see Barney and Souksakoun, 2020). Today, the state-owned enterprise China Southern Power Grid Company (CSG) holds a controlling stake in Laos's national grid, possibly up to 90 per cent of EdL-T. In exchange, CSG will complete construction of the transmission grid, thus allowing the Lao government to achieve its second core objective of regional energy connectivity. As Barney and Souksakoun (2020, 100) note, EdL 'has maintained that the sale does not represent a loss of sovereignty; that CSG is a large and professionally-run utility with deep pockets; and that the completion of the domestic grid through the CSG investment could allow for more efficient and profitable distribution of energy production within the country'. They also suggest that this may allow EdL and EdL-T increased bargaining power with 'external off-takers', indicating that CSG may disrupt regional competition over electricity costs and distribution. As in other cases of state restructuring for infrastructure, however, many details of deals and negotiation processes remain behind closed doors, leaving the infrastructures themselves as the main evidence of such processes.

## Conclusion

Critically examining the US–China rivalry in Laos as expressed through infrastructure reveals cracks in the notion of a 'new Cold War'. While geopolitical tension has undoubtedly intensified, bipolar rivalry is scarcely visible despite recent US overtures. Rather, other bilateral relations such as those with Vietnam, Japan, as well as Russia, remain pivotal for three reasons. They are not only investing in and building much of Laos's infrastructure, but they also have deep historical relations, and their capital and projects have become tools for the Lao government to balance Chinese investment, even as they lean into it.

While Vietnam and China are Laos's closest political allies, Moscow's linkages to Laos's local communist regime date back to the Cold War. As state restructuring demonstrates, Lao state agency allows certain countries to exert more influence than the US to balance China. For example, in July 2021 Russian Foreign Minister Sergey Lavrov visited Vientiane with special focus on the 'China challenge'. On the visit, Lao President Thongloun Sisoulith noted the close ties that Laos has maintained with its 'Russian brothers' over 60 years. Indeed, since the founding of Lao People's Democratic Republic in 1975, Russia has assisted by supplying food, goods, and vehicles. While Russia cannot compete with China at scale, the Lao government actively courted Russian support as a means of balancing China. The two counties are poised to engage in military infrastructure development as Moscow pledged to upgrade a Lao military airstrip,[1] supply military technology, and train Lao armed forces.

Still, Laos's integration to the global economy is largely and undeniably mediated by Chinese-backed projects. State restructuring for Chinese-backed projects hints at how Laos is increasingly shifting its geopolitical orbit on its own terms. As the Lao government embraces new spatial strategies, the BRI presents capital and momentum to achieve longstanding projects, but this necessitates state restructuring. The *raison d'être* for new governmental institutions and agendas is to design and produce an integrated territory, as a landlinked energy hub. This has not been without controversy – the social and environmental costs of the BRI have attracted scrutiny, especially since economic gains have yet to materialize. However, Lao Prime Minister Sisoulith assured members of the National Assembly that the economy will pick up and indeed grow beyond 2020 as BRI and mega-infrastructure projects – the railway, a 'smart' city, expressways, bridges, industrial farming, and processing facilities – 'kick into gear' (Phouthonesy, 2019).

**Note**

[1]  Unexploded ordnances dropped by the US from 1964 to 1973 still have to be cleared from this area.

**References**

Australian Government (2019) *Southeast Asia Economic Governance and Infrastructure Facility*, Canberra: Department of Foreign Affairs and Trade. Available from: www.dfat.gov.au/sites/default/files/seaegif-investment-design-document.pdf.

Barney, K.D. and Souksakoun, K. (2020) 'Credit crunch: Chinese infrastructure lending and Lao sovereign debt', *Asian and the Pacific Policy Studies*, 8(1): 94–113.

Brenner, N. (2004) *New State Spaces: Urban Governance and the Rescaling of Statehood*, Oxford: Oxford University Press.

Creak, S. (2011) 'Laos: Celebrations and development debates', *Southeast Asian Affairs*: 107–28.

DiCarlo, J. (2021) 'Grounding Global China in Northern Laos: the making of the infrastructure frontier', Doctoral dissertation, University of Colorado Boulder.

Dwyer, M.B. (2016) 'Upland geopolitics: finding Zomia in northern Laos c. 1875', *The Journal of Lao Studies*, 3: 37–57.

Dwyer, M.B., Ingalls, M.L., and Baird, I. (2016) 'The security exception: development and militarization in Laos's protected areas', *Geoforum*, 69: 207–17

*The Economist* (2021) 'The battle for China's backyard: the rivalry between America and China will hinge on South-East Asia', 27 February. Available from: www.economist.com/leaders/2021/02/27/the-rivalry-between-america-and-china-will-hinge-on-south-east-asia.

Evans, G. (2002) *A Short History of Laos: The Land in Between*, London: Allen & Unwin.

FMPRC (2016) 'Xi Jinping meets with President Bounnhang Vorachith of Laos', 2 September, Ministry of Foreign Affairs of the People's Republic of China. Available from: www.fmprc.gov.cn/ce/cgla/eng/topnews/t1395040.htm.

Glassman, J. (2010) *Bounding the Mekong: The Asian Development Bank, China, and Thailand*, Honolulu, HI: University of Hawai'i Press.

Goh, E. (2014) 'The modes of China's influence', *Asian Survey*, 54(5): 825–48.

Gunn, G.C. (1990) *Rebellion in Laos Peasant and Politics in a Colonial Backwater*, New York: Routledge

Hiebert, M. (2020) *Under Beijing's Shadow: Southeast Asia's China Challenge*, Washington, DC: Center for Strategic and International Studies.

Ivarsson, S. (2008) *Creating Laos: The Making of a Lao Space between Indochina and Siam, 1860–1945*, Copenhagen: Nordic Institute of Asian Studies.

Jennings, R. (2021) 'Japan, Thailand, Vietnam vie with China for influence in impoverished, landlocked Laos', *VOA News*, 23 April. Available from: www.voanews.com/east-asia-pacific/japan-thailand-vietnam-vie-china-influence-impoverished-landlocked-laos.

Kuik, C.-C. (2021) 'Asymmetry and authority: theorizing Southeast Asian responses to China's Belt and Road Initiative', *Asian Perspective* 45(2): 255–76.

Lampton, D.M., Ho, S., and Kuik, C.-C. (2020) *Rivers of Iron: Railways and Chinese Power in Southeast Asia*, Oakland, CA: University of California Press.

Menon, P.K. (1971) 'Financing the Lower Mekong River Basin development', *Pacific Affairs*, 44(4): 566–79.

Pholsena, V. and Banomyong, R. (2006) *Laos: From Buffer State to Crossroads?* Singapore: Silkworm Books.

Phouthonesy, E. (2019) 'Laos PM sees economic recovery in sight', *Vientiane Times*, 29 November.

Runde, D.F. and McKeown, S. (2019) 'The Asian Development Bank: a strategic asset for the United States', Center for Strategic and International Studies. Available from: https://www.csis.org/analysis/asian-development-bank-strategic-asset-united-states.

Sayalath, S. and Creak, S. (2017) 'Regime renewal in Laos: The Tenth Congress of the Lao People's Revolutionary Party', *Southeast Asian Affairs*: 179–200.

Schindler, S. and Kanai, J.K. (2021) 'Getting the territory right: infrastructure-led development and the re-emergence of spatial planning strategies', *Regional Studies*, 55(1): 40–51.

Schindler, S., DiCarlo, J., and Paudel, D. (2020) 'The new cold war and the rise of the 21st century infrastructure state', *Transactions of the Institute of British Geographers*, 47(2): 331–46.

Shambaugh, D. (2020) *Where Great Powers Meet: America and China in Southeast Asia*, Oxford: Oxford University Press.

Soong, J.J. (2016) 'The political economy of the GMS development between China and Southeast Asian countries: geo-economy and strategy nexus', *The Chinese Economy*, 49(6): 442–55.

Strangio, S. (2020) *In the Dragon's Shadow: Southeast Asia in the Chinese Century*, New Haven, CT: Yale University Press.

Stuart-Fox, M. (1995) 'The French in Laos, 1887–1945', *Modern Asian Studies*, 29(1): 111–39.

Stuart-Fox, M. (2009) 'The Chinese connection', *Southeast Asian Affairs*: 141–69.

Tan, D. (2012) '"Small is beautiful": lessons from Laos for the study of Chinese overseas', *Journal of Current Chinese Affairs*, 41(2): 61–94.

US Department of Defense (2019) 'Indo-Pacific strategy report: preparedness, partnerships, and promotion a networked region', US Department of Defense, 1 June. Available from: https://media.defense.gov/2019/Jul/01/2002152311/-1/-1/1/DEPARTMENT-OF-DEFENSE-INDO-PACIFIC-STRATEGY-REPORT-2019.PDF.

USAID (2018) 'Interview with administrator Mark Green', 26 November. Available from: https://www.usaid.gov/news-information/press-releases/nov-26-2018-administrator-mark-green-interview-cspan-newsmakers.

Wong, C. (2017) 'After summit, ASEAN remains divided on South China Sea', *The Diplomat*, 3 May. Available from: https://thediplomat.com/2017/05/after-summit-asean-remains-divided-on-south-china-sea/.

Yeh, E.T. (2016) 'Introduction: the geoeconomics and geopolitics of Chinese development and investment in Asia', *Eurasian Geography and Economics*, 57(3): 275–85.

Yellinek, R. and Chen, E. (2019) 'The "22 vs. 50" diplomatic split between the West and China over Xinjiang and human rights', *China Brief*, 19(22).

Yuvejwattana, S. (2019) 'Laos says it's set to benefit from the US–China trade war', Bloomberg, 25 November. Available from: https://www.bloomberg.com/news/articles/2019-11-25/laos-says-it-s-set-to-benefitfrom-the-u-s-china-trade-war.

# *Interlude*: Locating Host-Country Agency and Hedging in Infrastructure Cooperation

*Cheng-Chwee Kuik*

## Introduction

Since the 2013 launch of China's Belt and Road Initiative (BRI) – a multitrillion dollar infrastructure investment and lending programme connecting China with Asia, Africa, Europe, and Latin America – bilateral infrastructure cooperation has gained attention from policy and academic communities. This trend has accelerated as other powers launched efforts such as Japan's Partnership for Quality Infrastructure (PQI) in 2015, the EU–Asia Connectivity Strategy in September 2018, and the US's BUILD Act in 2018. More recently, at its June 2021 Summit, the Group of Seven (G7) announced the establishment of Build Back Better World (B3W), an initiative widely viewed as an alternative to rival the BRI in helping developing countries with their infrastructure, health, digital technology, and other matters. In July 2021, the EU launched 'A Globally Connected Europe', signalling its determination to pursue a geostrategic approach to connectivity at the global level.

In the field of International Relations, most scholarly works on infrastructure cooperation focus on China and other big powers, with attention to their motivations, and implementation progress and problems. Relatively few works concentrate on the smaller host countries where infrastructure is built. Some scholars portray countries hosting BRI projects as victims of Beijing's 'debt trap', describing them as passive, powerless, or even clueless actors being manipulated by 'predatory' actors from China via central or provincial authorities, state-owned enterprises, and private commercial entities. Very few studies have explored the phenomenon of host-country agency thematically or comparatively.

The chapters in this book – particularly those that follow in Part III – help to fill these gaps. Drawing on the findings in these chapters and related articles

(Kratz and Pavlićević, 2019; Camba, 2020; Lampton et al, 2020; Liao and Dang, 2020; Kastner and Pearson, 2021; Kuik, 2021a, 2021b; Schindler et al, 2021; Schneider, 2021), this interlude develops a typology of host-country agency and addresses key questions: What explains variations of host-state agency? Why do some countries welcome foreign-backed infrastructure cooperation more than others? How do some smaller states hedge more effectively in infrastructure cooperation?

Variations in patterns of host-country receptivity and agency are brought about by internal and external factors. Specifically, while elite legitimation is a key driver in determining how receptive the smaller states are towards foreign-backed infrastructure connectivity projects, the forms of host-country agency are conditioned and constrained by internal resilience and external alternatives (or lack thereof). I further contend that host states responding to and leveraging US–China rivalry as well as competitions involving regional powers such as Japan and India (see Chapters 11, 13 and 15 in this volume). These multi-level power dynamics prompt host countries to pursue and connect their respective spatial objectives the ways they do.

To support these arguments, this interlude first draws on empirical observations across countries and regions to illuminate the different types of host-state agency in infrastructure cooperation. The second and third sections unpack internal and external dynamics influencing agency, to examine how and why certain factors both drive and limit host-country agency in foreign-funded infrastructure partnerships. It concludes by connecting these findings to chapters in this volume.

## Defining host-country agency and small-state hedging

Agency is conceived of here as an intermediary between micro-processes and macro-structures (Parsons, 1951; Giddens, 1984; Sewell, 1992; Wight, 2006). When applied to international relations, macro-structures refer to such *structural*, top-down conditions as asymmetrical power relations, power rivalries, and systemic uncertainties, while micro-processes refer to the *unit-level* action–reaction along the key stages of inter-state cooperation, initiation, negotiation, and implementation. Accordingly, host-country agency is defined as the capacity of a sovereign actor hosting a foreign-backed venture in making its own decisions, shaping the circumstances, and pursuing its desired outcomes during the micro-processes, despite power asymmetry, rivalry, and uncertainty.

Host-country agency in infrastructure cooperation manifests in multiple forms. These include: (a) proactive initiation (proposing to a foreign power to forge an infrastructure connectivity partnership); (b) active involvement (enthusiastically participating in and partnering with an external power on an infrastructure construction venture); (c) active renegotiation (advocating amendments to the previously agreed-upon terms of an infrastructure partnership); and/or (d) passive resistance (denying, delaying, or distancing from a stronger power's initiative). All four patterns of host-country agency have been observed in smaller states' engagement with and response to Chinese-backed projects across Asia, Africa, Europe, and beyond (Goh, 2016; Oh, 2018; Carrai et al, 2020; Lampton et al, 2020; DiCarlo, 2021; Kuik, 2021b, 2021c; Schneider, 2021). Schneider (2021, 25) notes that many outcomes depend on host countries: 'the BRI is shaped by diverse local actors who exercise their agency for their own developmental ends or other goals.'

The cases in this volume and other studies vividly illustrate these various manifestations of agency. Vietnam, for instance, displayed a persistent tendency to distance itself from the BRI, as it delays or declines several China-related projects. As Liao (Chapter 15 in this volume) points out, driven primarily by deep-seated anti-China public sentiment, Hanoi has deliberately kept a low profile regarding its connectivity cooperation with China, cautiously controlling the speed and scope of cooperation, while strengthening and widening its partnerships with Japan. Since the early 2010s, following China's increasingly assertive actions in the South China Sea, Hanoi has further demonstrated its agency. After the 2014 HD-981 oil rig standoff between China and Vietnam, Hanoi pivoted further away from infrastructure cooperation with China, and instead collaborated on a thermal power project and various ventures with Japan and other partners.

Other cases highlighted in this part and the wider volume – Indonesia, Kazakhstan, Laos, Nepal, Ethiopia, and Hungary – display differing forms of host-country agency. They have all actively partnered with China on infrastructure connectivity projects in some form. Indonesia collaborated to construct the Jakarta-Bandung High-Speed Rail (the first high-speed railway, HSR, in Southeast Asia, see Chapter 14), Kazakhstan on the Khorgos dry port (the largest dry port in Central Asia, see Chapter 17), Laos on the Laos–China Railway (the first rail from Southeast Asia to connect with China's vast railway network, see Chapter 12), Nepal on the Kerung-Kathmandu Trans-Himalayan Railway (see Chapter 13), Ethiopia on the 756-km Addis Ababa–Djibouti Railway (the first Chinese-built and Chinese-operated electrified railway in Africa) and the Hawassa Industrial Park (the largest industrial park in Africa, see Chapter 8), and Hungary on the Hungarian section of the Budapest–Belgrade Railway (see Chapter 16). In a study on African agency through Ethiopia's engagement with the Digital Silk Road, van der Lugt (2021, 339) concludes that contrary to the popular depiction in Western media that attributes Ethiopia's embrace of Chinese surveillance technology to the agency of Chinese firms and the Chinese state, her study has found that 'the Ethiopian government has the agency to independently choose what technology it acquires and from where'.

Some host countries have been more receptive and proactive in embracing China as a developmental partner, with several projects predating the announcement of the BRI. Malaysia under Abdullah Badawi (2003–09) and Najib Razak (2009–18), for instance, took the initiative to partner with China on infrastructure construction, including on the Second Penang Bridge and the Kuantan Industrial Park. After Mahathir Mohamad returned to power following the electoral victory of the Pakatan Harapan (PH) coalition in May 2018, however, Malaysia suspended the controversial East Coast Rail Link (ECRL) and two other China-backed projects. Intense renegotiation ensued, leading to the resumption of the ECRL at a lower cost and terms more favourable to the host country (Kuik, 2021d). Malaysia's fluctuating patterns of BRI engagement indicate that BRI projects are not always about big-power push (China pushing the cooperation envelope); sometimes smaller states pull as well and at times even push back, despite their asymmetrical power structures vis-à-vis China.

Regardless of its form, some scholars consider such agency a manifestation of small-state 'hedging', an approach where a state seeks to maximize benefits from competing powers, while simultaneously mitigating risks and cultivating options to preserve autonomy and prepare for contingencies (Kuik, 2020; Lampton et al, 2020, 16–17). Hedging is particularly prevalent in asymmetrical partnerships involving a weaker state and a much more powerful partner. To mitigate the risks of over-dependence on and unbalanced influence from any single power, weaker states tend to diversify their strategic and developmental ties with multiple powers, exploring ways to benefit from competitive dynamics at various levels (Liao and Dang, 2020; Pitakdumrongkit, 2020; Liao, Chapter 15 in this volume).

In their chapter on Indonesia's Jakarta-Bandung HSR, Tritto et al discuss how the Jokowi government hedges vis-à-vis competing powers by exploiting the eagerness of Chinese companies and policy banks and 'making them compete with Japanese institutions'. They cite Luhut Panjaitan, the Coordinating Minister of Maritime Affairs, who said: 'Let them race to invest in Indonesia. It is good for us', adding that it is like 'a girl wanted by many guys; the girl then can pick whomever she likes'. The Indonesian authorities initially favoured Japan but ultimately selected China. Jakarta's 'project-specific' hedging approach enabled it to secure its desired outcomes from the China deal: an uncommon financing mode (a loan that requires no government guarantees, with low interest rate), favourable joint venture terms (Kereta Cepat Indonesia China, KCIC, comprising Chinese 40 per cent and Indonesian 60 per cent ownership), concessions in terms of employment and training, resources, manufacturing of rolling stock, as well as financial disbursement prior to the finalization of land acquisition processes (Negara and Suryadinata, 2018).

Hedging is dynamic. It necessarily involves *adaptive, selective,* and even *contradictory* approaches (Kuik, 2020), as the Indonesian case illustrates. Even as Jakarta actively engages with China on the Jakarta–Bandung HSR and other infrastructure projects, it does so cautiously and selectively, while simultaneously pursuing seemingly contradictory measures aimed at keeping its distance and limiting Beijing's influence. In addition to pushing back China's growing maritime assertiveness by conducting a series of high-profile military exercises near its Natuna Islands, Indonesia has also stepped up its defence partnerships with the US, Australia, Japan, and other powers (Tritto et al, Chapter 14 in this volume; Putten and Petkova, 2021).

Dynamic hedging approaches are observable in other countries. Vietnam has leveraged Japan–China rivalry to maximize gains and minimize risks across security and economic domains. Liao and Dang (2020) observe that over the past three decades Vietnam has taken a nuanced approach to optimize its interests and achieve its developmental priorities from Japan and China's infrastructure financing programmes, embracing both programmes at earlier periods but recently pivoting away from risky China-backed projects while enhancing collaboration with Japan, an increasingly important strategic partner.

Beyond Southeast Asia, Nepal, Kazakhstan, and Ethiopia exemplify how countries considered peripheral or small have also endeavoured to broker better deals by hedging and being pragmatic towards big-power competition. Nepal cautiously hedges towards India, the US and China's infrastructure courtships, seeing the BRI as 'a historic opportunity to challenge Indian market monopolies and influence in its internal political and economic spheres' (Paudel and Rankin, Chapter 13 in this volume; Ghiasy, 2021). Taking advantage of Kazakhstan's strategic location at the crossroads of Eurasia, Nur-Sultan hedges between the US and China's economic statecraft to maximize support from both powers, while pursuing its spatial projects under the banner of *Nurly Zhol* ('bright path'), aimed at developing Kazakhstan as a central hub for trade, goods transportation, and engine of economic growth for Central Asia (Neafie, Chapter 17 in this volume). Successive governments in Ethiopia – from the Ethiopia People's Revolution Democratic Front (EPRDF, once Africa's largest political party) to the present Prosperity Party (PP) – have both capitalized on the US–China rivalry and hedged to maximize benefits and offset risks. According to Huang and Goodfellow, the inflow of Chinese finance grew steadily under EPRDF rule (1991–2019), with infrastructure as the key area of investment and state-led economic transformation. For the EPRDF regime, a close partnership with China was crucial not only for maximizing economic returns, but also mitigating external risks, for example, neutralizing external pressure from Western partners over ideological issues and the pace of political liberalization. In 2019, Beijing courted Prime Minister Abiy Ahmed of the

PP government with US$1.8 billion in funding for Ethiopia's electricity transmission and distribution network. China also facilitated the launch of Ethiopia's first satellite into space. In 2020, Washington pledged to invest US$5 billion in Ethiopia through its new International Development Finance Corporation (IDFC) in such sectors as telecommunications, geothermal energy, logistics, and sugar production (Huang and Goodfellow, Chapter 8 in this volume).

Small-state hedging in infrastructure cooperation is likely to deepen and widen across regions in the years to come. Indeed, the scope and scale of such hedging acts are likely to evolve in tandem with the intensifying US–China rivalry on both the high- and low-politics chessboards: the military and non-military domains (Schindler et al, 2020; Kuik 2021e). Under the current situation, big-power competition on the twin chessboards – involving not only the US and China, but also the EU, Japan, and other second-tier powers – is arguably providing more manoeuvring space for smaller states. This is because the two chessboards, in effect, constitute two marketplaces where multiple sets of geopolitical and geoeconomic supply and demand intersect. When more powers compete to court and supply the regional public goods and private goods (security, prosperity, and autonomy) required by regional countries, this accords *agency* to the consumers (that is, the smaller states in the region; Quah, 2019). Hence, as long as big-power competition does not deteriorate into polarized camps, the space for small-state hedging is likely to increase in the decades ahead.

## Drivers: elite legitimation and patronage

Why are some host countries more receptive and enthusiastic than others when embracing foreign-funded developmental partnerships such as BRI-related projects? There are numerous drivers, but the primary ones are the ruling elites' pathways of legitimation which often intersect with factors like political patronage and special interests, leading to different degrees and patterns of host-country receptivity.

This interlude focuses on governing elites or political classes, that is, the small group of actors who exercise disproportionate power and influence over a particular society and country on the grounds of governance authority, coercive capacity, and ideology (Bottomore, 1964; Lipset and Solari, 1967; Parry, 1969/2005; Axelrod, 2015). Regardless of their countries' political systems, elites claim their 'right' to rule by appealing to certain ideals, constructing substantial, or substantiated narratives, and resorting to corresponding pathways to justify, enhance, and consolidate their domestic authority vis-à-vis other contesting elites and society at large. There are three major ideals or pathways: (a) *procedural* virtues (such as democratic values, social justice); (b) *particularistic* narratives (all forms of identity politics, including nationalist sentiments, ethnic and religious appeals, personal charisma); and (c) *performance*-related rationales (including ensuring growth and delivering development fruits, managing well nationwide problems such as maintaining internal order or quelling a pandemic). No elites rely on a sole pathway to rule. In practice, all elites resort to a combination of legitimation pathways – with different degrees of emphasis and mobilization – for their inner justification (Lampton et al, 2020; Kuik, 2021a).

As I have argued elsewhere, these pathways of legitimation drive the direction and prioritization of state policies, including how receptive the host countries are towards the BRI or, for that matter, any foreign-backed developmental initiative (Kuik, 2021b, 2021c). These drivers explain the following three broad patterns of host-country receptivity.

*Highly receptive*

A host country is likely to embrace foreign-backed developmental initiatives when close cooperation with the external power serves to boost the ruling elites' growth-based performance and bolster other pathways of inner justification, such as managing internal conflict, neutralizing external pressure, and promoting national identity. The host country is likely to be more receptive if and when the converging pathways of legitimation are also accompanied by opportunities for patronage and state-capital congruence (among elites and their transnational allies; see, for instance, Wijaya and Camba, 2021).

In Laos, for instance, the Lao People's Revolutionary Party's (LPRP) government has enthusiastically embraced the BRI for numerous reasons: China's capital and knowhow is crucially needed to boost Laos' infrastructure development and economic growth. When the construction of the Laos–China Railway (DiCarlo and Ingalls, Chapter 12 in this volume) is completed, Laos will be transformed from a landlocked to a landlinked nation, thereby enhancing the ruling elites' performance legitimation (Lampton et al, 2020). In addition, a stronger partnership with China will reduce Western political pressure and enable Laos to deal more effectively with its two larger neighbours, Thailand, and Vietnam, enhancing the LPRP's identity-based ethno-nationalistic justification. As the LPRP struggles to maintain its ideological appeal, enhancing performance and particularistic legitimation are keys to preserving its political relevance and authority domestically. Moreover, the developmental projects provide the ruling elites with patronage resources.

Similar drivers explain other countries' relatively high receptivity of the BRI. In Kazakhstan, the Nur Otan government views Chinese finance as a catalyst required to actualize its 'Kazakhstan 2050' vision of diversifying the national economy and transforming Kazakhstan into an 'Eurasian

Land Bridge', a central hub for national and transnational infrastructure development, the transportation of goods, and sustainable regional economic growth (Neafie, Chapter 17 in this volume). In Nepal, both elected politicians and bureaucrats want to leverage on the BRI and the wider geopolitical competition on infrastructure development for 'national integration and modernization' while 'asserting self-reliance and sovereignty' vis-à-vis India's regional hegemony (Paudel and Rankin, Chapter 13 in this volume). In Ethiopia, the Prosperity Party (PP) government – despite its ideological reorientation and commitment to diversify sources of foreign investment – wants to capitalize on Chinese-funded mega-infrastructure projects to promote 'a narrative of growth and development' and pursue structural transformation in accordance with the country's state centralization (Huang and Goodfellow, Chapter 8 in this volume). This approach effectively continues the previous regime's receptivity of the BRI, even as the PP government welcomes infrastructure finance from other sources, including the US. In Hungary, since the Viktor Orbán government came to power in 2010, it has actively sought Chinese investment to enhance its governing elites' growth-based performance, in the wake of 'persistently sluggish growth and the EU's bureaucratic inertia' following the 2008 global financial crisis (Gyuris, Chapter 16 in this volume).

For the ruling elites in these and many other countries, growth and developmental problems are not only economic but also *political* issues. Infrastructure building, a prerequisite for economic growth, is key to the performance legitimation of governing elites. While important in its own right for all governments, development performance is especially critical for authoritarian and semi-democratic regimes which lack democracy-based procedural legitimacy (Kuik 2021a). These regimes then have to rely much more heavily on performance- and identity-based nationalist legitimation to maintain and justify their right to rule. Other sources of performance legitimacy exist but economic performance is often the most crucial one as it creates jobs, ensures socio-political stability, boosts the elites' electability, and even provides opportunities for special interest and patronage. For some or all of these reasons, elites in Laos, Kazakhstan, Nepal, Ethiopia, and Hungary have thus turned to China, with hopes that close and cordial ties with Beijing would facilitate flows of Chinese capital and technology, infrastructure, and transport connectivity construction, stimulate broader economic opportunities, and/or fend off political pressure from certain external actors.

## Moderately and selectively receptive

A host state is likely to collaborate with the BRI on a *selective* and *cautious* basis if and when the collaboration is expected to have *mixed* impact on the elites' legitimation efforts, such as enhancing performance justification but harming identity-based particularistic legitimation, eroding electoral support, and/or undermining national identity. Indonesia illustrates such selective receptivity. While potential synergies exist between Jakarta's Global Maritime Fulcrum and Beijing's BRI, President Joko Widodo (Jokowi)'s government has engaged China-funded projects selectively and gradually, insisting on Indonesia's longstanding 'free and active' foreign policy, while cultivating close partnerships with Japan, the US, and other powers (Tritto et al, this volume).

Such a delicate balancing act is rooted in the need to optimize the elites' bases of legitimation. As Yeremia (2021) observed, caution arises as a result of the incumbent coalition's need to strike a balance between two pathways of legitimation: growth-based performance and nationalist legitimation. While Jokowi's developmental agenda and external outlook have pushed him to expand Indonesia's engagement with China, his pragmatic push has been limited by nationalism, Islamism, and anti-Chinese sentiment mobilized by the hardliners and opposing socio-political groups. Should Jokowi fully engage China's BRI, he runs the risk of being perceived as a pro-China leader, which would threaten his authority in Indonesia, the world's largest Muslim country.

## Coolly receptive

Finally, a smaller state is likely to keep its distance when close cooperation is expected to hurt the ruling elites' main basis of authority and legitimacy (nationalism, electoral pledge, reformist agenda), and when the perceived developmental benefits would be gained at the expense of security and territorial interests, autonomy, and/or sustainable development. Vietnam offers a clear example. Despite its geographical proximity, ideological ties, and growing commercial links with China, the ruling Communist Party of Vietnam (CPV) has persistently kept its distance from Beijing-backed infrastructure and connectivity ventures, limiting its BRI cooperation primarily to the construction of the Cat Linh–Ha Dong Elevated Railway, an urban rail project in Hanoi. Scholars have attributed Vietnam's lukewarm and low-profile response to strong anti-China sentiments in Vietnamese society (Liao, this volume; Liao and Dang, 2020). This identity-based particularistic legitimation is arguably a more important pathway of inner justification than development performance for the CPV elites. Considering anti-China nationalism in Vietnam, overly embracing the BRI would be politically risky for the ruling CPV elites. In contrast, distancing from but expressing verbal support for the BRI avoids a political landmine, while earning diplomatic mileage and keeping open opportunities for economic cooperation (Pham and Ba, 2021). While development-based performance legitimation is the principal driver that explains why many countries have receptively embraced the BRI, other pathways of inner justification – most notably identity-based nationalist legitimation – have constrained the extent to which elites in some countries can and would forge a close collaboration with China.

## Determinants: external alternatives and internal resilience

While elite legitimation and patronage politics explain varying degrees of host-country receptivity, they do not explain why different countries show different *patterns* of agency and hedging tendencies in infrastructure cooperation. The latter is determined by the presence or absence of *external* alternative partnership(s), as well as the respective country' degree of *internal* resilience. Resilience – the ability to manage, withstand, and respond to particular challenges – involves multiple internal attributes and mechanisms for ensuring the survival, stability, and societal cohesion of a given country (Brassett and Vaughan-Williams, 2015; Cavelty et al, 2015). The greater a host country's internal resilience and external alternatives, the greater its agency and ability to hedge. These internal and external factors determine the host country's capacity to promote its interests and offset risks amid power asymmetry and uncertainty.

The availability of external alternative partnerships is a *sine qua non* of host-country agency and hedging. A host country can pursue economic hedging only if and when two or more powers vying for influence are courting the country. As discussed in this volume, the governments of Indonesia, Vietnam, Kazakhstan, Nepal, and Ethiopia are able to hedge and exercise agency primarily because competing powers such as the US and Japan (and India in South Asia) provide alternative partnership opportunities, which increase the smaller states' leverage when bargaining with China to maximize their benefits from BRI-related projects. Conversely, when there is little alternative or when China is the only available provider of infrastructure capital and technology, there is little room for smaller states to hedge and bargain for better deals as equally. Laos is an example in point: despite a preference for balanced connectivity, Chinese investment remains the dominant source of funding for infrastructure.

However, alternative partnerships, in and of themselves, only provide external space for small-state hedging; they do not translate automatically into host-country agency. A more fundamental determinant is the host country's internal resilience, which determines the extent to which it can actively promote its interests from a partnership. Different patterns of internal resilience are the result of numerous internal attributes, including political systems, inter-elite contestation, bottom-up activism, state restructuring, and actual implementation. These themes play out across countries and across time. Take Cambodia, Laos, and Malaysia, three enthusiastic embracers of China's BRI in Southeast Asia. While all three countries have actively partnered with China on infrastructure cooperation, they have demonstrated different patterns of host-country agency which are attributable to their different political systems and other attributes. Malaysia is a quasi- or semi-democracy; Cambodia a hegemonic authoritarian regime; and Laos a Marxist-Leninist party state. In the three countries, leaderships intersect with inter-elite dynamics, shaping the extent and manner in which societal and bottom-up sentiments vis-à-vis China-backed projects are addressed. Different political systems, however, dictate different patterns of processes and outcomes.

Malaysia's fluctuating engagement with the BRI under Najib Razak (2009–18) and Mahathir Mohamad (May 2018–February 2020) illustrates that democratic institutions, even flawed ones, are sources of internal resilience that enable inter-elite struggle to constrain and 'correct' decisions made by a previous leadership (Kuik, 2021d). When Najib was still in power from 2016 to 2018, several China-funded projects in Malaysia were highly contested at both the inter-elite and societal levels. The East Coast Rail Link (ECRL) and the two pipeline projects were viewed as *quid pro quos* for Beijing's help in bailing out Najib's 1MDB, a scandal-plagued sovereign fund. The three projects were highly controversial because of their perceived links with the troubled 1MDB, their lack of transparency, questionable financing arrangements, and unfavourable terms for Malaysians. The ECRL was criticized by opposition groups, non-governmental organizations, and members of parliament from different political parties. Mahathir was the harshest critic. In 2016, after quitting the ruling United Malays National Organization following his fallout with Najib over the 1MDB and other issues, Mahathir criticized the ECRL and other China-related projects in Malaysia. Popular discontent about Chinese investment converged with inter-elite competition, leading Mahathir and other opposition elites to intensify their attacks on China-backed projects during the election campaign. Mahathir vowed to review the contracts for these projects should his PH coalition win.

Upon winning the 2018 elections, the 93-year-old leader leveraged his democratic mandate and grass roots sentiments and suspended the three controversial projects and requested renegotiation with China. Through public statements and media interviews, Mahathir made his goals clear: reduce costs significantly, reset terms, and ensure that implementation of the projects would be favourable to Malaysia. In a media interview, he said 'if we get better terms, then of course we will continue' (Azim, 2018). Democratic institutions and the electoral process provided the PH government with strong bases to leverage and bargain with China, allowing Malaysia, a smaller host country, to push back.

This pattern of host-country agency and internal resilience has not occurred in Cambodia and Laos. In the absence of democratic space, bottom-up discontent and sentiments against controversial projects are handled differently. In Cambodia, these sentiments have been selectively co-opted and partially suppressed. In Laos, they have been ignored and largely denied. The extent to which political power is diffused and distributed across state elites and society help shape a country's internal resilience, which in turn determines host-country agency.

Cambodian authorities have displayed agency by increasing efforts to hedge and diversify its external policy beyond China, for example, by developing developmental, diplomatic, and strategic ties with Japan (Chheang, 2018; Luo and Un, 2021). This agency, however, is restricted. Indicators such as the unequal partnership and the dominant involvement of China in energy and other key sectors (many of which do not meet international standards and best practices in social and environmental safeguards, see Heng, 2016); the massive influx of Chinese nationals, the presence of unproductive and unsustainable ventures (particularly casinos and related entertainment venues) especially in Sihanoukville (Po and Heng, 2019; Luo and Un, 2020) point to the limits of Cambodia's internal resilience and external agency. Consequently, public resentment has been on the rise.

Laos' limited host-country agency is manifested in its disadvantaged position in the negotiation and implementation of China-funded projects (DiCarlo and Ingalls, Chapter 12 in this volume). The loan for the Laos-China rail project, for instance, is guaranteed by underground mineral resources in Laos. China was granted mining concessions as collateral – should revenues from the railway project be too low to service the debt, the lender, China, would have the right to extract said minerals as an additional mean to secure repayment of its loan (Tappe, 2018; Freeman, 2019). The Lao government has also made tax concessions, including the waiving of import duties on imported Chinese equipment associated with the project. These moves are likely to reduce the benefits accruing to Laos (Morris, 2019). Laos' low internal resilience is reflected in the authorities' limited ability in performing the governance functions and obtaining strong commitment from its Chinese partners to safeguard its environment and other issues (DiCarlo, 2020).

A host country's internal resilience also depends on the restructuring of state institutions and actual implementation of state regulations, as illustrated by the Ethiopian case. According to Huang and Goodfellow (this volume), to manage the inflow of Chinese capital and other foreign resources more effectively, while channelling them towards the state's ambitious infrastructure vision, the Ethiopian authorities created several semi-autonomous state agencies, most notably the Ethiopian Railway Corporation (FRC) and the Industrial Parks Development Corporation (IPDC). These new agencies have enabled the Ethiopian government to deliver major infrastructures, resolve bureaucratic issues, and further consolidate its control over developmental and spatial projects. In addition, these agencies have also allowed Ethiopia to draw on China's experience with large-scale infrastructure governance while executing its regulation policies vis-à-vis Beijing and China's State-Owned-Enterprises (SOEs). However, implementation remains a key challenge. This is evident in the Addis Ababa–Djibouti Railway (ADR), a project financed by China EXIM Bank, built by China Railway Engineer Corporation (CREC) and China Civil Engineering Construction Corporation, and operational from January 2018. Despite Ethiopia's relatively well-developed legislation on environment, development, and related matters, there has been 'legal and implementation lacunae' on various fronts, alongside a practice in 'disregarding its policies and medium-term debt strategy' to secure infrastructure financing, presumably to speed up the ADR's negotiation, preparation, and operation (Carrai, 2021). A lack of civil society engagement can further constrain host country agency.

## Conclusion

Internal and external factors affect the degrees and manifestations of host-country agency and hedging in infrastructure cooperation between great and regional powers. Specifically, while a host country's receptivity (or lack thereof) towards foreign-funded infrastructure partnerships (in this case, China's BRI projects) is driven primarily by its ruling elites' pathways of legitimation, its capability to actively maximize its own interests and to hedge against risks is determined by the availability of external alternatives and its internal resilience. Elite legitimation matters because the elites' principal pathways of inner justification motivate and limit the extent to which the host government engages with such foreign-backed projects as the BRI-related ventures. The availability of external alternatives matters because it determines the extent and space through which the host country bargains and hedges vis-à-vis an asymmetric partnership. Internal resilience matters the most because it determines the capability of host countries in pursuing desired outcomes (including pushing back when necessary), while offsetting and mitigating risks despite power asymmetry and uncertainty.

These findings – lucidly illustrated in this book – make three important contributions. First, infrastructure and connectivity cooperation – an increasingly salient aspect of international politics due chiefly to the growing US–China competition – are not just about big-power activism, but also about small-state agency. Despite the asymmetry in strength and status, smaller host countries are not necessarily passive partners or powerless recipients; rather, some of them have proven to be active actors capable of initiating, shaping, and reshaping their partnerships with stronger powers.

Small-state agency in infrastructure cooperation takes many forms, ranging from proactive initiation to active involvement, active renegotiation, and passive resistance. Second, hedging is increasingly prevalent in international cooperation. This, however, remains understudied and should be examined in light of how economic hedging overlaps with and relates to strategic hedging in international politics. Third, in the context of longstanding debate on internal–external linkages, the chapters here reveal that while power competition at global and regional levels provides the space and leverage for smaller states to hedge, exactly how individual states engage is necessarily conditioned by the domestic attributes, most notably elite legitimation and internal resilience, pointing to the importance of drilling down through multiple scales of internal sources of statecraft under conditions of uncertainty.

## Acknowledgements

The author gratefully acknowledges support from *Skim Geran Penyelidikan Dana Khas Universiti Malaysia Sabah* (Skim Dana Khas, grant number SDK0126-2020). He thanks Jessica DiCarlo, Fong Chin Wei, and Lai Yew Meng for their valuable feedback on earlier drafts. He also thanks Fikry A Rahman and Intan Baizura Jailani for superb research assistantship. The usual caveats apply.

## References

Axelrod, R. (ed) (2015) *Structure of Decision: The Cognitive Maps of Political Elites*, Princeton, NJ: Princeton University Press.

Azim, A. (2018) 'ECRL can proceed if better terms can be obtained, says PM', *The Edge Markets*, 25 June. Available from: www.theedgemarkets.com/article/ecrl-can-pro-ceed-ifbetter-terms-can-be-obtained-says-pm.

Bottomore, T.B. (1964) *Elites and Society*, Harmondsworth: Penguin Books.

Brassett, J. and Vaughan-Williams, N. (2015) 'Security and the performative politics of resilience: critical infrastructure protection and humanitarian emergency preparedness', *Security Dialogue*, 46(1): 32–50.

Camba, A. (2020) 'Derailing development: China's railway projects and financing coalitions in Indonesia, Malaysia, and the Philippines', GCI Working Paper 008.

Carrai, M.A. (2021) 'Adaptive governance along Chinese-financed BRI railroad megaprojects in East Africa', *World Development*, 141: 105388.

Carrai, M.A., Defraigne, J., and Wouters, J. (eds) (2020) *The Belt and Road Initiative and Global Governance*, Leuven: Institute for International Law and Leuven Centre for Global Governance Studies.

Cavelty, M.D., Kaufman, M., and Kristensen, K.S. (2015) 'Resilience and (in) security: practices, subjects, and temporalities', *Security Dialogue*, 46(1): 3–14.

Chheang, V. (2018) 'Lancang-Mekong cooperation: a Cambodian perspective', *ISEAS Perspective*, 70: 1–9.

DiCarlo, J. (2020) 'Mind the gap: grounding development finance and safeguards through land compensation on the Laos–China Belt and Road Corridor', GCI Working Paper, no 13.

DiCarlo, J. (2021) 'Grounding global China in Northern Laos: the making of the infrastructure frontier', Doctoral dissertation, University of Colorado Boulder.

Freeman, N. (2019) 'Laos's high-speed railway coming round the bend', *ISEAS Perspective*, 101: 1–7.

Ghiasy, R. (2021). 'The Belt and Road Initiative in South Asia: regional impact and the evolution of perceptions and policy responses', in F. Schneider (ed) *Global Perspectives on China's Belt and Road Initiative: Asserting Agency through Regional Connectivity*, Amsterdam: Amsterdam University Press, pp 265–290.

Giddens, A. (1984) *The Constitution of Society: Outline of the Theory of Structuration*, Berkeley, CA: University of California Press.

Goh, E. (ed) (2016) *Rising China's Influence in Developing Asia*, Oxford: Oxford University Press.

Heng, P. (2016) 'Chinese investment and aid in Cambodia's energy sector: impact and policy implications', PhD dissertation, VU Amsterdam.

Kastner, S.L. and Pearson, M.M. (2021) 'Exploring the parameters of China's economic influence', *Studies in Comparative International Development*, 56: 18–44.

Kratz, A. and Pavlićević, D. (2019) 'Norm-making, norm-taking or norm-shifting? A case study of Sino–Japanese competition in the Jakarta–Bandung high-speed rail project', *Third World Quarterly*, 40(6): 1107–26.

Kuik, C.C. (2020) 'Hedging in post-pandemic Asia: what, how, and why?', *The Asan Forum*, 6 June. Available from: www.theasanforum.org/hedging-in-post-pandemic-asia-what-how-and-why/.

Kuik, C.C. (2021a) 'Irresistible inducement? Assessing China's Belt and Road Initiative in Southeast Asia', *Council on Foreign Relations*, 15 June. Available from: www.cfr.org/sites/default/files/pdf/kuik_irresistible-inducement-assessing-bri-in-southeast-asia_june-2021.pdf.

Kuik, C.C. (2021b) 'Asymmetry and authority: theorizing Southeast Asian responses to China's Belt and Road Initiative', *Asian Perspective*, 45(2): 255–76.

Kuik, C.C. (2021c) 'Elite legitimation and the agency of the host country: evidence from Laos, Malaysia, and Thailand's BRI engagement', in F. Schneider (ed) *Global Perspectives on China's Belt and Road Initiative: Asserting Agency through Regional Connectivity*, Amsterdam: Amsterdam University Press, pp 217–44.

Kuik, C.C. (2021d) 'Malaysia's fluctuating engagement with China's Belt and Road Initiative: leveraging asymmetry, legitimizing authority', *Asian Perspective*, 45(2): 421–44.

Kuik, C.C. (2021e) 'The twin chessboards of US–China Rivalry: impact on the geostrategic supply and demand in post-pandemic Asia', *Asian Perspective*, 45(1): 157–76.

Lampton, D.M., Ho, S., and Kuik, C.C. (2020) *Rivers of Iron: Railroads and Chinese Power in Southeast Asia*, Oakland, CA: University of California Press.

Liao, J.C. and Dang, N. (2020) 'The nexus of security and economic hedging: Vietnam's strategic response to Japan–China infrastructure financing competition', *The Pacific Review*, 33(3–4): 669–96.

Lipset, S.M. and Solari, A. (1967) *Elites in Latin America*, New York: Oxford University Press.

Luo, J.J. and Un, K. (2020) 'Cambodia: hard landing for China's soft power?', *ISEAS Perspective*, 111: 1–8.

Luo, J.J. and Un, K. (2021) 'Japan passes China in the sprint to win Cambodian hearts and minds', *ISEAS Perspective*, 59: 1–10.

Morris, S. (2019) 'The Kunming–Vientiane railway: the economic, procurement, labor, and safeguard dimensions of a Chinese Belt and Road Project', CGD Policy Paper 142, Washington, DC: Center for Global Development.

Oh, Y.A. (2018) 'Power asymmetry and threat points: negotiating China's infrastructure development in Southeast Asia', *Review of International Political Economy*, 25(4): 530–52.

Parry, G. (1969/2005) *Political Elites*, Colchester: ECPR Press.

Parsons, T. (1951) *The Social System*, New York: Free Press.

Pham, S.T. and Ba, A.D. (2021) 'Vietnam's cautious response to China's Belt and Road Initiative: the imperatives of domestic legitimation', *Asian Perspective*, 45(4): 683–708.

Pitakdumrongkit, K. (2020) 'What causes changes in international governance details? An economic security perspective', *Review of International Political Economy*. doi.org/10.1080/09692290.2020.1819371.

Po, S. and Heng K. (2019) 'Assessing the impacts of Chinese investments in Cambodia: the case of Preah Sihanoukville Province', *Issues and Insights Working Paper*, 19(4), Honolulu, HI: Pacific Forum.

Putten, F.P.V. and Petkova, M. (2021) 'The geopolitical relevance of the BRI', in F. Schneider (ed) *Global Perspectives on China's Belt and Road Initiative: Asserting Agency through Regional Connectivity*, Amsterdam: Amsterdam University Press, pp 197–215.

Quah, D. (2019) 'Great power competition in the marketplace for world order', Lee Kuan Yew School of Public Policy Working Paper, November. Available from: www.DannyQuah.com/Quilled/Output/Quah-D 2019-Great-Power-Competition-Marketplace-World-Order.pdf.

Schindler, S., Jepson, N., and Cui, W. (2020) 'COVID-19, China and the future of global development', *Research in Globalization*, 2: 1–7.

Schindler, S., DiCarlo, J., and Paudel, D. (2021) 'The new cold war and the rise of the 21st-century infrastructure state', *Transactions of the Institute of British Geographers*. https://doi.org/10.1111/tran.12480.

Schneider, F. (ed) (2021) *Global Perspectives on China's Belt and Road Initiative: Asserting Agency through Regional Connectivity*, Amsterdam: Amsterdam University Press.

Sewell, Jr., W.H. (1992) 'A theory of structure: duality, agency, and transformation', *American Journal of Sociology*, 98(1): 1–29.

Tappe, O. (2018) 'On the right track? The Lao People's Democratic Republic in 2017', *Southeast Asian Affairs*: 169–184.

van der Lugt, S. (2021) 'Exploring the political, economic, and social implications of the Digital Silk Road into East Africa', in F. Schneider (ed) *Global Perspectives on China's Belt and Road Initiative: Asserting Agency through Regional Connectivity*, Amsterdam: Amsterdam University Press, pp 315–46.

Wight, C. (2006) *Agents, Structures and International Relations: Politics and Ontology*, Cambridge: Cambridge University Press.

Wijaya, T. and Camba, A. (2021) 'The politics of public-private partnerships: state-capital relations and spatial fixes in Indonesia and the Philippines', *Territory, Politics, Governance*. doi.org/10.1080/21622671.2021.1945484.

Yeremia, A.E. (2021) 'Explaining Indonesia's constrained engagement with the Belt and Road Initiative: balancing developmentalism against nationalism and islamism', *Asian Perspective*, 45(2): 325–47.

PART III

# Geopolitics and State Spatial Strategies

# 13

# Himalayan Geopolitical Competition and the Agency of the Infrastructure State in Nepal

*Dinesh Paudel and Katharine Rankin*

## Introduction

The Himalaya is witnessing new rounds of competition in infrastructure development among global powers seeking to shore up allies and expand value chains. This competition has shifted the orientation and priorities of development in Himalayan countries whose sovereign territory has become a battleground within this mode of geopolitical competition; building infrastructure has now, once again, come to define development after a period focused on poverty alleviation, empowerment and human development. The incorporation of Nepal into China's Belt and Road Initiative (BRI) since 2017 (within a context of expanding Chinese financial and technical support more generally) and subsequent counter-BRI projects from the US and its allies have catalysed vigorous new rounds of infrastructure development on the Himalayan landscape. In this chapter we suggest that the current conjuncture manifests traces of Cold War modalities, with competition in infrastructure development serving as a geopolitical and geoeconomic tool for global and regional hegemonic powers as well as for a Nepali state seeking to leverage opportunity for development.

China is advancing Sino-centric connectivity throughout the Himalaya, and India and the US (via the 'Pivot to Asia' strategy of the Obama and Biden Administrations (Madhani and Lemire, 2021) and more recently the Indo-Pacific Alliance with the UK and Australia) are pursuing a containment strategy against the Chinese footprint by building infrastructure-oriented market networks and commodity production in the Himalaya that could renew connections with India and its shipping ports. In this sense, the current conjuncture presents striking resonances to an earlier round of infrastructural competition between China (and Russia) and India (and the US) on Himalayan territory, where the aim had been to contain communism.

At the same time, now, as in the 1950s, countries commonly recognized as at the 'periphery' of the world system or 'buffer' to hostile, hegemonic blocs have historically complicated the power dynamics of imperialist geopolitics and geoeconomics by exercising their sovereign agency to safeguard their needs, priorities, and interests. Today the Nepali state engages the current conjuncture, and the thrust toward BRI connectivity, in particular, as a historic opportunity to diminish Nepal's dependence on India and challenge Indian influence in its internal political and economic spheres, while also strategically mobilizing infrastructure supports from the US, India and others to rebuild after the 2015 earthquakes and enhance transnational connectivity. Through negotiations over how infrastructure is developed materially on the landscape, small Himalayan states are thus articulating a strategic Himalaya-centric infrastructuralism to pursue their own spatial projects. In this chapter, we ask: In what ways is competition in infrastructure development becoming a key domain of geopolitics and geoeconomics on the Himalayan landscape? How has the Nepali state, as an 'infrastructure state' in the sense developed in this volume, leveraged the opportunity to mobilize foreign capital for infrastructure projects serving national interests?

## Geopolitics and Himalaya-centric infrastructuralism

Competition in infrastructure development became an everyday reality in Nepal after several recent events, which marked an increasing trajectory of Chinese and Indian (and US and other allies') mutual strategy of containing and constraining one another's long and ongoing integration and influence, infrastructurally and otherwise.[1] First, Nepal joined the BRI on 12 May 2017 by signing various agreements to build a railway line, several ports and three different highways through the Himalaya (Paudel and Le Billon, 2018; Murton and Plachta, 2021). Second, on 14 September 2017 Nepal and the US signed an agreement to construct a high-voltage transmission line from Kathmandu to India – a half billion-dollar project funded by the Millennium Challenge Corporation (MCC), an agency created by the US government to facilitate infrastructure development in 'less-developed countries' (MCC, 2017). Third, during an unprecedented visit by the Indian prime minister, Narendra Modi, in May 2018, Nepal and India signed various infrastructure agreements including a railway line between Raxaul in India to Kathmandu, dams and hydro-projects, and several highways across the India–Nepal border (*Times of India*, 2018). The competitive nature of these developments became more concrete after President Xi Jinping visited Nepal in October 2019. It was the first time a Chinese President had visited Nepal in 23 years, and in the context of the BRI, he reaffirmed China's commitments to provide support to build the much-anticipated Kerung–Kathmandu Trans-Himalayan Railway, several cross-border highways, trans-Himalayan economic corridors, dry ports, the Kathmandu ring road and a university. He emphasized that these projects would integrate the landlocked Himalayan nation with China and beyond, and presented a vision of shared prosperity. This visit came on the heels of a second virtual meeting between Modi and K.P. Oli in September 2019, which inaugurated a 69-km cross-border oil pipeline between India and Nepal, the first of its kind in South Asia. These visits were followed by numerous exchange visits between Nepal and US in which Nepal was invited to be 'at the center of the Indo-Pacific Alliance' and timely implementation of the MCC-sponsored transmission line project and other infrastructure projects (Department of Defense, 2019).

As the most recent manifestation of the US 'Asia Pivot', the Indo-Pacific Alliance seeks to expand the US's regional partnership in Asia to contain the growing influence of China in the regional trade and security. As part of this initiative, the MCC funded infrastructure development projects aim to build an India-centric connectivity and counter the potential growth of the influence of the Chinese BRI in Nepal and South Asia. By competing with China to integrate Nepali infrastructure networks, India's interests align with the American objectives, most notably the US government's various infrastructure initiatives especially the MCC funded infrastructure. These new infrastructure conjunctures are also facilitated by the Nepal's priority of infrastructure development in the context of post-earthquake reconstruction and the significant interests shown by global power centres after the earthquake in Nepal. The unprecedented level of support from China for the post-earthquake rescue and reconstruction prompted similar initiatives from India, US and others. Consequently, the 2015 earthquakes and subsequent Indian economic blockades reshaped Nepal's geopolitical and financial relationships with China and others but most importantly with India (Chand, 2017; Murton and Lord, 2020). The growing involvement of India, a regional power with far-reaching geoeconomic aspirations and a US ally in the Indo-Pacific against BRI, thus marks a significant intensification in ongoing geopolitical 'infrastructural relations' through which regional hegemonic powers have sought to register their territorial influence within the Himalaya (Murton and Plachta, 2021). Given the positioning of China and India on either side of these relations, we find it apt to evoke Cold War modalities of infrastructural competition, while also recognizing the distinction between motivations of containing communism and containing geoeconomic as well as geopolitical ambitions.

However, through various strategic practices, the Nepali state has refused the positioning of the Nepal Himalaya as a mere arena for the expression of dominant forces in the region (see also Murton et al, 2017). Nepali state actors have engaged the arrival of BRI and the ongoing competition in infrastructure development first to mobilize assistance in rebuilding after the 2015 earthquakes and second to achieve greater autonomy after the 2015 Indian economic blockade of Nepal, which catalysed powerful public opinion about the danger of vulnerability to regional hegemonic power. At the time, India had deemed Nepal's newly ratified constitution unsatisfactory because it had omitted Indian suggestions for how to craft provincial boundaries and federal systems. The Indian government seized the opportunity to support the protests against provincial demarcations in the southern Terai, which had sought to block border crossings with India and thus also import of essential goods.

The blockade fomented broad-based resentment towards India from a cross-section of Nepali society and created a national debate and consciousness toward developing alternative connectivity methods to international markets. In this context, while Nepal remains dependent on transport and trade with India in many ways, the BRI (as well as other forms of Chinese infrastructure support, and in fact a largely symbolic donation of fuel through the Northern border at Rasuwagadhi) marked 'an important turning point in a broader geopolitical reconfiguration between Nepal, India and China' (Murton and Plachta, 2021).

Increasing infrastructural integration with China, Nepal's own 'pivot to China', now appears as a spatial strategy that will allow Nepal to challenge India's longstanding domination in favour of a precarious balance of alliances between China and India that will yield infrastructure development, while thwarting one-way dependency. The intensifying tension and competition between China and India (with its US and other allies) have thus provided a political opportunity structure in the Himalaya where a small 'buffer' state like Nepal can exercise its agency and mobilize geopolitical power through infrastructure construction.

## History of geopolitical infrastructure and state agency in Nepal

This strategic geographic, political and economic positioning in the context of Himalayan infrastructure competition must be contextualized in a history of state spatial strategies for infrastructure development. Infrastructure has been deployed by the state as an organizing force since the unification of Nepal in the 18th century (Gurung, 1969; Murton, 2017; Rankin et al, 2017). Nepali rulers symbolized their divine power through traditional infrastructures such as temples, monasteries, horse cart trails and statues. After Prime Minister Janga Bahadur Rana's visit to Europe in 1850, emphasis shifted to modern infrastructure such as roads, hydro-management projects and urban planning, in the context of a hereditary system of governance no less autocratic than the foregoing monarchy (Isaacson et al, 2001). These priorities were soon institutionalized by a series of democratically elected governments that sought to establish a modern bureaucratic state apparatus, with roads identified as the cornerstone for forging national unity and progress. From the 1950s to the 1980s infrastructure was deployed by king and panchayat as a tool for national integration; expansion promised to deepen national unity and modernize backward places and societies (Rose, 1971; Blaikie et al, 1977; Isaacson et al, 2001). Key to these initiatives was the consolidation of a national highway network, anchored by the East–West Highway across Nepal's southern Terai belt. The East–West Highway was in turn cross-cut by several major north–south axes oriented to promoting a regional approach to development allowing for internal migration and trade within planned development regions (Gurung, 1969; Hagen, 1994).

State institutions and projects were established to realize these goals: the Ministry of Planning and Development (1951), the National Planning Commission (1955), a Regional Transportation Office (1958), and the institutionalization of five-year plans (1956). Planners divided Nepal into developmental zones and the population was categorized based on accessibility to roads. This period also marked the Nepali state's earliest engagements with international diplomacy via post-Bretton Woods multilateral and bilateral mechanisms. Nepal joined the UN in 1955, opened and a US embassy in 1959. Within this institutional framework, King Mahendra quickly became adept at leveraging geopolitical dynamics to secure funding and technical assistance for major infrastructure projects like the East–West Highway. He catalysed Cold War politics into an earlier round of infrastructure competition that allowed Nepal to play competing superpowers against one another to build roads in a perceived sensitive border zone between global ideological blocs (Rose, 1971; Rankin et al, 2017). Over the 1970s and 1980s, debates raged within Nepal about the potential for large-scale connectivity projects to catalyse import-substituting industrial production and improve social welfare across mountains, hills, and Terai.

By the 1990s, however, like many low-income countries, Nepal was subject to neoliberalization via conditionalities from donors. Infrastructure development, along with industry and service provision, was significantly privatized. During this period road development was largely orchestrated by private sector actors; the public sector was gutted of engineers – or they were reduced to the functions of evaluation and regulation – as multilateral donors sought to establish a viable contracting sector and markets for procurement (see World Bank, 2019). Privatization of road building manifested in rural areas as a kind of 'dozer terrorism', in which contractors and suppliers became the decisive force in infrastructure planning with little recourse to engineering standards and scant attention to the long term (Paudel, 2021). Meanwhile, the neoliberal reform especially in the economic sector started in the early 1990s in Nepal and the privatization of the industrial sector became an important aspect of development aid in the country. Most of the state-owned industries were sold to Indian firms (Murton, 2017), while the improvement of Human Development Indicators became the domain of an emergent non-governmental organization (NGO) sector (Paudel, 2016).

As a result, massive deindustrialization began in the 1990s, leading to diminished economic productivity, collapsed national industrial outputs, and migration of labourers to the Middle East. In parallel to these dynamics, the Maoist revolution erupted in the late 1990s, which was perceived as a response to the precarity and perceived growing levels of dependence on imperial powers, especially the US and India (Hutt, 2004). However, at its core of the revolutionary idea and subsequent political changes was the renewal of a modern Nepal. Infrastructure was key to this vision, yet after a decade of civil war, Nepali infrastructure systems were in tatters. The Federal Democratic Republic of Nepal was established in 2006, but Nepali institutions remained fragmented, and peace was mired with ongoing political gridlock. It was not until 2015 that the government ratified a new constitution, establishing a federal republic established and dismantling a two-centuries-old monarchy.

The 2015 constitution revived grand visions of infrastructure to stimulate a new 'economic' Nepal to match the New [political] Nepal that was struggled for in the Maoist revolution. For many, the constitution signalled a welcome end to decades of sluggish growth, violence, and political instability. In Delhi's corridors of power, however, Nepal's constitution was met with disapproval, and from September 2015 to February 2016 India imposed an economic blockade to punish Nepal for promulgating the constitution without the consent of the Indian government (Paudel and Le Billon, 2018). This aggressive move provoked outrage in Nepal, which served to unify parties across the political spectrum, with a growing consensus to reduce dependence on India by cultivating connectivity with China (Paudel and Le Billon, 2018; Murton, 2020).

## Infrastructure competition, and unfolding geopolitics and geoeconomics in Nepal

Three events in 2015 consolidated infrastructure competition in Nepal: (1) the 2015 earthquakes and subsequent processes of post-earthquake reconstruction that created ample opportunities for disaster capitalism to manifest (Le Billon et al, 2020; Murton, 2020; Paudel et al, 2020); (2) the promulgation of a new constitution paving the way for so-called economic prosperity 'toward socialism' through infrastructure (Government of Nepal, 2015) after the settlement of political transitions in the country; and (3) the Indian economic blockade and resulting orientation to transborder connectivity with China as a means for preserving Nepali sovereignty (Paudel and Le Billon, 2018). The arrival of the BRI offered a readymade framework for pivoting infrastructure development in a more internationally competitive direction. The Nepali state engaged Cold War geopolitics to catalyse infrastructure development in the mid-20th century; the correlation with contemporary dynamics of competition in infrastructure development are striking. Once again, Nepali bureaucrats and elected politicians alike recognize the opportunities to leverage infrastructure development for national integration and modernization out of an internationally competitive situation.

What distinguishes the current conjuncture is how infrastructure development has become not only a source of geopolitical competition, but also a means for asserting self-reliance and sovereignty vis-à-vis a long-time *regional* hegemon. The 1950s Cold War dynamics were premised on a depiction of the Himalaya as a physical barrier containing China and communism more broadly. By contrast, in the 2010s, connectivity with China represents an alternative within Nepal to dependence on India and decades of subjection to the neoliberal conditionalities on assistance from the 'free world'.

BRI developments are materializing on the ground swiftly. Detailed project reports are being prepared for trans-Himalayan railways, highways, transmission lines, cyber connectivity and economic corridors. China completed two inland ports in Kodari in December 2019 and five others are under construction. Three major highways connecting important ports in northern Nepal, namely Rasuwagadhi, Kodari, and Korala, are in the process of major upgrading. Various tunnels, dams and corridors and many BRI-related infrastructure projects are included in the flagship Trans-Himalayan Multi-Dimensional Connectivity Network between Nepal and China.

Of course, the risks of subjection to imperialist influence remain in the context of Sino-centric infrastructure development via BRI. However, BRI has two distinctive qualities that align with Nepali infrastructural nationalism relations: first, the explicitly transnational character of connectivity promised through BRI; and second, the seeming emphasis on the autonomy of partnering states. These paired qualities manifest in China's frequent assertions that BRI-related infrastructure decisions allow Nepal to stake control over the Himalayan economy. During a visit to India, Chinese State Councillor and Foreign Minister Wang Yi stated:

All walks of life in Nepal believe that strengthening connectivity under the framework of the BRI gives the Nepali side the hope of transforming itself from a landlocked to a land-linked country. The Trans-Himalayan Multi-Dimensional Connectivity Network will tighten the bonds between both countries and help the Nepali side play the role of a bridge in economic development in the region, which will be convenient for both countries and beneficial to the region. (Chinese Embassy in India, 2019)

The statement resonates with Nepal's ambitions for developing roads, railways, and hydropower as national priority projects. Nepal expects these projects to promote both internal connectivity and Nepal's capacity to bridge the two giant economies of India and China in a way that fosters autonomy and agency rather than diminishes Nepal to a battleground of geoeconomics.

The carefully crafted statement was made in India, it must be noted, where emphasizing Chinese unity with Nepal in the context of infrastructure development was received as a provocation; but from a Nepali perspective, the suggestion of Nepal's potential to serve as a bridge has important implications for its own ambitions to situate itself as an increasingly autonomous and agentic player in the region. Indeed, BRI developments correlate with various signs of assertion on the part of the Nepali state – including the recent move to include a large swath of India-occupied land in the north-west tip of Nepal, concomitant demands that India must remove its military from Nepali land (The Kathmandu Post, 2020), hosting of two international conferences on Himalayan climate change and culture (Government of Nepal, 2019), and claims of Nepal being a centre of Eastern or Himalayan civilization (Outlook, 2021). These events and assertions, which stake a leadership role for Nepal in the Himalaya, would have been formerly unthinkable and they signal its growing desire for autonomy and negotiating power that correlates well the resurgence of competition for infrastructure development in the region.

Not surprisingly, a counter-BRI initiative is evolving, that expresses a military alliance among US, Japan, India, and Australia, dubbed an Asian NATO (Hindustan Times, 2020). India has undertaken a series of infrastructure projects – the Indian and Nepali Prime Ministers jointly inaugurated two inland ports near their border, and a 65-km railway between the Indian border town of Jainagar in Bihar and Nepal's foothill town of Bardibas. India has committed to building five cross-border railway lines and Nepal is included in its initiative to build roads across much of South Asia (the South Asia Subregional Economic Cooperation Road Connectivity Investment Program). The Indian rush to develop enhanced connectivity infrastructure manifests a fear of losing Nepal from India's sphere of influence if it cannot compete with BRI connectivity in the north.

To complicate this tug-of-war, the US increased its interest in infrastructure projects in Nepal. The US Millennium Challenge Corporation (MCC) was created in 2004 to build infrastructure in strategically important developing countries, and it is currently funding five-year projects to construct a 100-km highway in western Nepal and a high-voltage transmission line to export hydroelectric power from Nepal to India and Bangladesh. This

US$500 million initiative serves to integrate South Asia in an India-centric manner. A US official sparked controversy when, during a visit to Nepal, he stated explicitly that this initiative was part of the Indo-Pacific Strategy designed to counter Chinese influence (Ghimire, 2020). Similarly, the US government initiated the Blue Dot Network in 2019, in collaboration with the Australian Department of Foreign Affairs and the Japan Bank for International Cooperation to promote 'high-quality, trusted standards for global infrastructure development in an open and inclusive framework' (DFC, 2019). Japan has increased aid to Nepal for tunnels, urban roads and rural connectivity, most notably through a US$157 million project to develop a tunnel road to Kathmandu from Naubise. The World Bank and the ADB have loaned millions of dollars to Nepal for infrastructure, primarily roads and urban development. These projects, too, aim to strengthen domestic supply chains and they orient Nepal's international connectivity towards India.

The Nepali state is seeking to navigate these competing forces as it pursues its spatial and economic agenda. Growing global narratives of a new cold war, and an Asian pivot policy of the US and related initiatives of the global power centres of course have some traction within Nepali discourse and policy making. And yet there are signs of a distinctively Nepali response designed to subvert global competition in infrastructure development to national priorities. To coordinate the acceleration of the geopolitical competition to build infrastructure, Nepal has already initiated restructuring of state institutions at multiple scales (Schindler et al, 2021). The National Planning Commission and several line ministries have historically coordinated large-scale infrastructure projects at the central level. To specifically coordinate the infrastructure initiatives, a new entity called the National Infrastructure Development and Investment Board was established to serve as an umbrella organization coordinating central government bureaucracies. This powerful national institution hosts annual infrastructure summits with investors and developers. Several new national institutions have been created since 2015: the Nepal Infrastructure Development Company Limited was established to build large-scale infrastructure; the Department of Shipping and Waterways was created to enhance connectivity with India through Himalayan rivers; the Department of the Railways was revitalized and strengthened to facilitate construction of lines from China and India; the Millennium Challenge Account Nepal (MCA-Nepal) was formed under the Ministry of Finance to implement US-supported infrastructure projects. Joint investment cooperation groups have been formed with both China and India, and several laws and regulations have been revised or promulgated to simplify land acquisitions, environmental impact assessments and forest clearance for priority infrastructure projects. Finally, national-level institutions have been established to implement at least 20 different BRI-related agreements.

One sign that the function of these institutions is not simply to do the bidding of regional hegemons is evidenced in recent initiatives to establish a coordinated negotiating platform among buffer states in the Himalaya – reminiscent of the objectives of the Non-Aligned Movement consolidated among so-called Third World countries in the 1950s to refuse being dragged as pawns into the struggles among major capitalist and communist powers during the Cold War. Nepal's prime minister visited Cambodia and Vietnam in 2019, and the president of Nepal visited Myanmar and Bangladesh subsequently, with the goal of fostering regional collaboration vis-à-vis and India-centric competition in infrastructure development. No formal proposal has yet been brought forward, in large part due to the disruptions of COVID-19, but a strong desire is clearly emerging to develop a collaborative mechanism among smaller economies in the Himalayas and nearby, which are directly experiencing the growing geopolitical and infrastructural competition.

These changes are not only at the government level, with the exchange of such ideas among intellectuals, organizations, and policymakers also growing. The COVID-19 pandemic might intensify such collaborative efforts as the smaller countries in the region struggle to obtain vaccines while geopolitical power centres either hoard or use the vaccines as a geopolitical tool (Pratt and Levin, 2021). At the very least, it is clear that the combination of institutional changes within Nepal and the heightened orientation to collaboration among smaller states within the region promote debates among intellectuals, civil society organizations and policymakers – debates that foreground how the current wave of competition in infrastructure development could function as modes of imperialism and neo-colonialism and that recognize the need for collaborative action to achieve moderation in relations with the major powers.

## Conclusion

Like many countries within the ambit of the BRI and other regimes of infrastructure-led development, Nepal is refashioning itself as an infrastructure state – driving development priorities with large infrastructure and imagining that not only economic growth but also that poverty alleviation and empowerment will follow. This orientation animates its domestic and international aspirations – for integration and unity after long periods of civil war, economic turmoil, and political gridlock, on the one hand, and greater transnational connectivity promising steadier trade and migration opportunities on the other hand. The transformation is facilitated by a growing competition in infrastructure development among geopolitical power centres in Asia. The recent round of competition is catalyzing massive investments in physical infrastructures that are rapidly becoming territorialized across Nepal's complex geographies – in mountains, rivers, flatlands, borders and interiors – potentially creating new waves of capital accumulation as well as of course articulating with ongoing caste, gender and feudal dynamics that still characterize agrarian livelihoods in the Himalaya.

This conjuncture resonates with a previous round of competition in infrastructure development during post-World War II period, where Cold War logics staged the Himalaya as an ideological battleground. In the earlier rounds of donor-financed, state-led infrastructure development in Nepal, infrastructure served as a symbol of the state's legitimacy, and a means of achieving economic growth and the transformation of 'backwards' populations. In the current conjuncture in Nepal, infrastructure development is portrayed as holding the key to national renewal, both in terms of economic recovery, and in terms of reduced dependence on India. The primary protagonists of the contemporary competition for infrastructure development in Nepal are China and India, while the US and other Indo-pacific partner countries, such as Japan and Australia, and multilateral institutions, such as the Asian Development Bank and the World Bank, participate by contributing to spatial projects that augment Nepal's India-centric orientation. The rationalities are not only geopolitical, but also geoeconomic.

Thus, the infrastructure competition among geopolitical powers in the Himalaya unwittingly creates opportunities for assertion of agency and autonomy by small states like Nepal. First, the plethora of opportunities for financing investment catalysed by BRI and the ensuing competition from India and its liberal-democratic allies creates opportunities for reducing bilateral dependencies such as Nepal's with India. Second, it has created opportunities to foster debate and regional collaboration in a manner that weighs perspective of 'non-aligned' participants (for example via topical conferences convened in Nepal and other small states to deliberate climate change – an issue alarmingly absent in the rush for transportation relying on fossil fuels); such issue-focus deliberation has the potential to develop understandings of infrastructure and connectivity that pose alternatives to those advanced by dominant regional forces. Third, signs of growing collaboration among smaller recipient countries might foster solidarity and allow them to create blocs vis-à-vis the major powers. This, in turn, may augment their negotiating position in a manner that could genuinely disturb longstanding dynamics of power, and especially neo-colonial dynamics.

Finally, as Nepal's case demonstrates, competition in infrastructure development also reveals the limits of hegemony, and indeed the vulnerability of the major geopolitical powers. Powerful countries are not always able to control how receiving countries articulate and implement their geopolitical and geoeconomic infrastructure initiatives. Indeed, the capacity of powerful states – in this case China, India, the US and Japan – to realize their territorial visions in ways that underpin their hegemony is highly uncertain. These visions can be undermined by determined national actors as well as solidarity networks among states typically deemed dependent and powerless. If the US–China rivalry continues to intensify, these solidarity networks and the opportunities they present small states will become an important area for academic scholarship and political organizing oriented towards achieving environmentally sustainable and socially just versions of infrastructure development.

## Note

1   We are grateful for Galen Murton's comment that Chinese involvement in Nepal is not limited to BRI; rather the BRI must be understood in relation to a wider range of Chinese development assistance (also involving ChinaAid and the Chinese Embassy in Kathmandu) outside BRI frameworks, including China's massive and unprecedented humanitarian support and assistance for infrastructure development after the 2015 earthquakes (see Murton and Plachta, 2021).

## References

Blaikie, P.M., Cameron, J., and Seddon, D. (1977) *The Effects of Roads in West Central Nepal: A Summary*, Norwich: University of East Anglia.

Chand, B. (2017) 'Disaster relief as a political tool: Analysing Indian and Chinese responses after the Nepal earthquakes', *Strategic Analysis*, 41(6): 535–45.

Chinese Embassy in India (2019) 'The road ahead is long and winding though, a start will bring an arrival', 17 October. Available from: https://www.mfa.gov.cn/ce/cezm//eng/zgxw/t1707868.htm

Department of Defense (2019) 'Indo-Pacific Strategy Report: preparedness, partnerships, and promoting a networked region'. Available from: https://media.defense.gov/2019/Jul/01/2002152311/-1/-1/1/DEPARTMENT-OF-DEFENSE-INDO-PACIFIC-STRATEGY-REPORT-2019.PDF.

DFC (US International Development Finance Corportation) (2019) 'The Launch of Multi-Stakeholder Blue Dot Network', press release, 4 November. Available from: https://www.dfc.gov/media/opic-press-relea ses/launch-multi-stakeholder-blue-dot-network.

Ghimire, B. (2020) 'Why the MCC compact courted controversy in Nepal', *The Kathmandu Post*, 9 January. Available from: https://kathmandupost. com/national/2020/01/09/why-the-mcc-compact-courted-controve rsy-in-nepal.

Government of Nepal (2015) *The Constitution of Nepal*, Kathmandu: The Government of Nepal

Government of Nepal (2019) 'Nepal government plans to launch a regional international forum every year', AIDIA, 12 April. Available from: www.aidiaasia.org/news/nepal-government-plans-to-launch-a-regional-international-forum-every-year.

Gurung, H.B. (1969) 'Regional development planning for Nepal (No. 1)', National Planning Commission, His Majesty's Government, Kathmandu.

Hagen, T. (1994) *Building Bridges to the Third World*, Delhi: Book Faith India.

Hindustan Times (2020) 'India building 'Asian NATO' through Quad?' S Jaishankar answers', 23 October. Available from: www.hindustantimes.com/india-news/india-building-asian-nato-through-quad-s-jaishankar-answers/story-zBVAmTRIeXhnukEcHb8DbJ.html.

Hutt, M. (2004) *Himalayan People's War: Nepal's Maoist Rebellion*, Bloomington, IN: Indiana University Press.

Isaacson, J.M., Skerry, C.A., Moran, K., and Kalavan, K.M. (2001) *Half-a Century of Development: The History of US Assistance to Nepal, 1951–2001*, Kathmandu: United States Agency for International Development.

*Kathmandu Post* (2020) 'Government unveils new political map including Kalapani, Lipulekh and Limpiyadhura inside Nepal borders', 20 May. Available from: https://kathmandupost.com/national/2020/05/20/government-unveils-new-political-map-including-kalapani-lipulekh-and-limpiyadhura-inside-nepal-borders.

Le Billon, P., Suji, M., Baniya, J., Limbu, B., Paudel, D., Rankin, K., Rawal, N., and Sheneiderman, S. (2020) 'Disaster financialization: earthquakes, cashflows and shifting household economies in Nepal', *Development and Change*, 51(4): 939–69.

Madhani, A., and Lemire, J. (2021) 'Biden announces Indo-Pacific alliance with UK, Australia', *The Diplomat*, 16 September. Available from: https://thediplomat.com/2021/09/biden-announces-indo-pacific-alliance-with-uk-australia/.

Millennium Challenge Corporation (MCC) (2017) 'Nepal Compact'. Available from: www.mcc.gov/where-we-work/program/nepal-compact.

Murton, G. (2017) 'Making mountain places into state space: infrastructure, consumption, and territorial practice in a Himalayan borderland', *Annals of the American Association of Geographers*, 107(2): 536–45.

Murton, G. (2020) 'Roads to China and infrastructural relations in Nepal', *Environment and Planning C: Politics and Space*, 38(5): 840–47.

Murton, G. and Lord, A. (2020) 'Trans-Himalayan power corridors: Infrastructural politics and China's belt and road initiative in Nepal', *Political Geography*, 77: 102100.

Murton, G. and Plachta, N. (2021) 'China in Nepal: On the politics of the Belt and Road Initiative development in South Asia', in *Research Handbook on the Belt and Road Initiative*, Cheltenham: Edward Elgar Publishing.

Murton, G., Lord, A. and Beazley, R. (2016) '"A handshake across the Himalayas": Chinese investment, hydropower development, and state formation in Nepal', *Eurasian Geography and Economics*, 57(3): 403–32.

Outlook (2021) 'After Lord Ram claim, Nepal PM now says "yoga didn't originate in India"', Outlook India, 22 June. Available from: www. outlookindia.com/website/story/world-news-india-didnt-exist-when-yoga-was-discovered-kp-sharma-oli-claims-yoga-originated-in-nepal/ 385815.

Paudel, D. (2016) 'The double life of development: USAID, empowerment, and the Maoist uprising in Nepal', Development and Change, 47(5): 1025–50.

Paudel, D. (2021) डोजर आतंकः विकासको नाममा विनाशको बाढी, Dozer terrorism: destruction on the name of development in Nepal, Nepal Press, 17 June. Available from: www.nepalpress.com/2021/06/18/68294/.

Paudel, D. and Le Billon, P. (2018) 'Geo-logics of power: disaster capitalism, Himalayan materialities, and the geopolitical economy of reconstruction in post-earthquake Nepal', Geopolitics, 25(4): 838–66.

Paudel, D., Rankin, K., and Le Billon, P. (2020) 'Lucrative disaster: financialization, accumulation and post-earthquake reconstruction in Nepal', Economic Geography, 96(2): 137–60.

Pratt, S.F. and Levin, J. (2021) 'Vaccines will shape the new geopolitical order: the gulf between haves and have-nots is only growing', Foreign Policy, 29 April. Available from: https://foreignpolicy.com/2021/04/29/ vaccine-geopolitics-diplomacy-israel-russia-china/.

Rankin, K.N., Sigdel, T.S., Rai, L., Kunwar, S., and Hamal, P. (2017). 'political economies and political rationalities of road building in Nepal', Studies in Nepali History and Society, 22(1): 43–84.

Rose, L.E. (1971) Nepal: Strategy for Survival, Berkeley, CA: University of California Press.

Schindler, S. DiCarlo, J., and Paudel, D. (2021) 'The new cold war and the rise of the 21st century infrastructure state', Transactions of the Institute of British Geographers. https://foreignpolicy.com/2021/04/29/vaccine-geop olitics-diplomacy-israel-russia-china/.

Times of India (2018) 'Why PM Modi's visit to Nepal is strategically important for both countries', 11 May. Available from: https://timesofindia.indiatimes. com/india/why-modis-visit-to-nepal-is-strategically-important-for-india/ articleshow/64120383.cms

World Bank (2019) Nepal Infrastructure Sector Assessment: Private Sector Solutions for Sustainable Infrastructure Development, Washington, DC: World Bank.

# 14

# Indonesia's 'Beauty Contest': China, Japan, the US, and Jakarta's Spatial Objectives

*Angela Tritto, Mary Silaban, and Alvin Camba*

## Introduction

> 'Let them race to invest in Indonesia. It is good for us. ... It is like a girl wanted by many guys; the girl then can pick whomever she likes.'
> Luhut Panjaitan, Coordinating Minister of Maritime Affairs
> (Agence France-Presse, 2015)

As the quote suggests, Indonesia has deliberately balanced its overseas entanglements in the context of US–China competition. The Indonesian government held a closed bid between a Chinese and Japanese consortium to build the country's high-speed rail (HSR), the first in Southeast Asia, which was billed as a 'beauty contest' (Suroyo, 2015). This, first, highlights how Indonesia operates in an international environment characterized by a high degree of multipolarity that includes powerful regional actors, most notably Japan. Second, it demonstrates that Indonesia employs a dynamic hedging strategy as it establishes configurations of alliances among great and regional powers that are unique to particular spatial projects.[1]

Given Indonesia's unique archipelagic geography, its overarching spatial objectives[2] include terrestrial and maritime projects. This chapter focuses on the expansion of Indonesia's HSR network and Jakarta's attempts to secure maritime borders and navigation routes. In the first case, Japan's willingness to export HSR technology allowed Indonesia to acquire a better infrastructure financing deal with China. Jakarta was thereby able to integrate territory and strengthen political support for the Joko 'Jokowi' Widodo administration. The second case examines international maritime affairs, in which Indonesia harnessed the US-led Freedom of Navigation Operations (FONOPs) to its advantage, in concert with Japan, without explicitly allying with the US or China. Jakarta was thus able to enhance maritime security and control while remaining somewhat aloof from the intensifying US–China rivalry. The pursuit of both spatial projects – the expansion of the HSR network and joining FONOPs – were only possible through varying degrees of state restructuring (Brenner, 2004).

Indonesia's dynamic hedging strategy is consistent with its longstanding ambivalence towards great power politics. The country's relationship with the People's Republic of China (PRC) was curtailed after the military seized power in 1965, which led to Suharto's 31-year presidency (Van der Kroef, 1976; Human Rights Watch, 2017; Wright, 2017). Throughout the Suharto era, Indonesia was a reliable US ally and supported American efforts to contain communism in the region (Simpson, 2008). In recent years, particularly since the Yudhoyono presidency (2004–14), Indonesia's troubled relations with China evolved into a strong economic partnership (Sukma, 1999). This relationship has flourished under President Widodo (2014–), who has sought to mobilize Chinese support – loans, foreign direct investments, and technology transfer – in support of Indonesia's developmental and infrastructural agenda (Nabila et al, 2018; Tritto, 2020).[3]

This opportunistic relationship has evolved despite the longstanding dominance of Japanese capital in the Indonesian economy. Indeed, Jakarta leveraged its relationship with Japan – itself a close US ally but not a proxy – to garner support from China in pursuit of its infrastructural ambitions. Meanwhile, Indonesia's maritime strategy has led Jakarta to pursue a closer relationship with Tokyo. Indonesia's nuanced policy towards Japan and China unfolded in the context of waning US influence under the Trump administration, and its infrastructural and maritime policies contributed to US anxiety about its declining regional influence. Washington initiated a series of bi- and multilateral agreements in response, and Joe Biden has sought to renew its influence in the region.

This chapter ultimately demonstrates that although US–China competition establishes parameters of action for other states, it does not determine outcomes. Thus, rather than a bilateral struggle whose endogenous dynamics impact third countries, domestic and regional politics can draw in great powers, reverberate to the global scale and influence 'big G' geopolitics. This insight puts this chapter in conversation with scholarship on great power competition. First, Lüthi's (2020) book on 'Cold Wars' illustrates how regional and national politics of third countries (such as Egypt and the Vatican) shaped the Soviet–US rivalry, thereby restoring the agency of small countries in this global struggle. Similarly, this chapter illustrates that while Indonesia's domestic politics are shaped by the US–China conflict, they also 'boomerang' back to the international scale and influence geopolitics (see

Gourevitch, 1978 on the relationship between domestic and international politics). Thus, this chapter situates influence beyond the domestic politics of great powers, and it thereby highlights the agency and influence of third countries, elites, and non-state actors in shaping the US–China rivalry.

## The evolution of Indonesia's 'free and active' policy

Since independence, Indonesia has upheld a 'free and active' doctrine on foreign policy, however, its meaning has evolved over time. Introduced by Vice President Hatta in 1948, it prevents the country from forging military alliances or hosting foreign military bases within its territory. As a new nation with a burdensome colonial past, Indonesia prioritized sovereignty and independence. President Sukarno believed the best way to safeguard Indonesia's newfound independence and sovereigny was to cultivate an axis of likeminded societies, independent from Soviet and American influence (Ciorciari, 2010; Hamilton-Hart and McRae, 2015). In 1955, he hosted the Bandung Conference, which was attended by Third World luminaries such as Jawaharlal Nehru, Zhou Enlai, and Gamal Abdel Nasser and marked the birth of the Non-Aligned Movement.

Jakarta was formally not aligned with the USSR or the US, but in the context of the Cold War skulduggery, observers searched for signs that its de facto political orientation was partisan. One thing that worried Washington was the Soviet aid programme. From 1945 to 1965, Indonesia was the second-largest recipient of Soviet aid (after Egypt) and in 1964 Sukarno gave an incendiary speech in which he led Indonesia out of the UN, telling Western member states to 'go to hell with your aid' (Wejak, 2000). Meanwhile, Sukarno strengthened bilateral ties with China (Sukma, 2009). China held a political observer status in and supported many of the positions of the Non-Aligned Movement, while it provided significant support to the Communist Party of Indonesia (Partai Komunis Indonesia), making it the largest communist party in the world in a non-communist country (Anwar, 2009). Events came to a head in September 1965, when General Suharto and his supporters in the military led an anti-communist and anti-Chinese purge that removed Sukarno from power.

General Suharto assumed leadership in 1967, and he refashioned the country's approach to foreign policy. Socialism and communism were banned, Indonesia re-joined the UN and cut diplomatic ties with China, and the Indonesian military enjoyed significant support from the US (Simpson, 2008). Japan and Indonesia established diplomatic relations in 1958 when the former agreed to pay US$220 million in war reparations, and it subsequently became a significant source of FDI (Anwar, 1990). From 1960 to 2016 Japan was Indonesia's largest source of overseas development assistance (ODA), channelling a total of US$49.5 billion toward infrastructure development and disaster relief. Thus, during Suharto's presidency Indonesia cooperated with the US on matters of security, while Japan provided much-needed aid and investment. It was in this context that China sought to improve relations under the leadership of Jiang Zemin in the late 1990s (Sukma, 1999; Kurlantzick, 2006).

Bilateral relations between Indonesia and China strengthened rapidly and in the 2000s Jakarta realized that China may be able to contribute to its ambitious infrastructural agenda. This insight was taken a step further by President Widodo, who linked China's Belt and Road Initiative (BRI) with his Global Maritime Fulcrum development agenda. For some Indonesian elites, the deepening relationship with China is driven more by realism than strategy. For example, Luhut Pandjaitan, a retired army general who currently serves as a Coordinating Minister of Maritime Affairs noted: 'Like it or not, China is a global powerhouse and Indonesia must support Chinese investments in the country' (Al Hikam, 2020). Nevertheless, other policy makers have sought to reassert a free and active foreign policy, and this has created an opening for Japan and the US, which have watched China's expanded presence in Indonesia with suspicion.

Joe Biden plans to focus on the ASEAN region and pursue a more consistent policy towards Beijing than his predecessor (Engel, 2020; Palma, 2020). He also referred to Indonesia's strategic importance in Southeast Asia on numerous occasions. Meanwhile, a month after his inauguration, Japan's Prime Minister Yoshihide Suga selected Indonesia and Vietnam as his first overseas diplomatic visits. Suga discussed cooperation in defence and announced a US$473 million loan to Indonesia to support economic recovery from the COVID-19 pandemic. In sum, the US and Japan currently seek to renew relations with Indonesia in response to China's growing influence. China supplanted the US as Indonesia's leading investor and trade partner, and it expanded economic ties when the US advanced an 'America First' foreign policy (Damuri et al, 2019). However, it would be a mistake to underestimate Jakarta's agency. First, it professes its neutrality. Second, its ties to American and Japanese firms remain important in certain sectors such as manufacturing and resource extraction. Finally, Indonesia's security strategy includes preventing Chinese vessels from encroaching on its sovereign territory surrounding the Natuna Islands. As the cases in the following sections demonstrate, Indonesia has complex relations with the US, Japan, and China, and it aptly reconfigures its international relations in accordance with particular spatial projects.

## Infrastructure development and the case of the Jakarta–Bandung HSR

The Jakarta–Bandung HSR illustrates Indonesia's pragmatic stance towards great power competition, exemplifying how the country managed to broker a better deal by leveraging the rivalry between China and Japan. Through a flashy 'beauty contest' broadcast on the headline news, President Widodo exploited the eagerness of Chinese companies and policy banks to establish their presence in the region and institute a flagship project for the newly announced Belt and Road Initiative by making them compete with Japanese institutions.

A pioneer in HSR technology and Indonesia's largest infrastructural capital exporter, Japan had sought to export its railway technology to the country since 2008 when it showcased Shinkansen trains at the Indonesia–Japan Expo (JICA, 2008). In 2013, Japanese authorities proposed the first plans to build a 142-km HSR line from Jakarta to Bandung. The Japanese Minister of Transportation met with his Indonesian counterpart and explained that Japan was 'keen to invest in Indonesia' by building two HSR lines connecting Jakarta to Bandung and Surabaya (Tempo, 2013). That same year, the Japan International Cooperation Agency (JICA) initiated a feasibility study (Parlina, 2013).

In March 2015, on his first official visit outside the country, Widodo travelled to Japan and China to seek out 'concrete trade and investment cooperation, with resulting capital inflows to Indonesia, particularly in areas related to infrastructure, seaports, airports, power plants, railways, and toll roads' (*Jakarta Post*, 2015). He and President Shinzo Abe signed the Japan–Indonesia Joint Statement: Towards Further Strengthening of the Strategic Partnership Underpinned by Sea and Democracy (Ministry of Foreign Affairs of Japan, 2015). In this statement, they agreed to strengthen partnerships in several areas including maritime security and defence, economic cooperation, people-to-people, and cultural exchange, as well as regional and international cooperation. Among many projects, Widodo and Japan spoke about the HSR; however, Widodo hoped that the private sector would finance it, and the meeting ended without the two reaching an arrangement on how Japan might engage. Widodo then visited China to meet President Xi Jinping, and they signed the Joint Statement on Strengthening Comprehensive Strategic Partnership, which was similar to the agreement with Japan. However, on this occasion the National Development and Reform Commission of the PRC and the Ministry of State-Owned Enterprises of the Republic of Indonesia also signed a Memorandum of Understanding for the Jakarta–Bandung HSR Project (Ministry of Foreign Affairs of China, 2015), signalling a clear shift towards China as the preferred infrastructure provider.

In April 2015, Xi and Widodo renewed their commitment to the HSR by signing another framework. Here, a heated competition between Japan and China began to unfold (Tritto, 2020). A closed bidding process that local headlines branded a 'beauty contest' (Suroyo, 2015) showcased the two proposals. It was in this context that Luhut Panjaitan was quoted saying: 'Let them race to invest in Indonesia. It is good for us. ... It is

**Table 14.1:** Comparison of the Chinese and Japanese proposals to build the Jakarta–Bandung high-speed railway

| Parameters | Chinese–Indonesian consortium (KCIC) | Japan |
| --- | --- | --- |
| Project value | US$5.13 billion | US$6.2 billion |
| Government commitment | - There is no government underwriting, funding, or tariff subsidy<br>- Cost overrun: KCIC's responsibility | - Government underwriting is required, funding from state's budget and tariff subsidy<br>- Cost overrun: government's responsibility |
| Business concept | The Joint Venture Company (i.e., KCIC) is responsible for construction and operations. Project risk: KCIC | Engineering, Procurement, and Construction, Financing (Regular Contractor). Risk/liability: government |
| Local content | 58.6% | 40.0% |
| New jobs creation | Construction period: 39,000 workers per year. The Chinese workers employed are limited to experts and supervisors. | Construction period: 35,000 workers per year. Expatriates from Japan are required. |
| Technology transfer | Through the opening of rolling stock factory in Indonesia. | There was no tangible technology transfer programme. |

Source: Jakarta-Bandung high-speed train informative booklet provided during interview with KCIC in 2017, in Tritto (2020).

like a girl wanted by many guys; the girl then can pick whomever she likes' (Agence France-Presse, 2015). The Indonesian government initially favoured the Japanese proposal, but ultimately turned it down (*Straits Times*, 2015). Instead, the Indonesian government selected a Chinese proposal that did not require government guarantees and established a consortium of Indonesian and Chinese companies to develop the project. Table 14.1 compares two proposals.

Through the negotiations and deal for the Jakarta–Bandung HSR, the Indonesian government leveraged the eagerness of both Chinese and Japanese institutions and enterprises to export their HSR technology (see Figure 14.1). The deal struck with China is unique and unprecedented in many ways (Tritto, 2020). The project is built by a joint venture company called Kereta Cepat Indonesia China (KCIC), comprising Chinese (40 per cent) and Indonesian (60 per cent) state-owned enterprises, and financed

**Figure 14.1:** Current and proposed railways, Indonesia

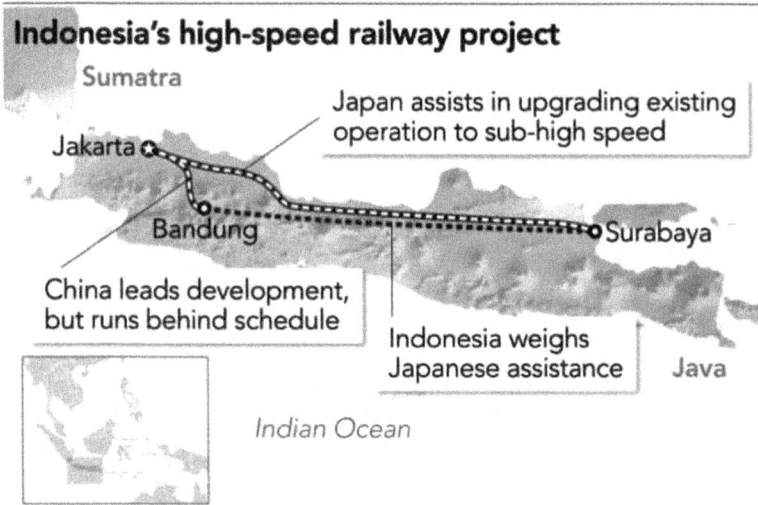

Source: Adapted from Jibiki (2020)

through a loan that requires no government guarantees, something quite uncommon for a project of this size and complexity. The deal also incorporates several contractual elements to ensure employment/training of local human resources and manufacturing of rolling stock, making it atypical from standard Engineering, Procurement, and Construction contracts. These concessions were followed by the renegotiation of railway deals by other ASEAN countries, such as Malaysia, and Thailand (*The Star*, 2019; Reuters, 2019) and led to new concessions from both sides to speed up project development. The project was financed by the China Development Bank, which previously required that all land should be acquired before funds could be disbursed for the project. In this case, however, it compromised on this condition and disbursed US$40 billion prior to the finalization of land acquisition processes (Negara and Suryadinata, 2018).

The project necessitated the Indonesian state to restructure and undertake several reforms. Widodo issued a series of presidential decrees and executive orders to pave the way for project implementation and managed to incorporate members of the political opposition into his new cabinet to ensure internal support. He also offered the extension of the line to Surabaya (the 'South line') to the Japanese consortium (Camba, 2020). This offer was rejected, however, as the Japanese consortium wanted to focus on the 'North line' (*Jakarta Post*, 2021).

On one hand, the Jakarta–Bandung HSR deal constitutes an important milestone for Indonesia, a country that has struggled to build large-scale infrastructure due to its cost and complexity, as well as challenges surrounding land acquisition and inter-departmental bureaucratic infighting. The HSR project has several innovative components that may make it economically viable, and it has significant political tailwind behind it. On the other hand, exploiting the Sino-Japanese rivalry has also led to a race that has, to many, bypassed important assessments related to the environmental and social impacts of the project. Finally, the absence of government underwriting of the loan heightened risks for China's financial institutions; it is unclear whether Beijing is willing to replicate this model.

## Maritime security and the US Freedom of Navigation Operations

Turning to Indonesia's maritime spatial objectives presents a different picture in how rivalry plays out. The primary maritime objective is safeguarding sovereignty and shipping lanes. To this end, Indonesia's strategy mirrors Japan's position on US-led FONOPs, thereby circumventing US–China competition over contending interpretations of maritime rights. Thus, the role of a third-party actor – in this case Japan – shapes Indonesia's spatial strategy more than the dynamics of great power rivalry.

The US FONOPs was established to uphold customary practices over the freedom of navigation by the United Nations Convention on the Law of the Sea (UNCLOS). This includes, for example, 'innocent passage' of ships through sovereign territorial and archipelagic waters, and it establishes forms of variegated sovereignty in maritime areas (Malone, 1983). By insisting on exercising the right of passage, FONOPs challenge nationalist and military claims in contested maritime territories such as the East China and the South China Sea (Hossain, 2013). Though the US has not signed the UNCLOS, it has adopted multiple practices related to FONOPs, regularly sending warships and navies to effectively contest maritime claims by states, particularly by China. Between 2016 and 2018, the US launched 15 FONOPs in contested areas of the South China Sea (Power, 2020), notably near the Spratly Islands, Paracel Islands, and Senkaku Islands.

China has vociferously objected to US-led FONOPs, arguing that they increase tensions in the region and that states should resolve disputes through bilateral negotiation.

Beijing insists that the US Navy should not be involved in FONOPs, and that FONOPs violate Chinese laws, specifically the PRC's law on the Territorial Sea and the Contiguous Zone and their declaration on the Baselines of the Territorial Sea (Ng, 2018). In an attempt to establish sovereign control over contested maritime spaces, China has constructed artificial islands and its coast guard has expanded operations (Camba and Magat, 2021). Additionally, provincial Chinese coast guard ships protect private fishing vessels in outlying maritime areas (Kennedy, 2018). These activities have spooked Southeast Asian states given the superior operational capabilities and size of the Chinese coast guard (Patalano, 2018).

Indonesia has rebuffed the US and China by establishing a policy on FONOPs in concert with Japan. The country supports US-led FONOPs and their principles on paper. However, the participation of its armed forces in these operations has been limited. According to a coast guard colonel, Jakarta seeks to maintain control of the deployment of its maritime forces and determine the pace of its engagement with US-led FONOPs. The US has consistently argued that East and Southeast Asian states should take the lead in pushing back against Chinese encroachment into contested maritime spaces, and it will 'support' their efforts in any disputes that arise if they abide by UNCLOS rules (Interview, US Vice Admiral, 1 June 2019). While Indonesia and Japan have objected to China's unilateral imposition of its maritime claims, they have done so with caution. In the case of Indonesia, this is due to the lack of elite consensus on how to deal with China (Laksmana, 2016), while Japan's caution stems primarily from its regional proximity (Bao, 2016). In 2015, Indonesia and Japan signed an agreement entitled Towards Further Strengthening of the Strategic Partnership Underpinned by Sea and Democracy, enhancing maritime cooperation, security linkages, and economic exchange. Both countries also affirm the Declaration of the Conduct of Parties in the South China Sea, which codifies dispute resolution and seeks to limit aggressive military options. Japan has also pushed for the implementation of a Regional Code of Conduct in the South China Sea (Ministry of Foreign Affairs of Japan, 2015).

Indonesia's support for Japan's more cautious approach – and wariness of participation in US-led FONOPs – is evident in the way it has sought to defend claims to the Natuna Islands in the South China Sea. According to Jakarta, these islands are within its sovereign maritime boundaries, and the country has persistently restricted Chinese fishing ships accordingly. In several cases, this has resulted in conflict and the destruction of these fishing vessels (Maulia, 2020). As an Indonesian Coastguard Colonel explained: "We chase them with both our navy and coast guard. We drive them out, warn them, and tell them to stop. We have to, if we give them an inch, they take a mile." (Interview, 8 September 2019) The US, he explained, pushed Indonesia to conduct FONOPs during the Obama and Trump administrations, but Indonesia instead opted to focus on tensions in the Natuna. Thus, Jakarta preferred to focus on a single issue, in which it enjoys strong claims, rather than general principles of maritime sovereignty and passage. This maritime spatial strategy may be about to change. In 2020, Japan joined a US-led FONOP, signalling support for a rules-based international order and opposition to China's maritime manoeuvres (Eckstein, 2020). Indonesia followed suit, and it remains to be seen whether this signals a fundamental evolution of Jakarta's spatial strategy.

## Conclusion

As this chapter has shown, Indonesia has carefully developed a complex foreign policy towards Japan, China, and the US, designed to advance its spatial objectives. Jakarta recognizes Indonesia's value as an economic and security partner to these more powerful states. This was captured by the notion that it is a contestant at a beauty contest who can choose from several potential suitors. Rather than committing to any one of them, however, Indonesia continues to pursue a 'free and active' foreign policy and it reconfigures partnerships in accordance with specific spatial projects. The Jakarta–Bandung HSR pitted Chinese and Japanese bidders against one another, allowing Jakarta to secure rather favourable terms. It further integrates the Chinese and Indonesian economies, as it offers improved connectivity for several Chinese companies on this line (such as a new industrial zone by China Fortune Land Development). Jakarta's primary maritime spatial objective is to secure its sovereignty and, in particular, fend off Chinese access to the waters surrounding the Natuna Islands. To this end, it has voiced support for the principles underlying US-led FONOPs, but it has embraced Japan's cautious approach towards China. Until recently, both countries resisted participating in FONOPs, but this strategy is evolving.

Joko Widodo has embraced infrastructure-led development (Schindler and Kanai, 2021), and his presidency has committed to terrestrial and maritime spatial objectives. Japan features prominently in both cases presented in this chapter. In the case of HSR, Jakarta balanced Tokyo and Beijing, ultimately siding with the latter. Indonesia followed Japan's lead when it came to FONOPs. This strategy represented an attempt to remain aloof from US–China rivalry and maintain narrowly focused on the dispute with China over the Natura Islands (rather than a general dispute on the principles of innocent passage), therefore safeguarding the established economic relations with the PRC. Jakarta's project-specific hedging strategies in the context of the US–China rivalry demonstrate that it enjoys a measure of agency that can be translated into the achievement of spatial objectives.

### Notes

[1] On hedging, see Kuik (2016).

[2] The notion of 'state spatial objectives', as well as state spatial strategies and projects, was introduced by Brenner (2004).

[3] The Partai Komunis Indonesia purportedly launched a coup against Sukarno, resulting in a series of counter coup operations by Suharto. Thereafter, Suharto with the support of the CIA purged the Partai Komunis Indonesia.

# References

Agence France-Presse (2015) 'Beijing, Tokyo in tussle for Indonesia rail project', *The Straits Times*, 25 August. Available from: https://www.straitstimes.com/asia/beijing-tokyo-in-tussle-for-indonesia-rail-project.

Al Hikam, H.A. (2020) 'Luhut Suka Tidak Suka Tiongkok Kekuatan Dunia', *Detik*, 5 June. Available from: https://finance.detik.com/berita-ekonomi-bisnis/d-5042263/luhut-suka-tidak-suka-tiongkok-kekuatan-dunia.

Anwar, D.F. (1990) 'Indonesia's relations with China and Japan: images, perception and realities', *Contemporary Southeast Asia*, 12(3): 225–46.

Anwar, D.F. (2009) 'A journey of change: Indonesia's foreign policy', *GlobalAsia*. Available from: www.globalasia.org/v4no3/cover/a-journey-of-change-indonesias-foreign-policy_dewi-fortuna-anwar.

Bao, L. (2016) 'Japan's naval chief rules out joint-US Freedom of Navigation Patrols', *Voa*, 28 September. Available from: https://www.voanews.com/a/japanese-naval-chief-rules-out-joint-us-freedom-of-navigation-patrol/3528783.html.

Brenner, N. (2004) *New State Spaces: Urban Governance and the Rescaling of Statehood*, Oxford: Oxford University Press.

Camba, A. (2020) 'Derailing development: China's railway projects and financing coalitions in Indonesia, Malaysia, and the Philippines', GCI Working paper 008, *Global China Initiative*. Available from: www.bu.edu/gdp/files/2020/02/WP8-Camba-Derailing-Development.pdf.

Camba, A. and Magat, J. (2021) 'How do investors respond to territorial disputes? Evidence from the South China sea and implications on Philippines economic strategy', *The Singapore Economic Review*, 66(1): 243–67.

Ciorciari, J.D. (2010) *The Limits of Alignment: Southeast Asia and the Great Powers since 1975*, Washington, DC: Georgetown University Press.

Damuri, Y.R., Perkasa, V., Atje, R., and Hirawan, F. (2019) 'Perceptions and readiness of Indonesia towards the Belt and Road Initiative', CSIS. Available from: https://mail.csis.or.id/uploads/attachments/post/2019/05/23/CSIS_BRI_Indonesia_r.pdf.

Eckstein, M. (2020) '2 Japan-based destroyers conduct second Taiwan Strait transit this month', *USNI News*, 30 December. Available from: https://news.usni.org/2020/12/30/2-japan-based-destroyers-conduct-second-taiwan-strait-transit-this-month.

Engel, D. (2020) 'How the Biden administration should manage US–Indonesia relations', *The Strategist*, 24 December. Available from: www.aspistrategist.org.au/how-the-biden-administration-should-manage-us-indonesia-relations/.

Gourevitch, P. (1978) 'The second image reversed: the international sources of domestic politics', *International Organization*, 32(4): 881–912.

Hamilton-Hart, N. and McRae, D. (2015) 'Indonesia: balancing the United States and China, aiming for independence', Sydney: The United States Studies Centre at the University Sydney. Available from: www.ussc.edu. au/analysis/indonesia-balancing-the-united-states-and-china-aiming-for-independence.

Hossain, K. (2013) 'The UNCLOS and the US–China hegemonic competition over the South China Sea', *Journal of East Asia and International Law*, 6(107).

Human Rights Watch (2017) 'Indonesia: US documents released on 1965–66 massacres', 18 October. Available from: www.hrw.org/news/2017/10/18/indonesia-us-documents-released-1965-66-massacres

*Jakarta Post* (2015) 'Jokowi begins first state visit to Japan', *The Jakarta Post*, 22 March. Available from: www.thejakartapost.com/news/2015/03/22/jokowi-begins-first-state-visit-japan.html.

*Jakarta Post* (2021) 'Luhut welcomes China's Wang Yi for investment talks', *The Jakarta Post*, 12 January. Available from: www.thejakartapost.com/paper/2021/01/12/luhut-welcomes-chinas-wang-yi-for-investment-talks.html.

Jibiki, K. (2020) 'Indonesia woos Japan as China led high speed rail project stalls', Nikkei Asia, 8 June. Available from: https://asia.nikkei.com/Business/Transportation/Indonesia-woos-Japan-as-China-led-high-speed-rail-project-stalls.

JICA (2008) 'JICA in Indonesia–Japan Expo 2008'. Available from: www.id.emb-japan.go.jp/oda/en/topics_200901_ijexpo.htm.

Kennedy, C. (2018) 'The struggle for blue territory: Chinese maritime militia grey-zone operations', *The RUSI Journal*, 163(5): 8–19.

Kuik, C.-C. (2016) 'How do weaker states hedge? Unpacking ASEAN states' alignment behavior towards China', *Journal of Contemporary China*, 25(100): 500–14.

Kurlantzick, J. (2006) 'China's charm offensive in Southeast Asia', Carnegie Endowment for International Peace, 1 September. Available from: https://carnegieendowment.org/2006/09/01/china-s-charm-offensive-in-southeast-asia-pub-18678.

Laksmana, E.A. (2016) 'Here's why Jakarta doesn't push back when China barges into Indonesian waters', *The Washington Post*, 28 April. Available from: www.washingtonpost.com/news/monkey-cage/wp/2016/04/28/heres-why-jakarta-doesnt-push-back-when-china-barges-into-indonesian-waters/.

Malone, J.L. (1983) 'The United States and the Law of the Sea after UNCLOS III', *Law and Contemporary Problems*, 46(29): 29–36.

Maulia, E. (2020) 'Jokowi vows "no compromise" on Natuna standoff with China', Asia Nikkei, 7 January. Available from: https://asia.nikkei.com/Politics/International-relations/Jokowi-vows-no-compromise-on-Natuna-standoff-with-China.

Ministry of Foreign Affairs of China (2015) 'Joint statement on strengthening comprehensive strategic partnership between the People's Republic of China and the Republic of Indonesia', 27 March. Available from: www.fmprc.gov.cn/mfa_eng/wjdt_665385/2649_665393/t1249201.shtml.

Ministry of Foreign Affairs of Japan (2015) 'Joint statement – towards further strengthening of the strategic partnership underpinned by sea and democracy', 23 March. Available from: https://www.mofa.go.jp/files/100002843.pdf.

Nabila, A., Fauri, A., Yusriza, B., Atmanta, D., Surdiasis, F., Tarmizi, H., Priyadi, L., Lingga, V., and Christianto, Y. (2018) Belt and Road Initiative: What's in It for Indonesia? Tenggara Strategics Briefing Paper. Available from: http://tenggara.id/assets/source/Insights/BRI-Briefing-Paper-English.pdf.

Negara, D.S. and Suryadinata, L. (2018) 'Jakarta–Bandung high speed rail project poses big challenge for Jokowi', Today Online, 12 January. Available from: www.todayonline.com/commentary/jakarta-bandung-high-speed-rail-project-poses-big-challenge-jokowi.

Ng, T. (2018) 'Chinese navy sent to confront USS Chancellorsville in latest South China Sea stand-off', South China Morning Post, 1 December. Available from: www.scmp.com/news/china/military/article/2175916/chinese-navy-sent-confront-uss-chancellorsville-latest-south.

Palma, S. (2020) 'Indonesia looks to Joe Biden for more "professional" stance on China', Financial Times, 6 December. Available from: www.ft.com/content/7f5eade4-03b2-40ec-beb3-a412903386e4

Parlina, I. (2013) 'Indonesia's creaky railways may get "Shinkansen" touch', Jakarta Post, 25 November. Available from: www.thejakartapost.com/news/2013/11/25/indonesia-s-creaky-railways-may-get-shinkansen-touch.html.

Patalano, A. (2018) 'When strategy is "hybrid" and not "grey": reviewing Chinese military and constabulary coercion at sea', The Pacific Review, 31(6): 811–39.

Power, J. (2020) 'US freedom of navigation patrols in South China Sea hit record high in 2019', South China Morning Post, 5 February. Available from: www.scmp.com/week-asia/politics/article/3048967/us-freedom-navigation-patrols-south-china-sea-hit-record-high.

Reuters (2019) 'Thailand gives green light for US$7.4 billion high-speed rail link between Bangkok and Pattaya', South China Morning Post, 24 October. Available from: www.scmp.com/news/asia/southeast-asia/article/3034467/thailand-gives-green-light-us74-billion-high-speed-rail

Schindler, S. and Kanai, J.M. (2021) 'Getting the territory right: infrastructure-led development and the re-emergence of spatial planning strategies', *Regional Studies*, 55(1): 40–51.

Simpson, B.R. (2008) *Economists with Guns: Authoritarian Development and US-Indonesian Relations, 1960–1968*, Redwood City, CA: Stanford University Press.

*Straits Times* (2015) 'Indonesia defends bidding process for high-speed rail project after Japan angered at being rejected', 1 May. Available from: www.straitstimes.com/asia/se-asia/indonesia-defends-bidding-process-for-high-speed-rail-project-after-japan-angered-at.

Sukma, R. (1999) *Indonesia and China: The Politics of a Troubled Relationship*, London: Routledge.

Sukma, R. (2009) 'Indonesia-China relations: the politics of re-engagement', *Asian Survey*, 49(4): 591–608.

Suroyo, G. (2015) 'Govt to hold "beauty contest" for high-speed train project', *MailOnline*, 14 July. Available from: www.dailymail.co.uk/wires/reuters/article-3160185/INDONESIA-PRESS-Govt-hold-beauty-contest-high-speed-train-project–Jakarta-Globe.html.

Tempo (2013) 'Japan eyes high-speed trains in Indonesia', *Tempo*, 17 September. Available from: https://en.tempo.co/read/513965/japan-eyes-high-speed-trains-in-indonesia.

*The Star* (2019) 'Daim: renegotiated ECRL roffers plenty of opportunities to local contractors', *The Star*, 19 April. Available from: www.thestar.com.my/news/nation/2019/04/19/daim-renegotiated-ecrl-offers-plenty-of-opportunities-to-local-contractors

Tritto, A. (2020) 'Contentious embeddedness: Chinese state capital and the Belt and Road Initiative in Indonesia', *Made in China Journal*, 5(1): 182–7.

Van der Kroef, J.M. (1976) 'The 1965 coup in Indonesia: the CIA's version', *Asian Affairs: An American Review*, 4(2): 117–31.

Wejak, J. (2000) 'Soekarno: his mannerism and method of communication', *Jurusan Sastra Inggris, Fakultas Sastra, Universitas Kristen Petra*, 2(2): 54–9.

Wright, S. (2017) 'Newly declassified files show how the US supported a mid-1960s extermination campaign in Indonesia', *Insider*, 19 October. Available from: www.businessinsider.com/newly-declassified-files-show-the-us-supported-killings-in-indonesia-2017-10?IR=T.

15

# Vietnam's Spatial and Hedging Strategies in Response to Chinese and Japanese Infrastructural Statecraft

*Jessica C. Liao*

## Introduction

*Doi Moi*, Vietnam's economic reform programme, was launched in 1986 on the heels of the country's most desperate times since reunification. Since then, infrastructure development has been a task of utmost importance within the government's reform agenda. Over the past three decades Vietnam has made tremendous accomplishments in building a variety of infrastructure facilities that support rapid socioeconomic development. Although Vietnam is today one of the fastest growing economies in the world, infrastructure development and upgrade continues to be a policy priority and a key component of the government's strategy to sustain economic growth, structural transformation, and global competitiveness.

Nevertheless, Vietnam's ambitious infrastructural objectives face many obstacles. Among them, financing and funding remain the most critical, especially given the country's ever-expanding infrastructure development plans, from power supply facilities and its energy grid system to toll roads and ports. More importantly, infrastructure development is not the sole priority for a country whose establishment came after three decades of wars fought against various foreign powers. Security and sovereignty remain top concerns, particularly with ongoing territorial disputes in the South China Sea (SCS) that threaten to spark tensions with its neighbours, particularly China. While both economic and security goals are prioritized, how to balance between them poses a challenge for Hanoi. This balancing act increasingly involves infrastructure.

This chapter illustrates Vietnam's political balancing act by examining the country's state spatial strategies (see Brenner, 2004) around its infrastructure policy, particularly as Vietnam collaborated with both Japan and China over the past three decades. When *Doi Moi* began, Vietnam relied on Japan as the main capital source for infrastructure development. Yet, the stagnation of Japan's economy in the 1990s led Hanoi to expand its sources of finance for infrastructure projects. It turned tentatively towards China, and alongside efforts to expand economic cooperation was a concerted attempt to resolve longstanding territorial disputes.

Since the 2000s, Japan and China have competed for political and economic supremacy and leadership status in East Asia. A distinct feature of this competition has been their distribution of economic incentives to other Asian countries in exchange for diplomatic and policy support. This competition has recently manifested in the form of infrastructure financing, with China leading the ambitious Belt and Road Initiative (BRI), followed by Japan's foreign aid campaign, the Partnership of Quality Infrastructure (PQI) for developing Asia. How has Vietnam responded to this competition? How has Vietnam balanced trade-offs between its economic and security goals, and whether to accept Japanese or Chinese infrastructure financing offers, especially given the ongoing SCS disputes? How do security concerns shape Vietnam's infrastructure policy choices? To what extent have Vietnam's efforts to hedge between China and Japan enabled it to achieve its state spatial objectives? Finally, how have such state spatial strategies affected places or people in Vietnam?

Drawing on literature of international relations of East Asia, the next section examines the concept of hedging to explain how small states calculate risks and rewards in their navigation of great power politics. The second section illustrates how hedging is practised, through an examination of Vietnam's state spatial strategy, which has rested on dynamic infrastructure partnerships with China and Japan over three time periods since *Doi Moi*. This strategy has become increasingly dynamic since the competition between China and Japan over infrastructure finance and construction intensified in the 2010s. The third section discusses challenges facing Vietnam's hedging strategy and connects the geopolitics of hedging with spatial strategies. The chapter concludes with a discussion on the theoretical and policy implications.

## Great power statecraft and small-state hedging

How small states make their stance in the face of great power rivalry is a good starting point to understand Vietnam's foreign policy in the post-Cold War era. At the turn of the 21st century, Southeast Asia faced a power shift from the decline of US hegemony and the rise of Chinese influence. This change, coupled with regional territorial disputes, historical complexity, and power asymmetry vis-à-vis China, has challenged the regional stability that fostered Southeast Asia's decades-long economic growth. To ensure stability, these states have improved their various degrees of security cooperation with the US. Nonetheless, they are also active in promoting cooperation with China, especially in the economic sphere, because of the potential to benefit from a rising China and of advantages from using economic exchanges to coax their giant northern neighbour into regional cooperation. Additionally, Southeast Asian states are active in strengthening relations with other extra-regional powers, including Japan, at both economic and security levels. Goh (2005, 3) conceptualizes this tactic as Southeast Asia's strategic 'hedging', that is, 'to cultivate a middle position that forestalls or avoids having to choose one side at the obvious expense of another'. Strategic hedging, Kuik (2016, 504) explains, entails 'a bundle of opposite and deliberately ambiguous policies'.

The literature on strategic hedging has increasing discussion on power competition through using economic policy devices. Liao and Dang (2020) stress that economic exchanges may be used either to promote peace among countries or as an instrument for great powers to exploit power asymmetry and expand their influence over small states. Thus, small states, especially those who perceive a great power counterpart as a security threat, ponder security implications of their foreign economic relations and maintain an extent of economic cooperation with the great power that allows them room to preserve political autonomy. Similarly, Pitakdumrongkit (2020) highlights the importance of 'economic security' to small states and their sensitivity to the great power's decisions in withdrawing economic incentives or sanctioning. Cognizant of such risk, small states' political elites often aim to diversify their economic relations with various powers to mitigate asymmetrical dependence and the influence from a particular power. In this sense, small states develop hedging strategies in pursuit of spatial objectives, which are inextricably linked with security strategy and policy.

The discussion of economic hedging has taken on particular salience against the backdrop of China's BRI. To be sure, China is not the first Asian country whose economic statecraft includes infrastructure finance. For decades, Japan has been a major financier for developing Asia's infrastructure, through both bilateral and multilateral channels including the Japan International Cooperation Agency (JICA) and the Asian Development Bank (ADB). In the 1980 and 1990s, Japan's Official Development Assistance (ODA) became the staple of East Asia's infrastructure development and it expanded so rapidly that it made Japan the world's largest ODA donor. However, the burst of the national economic bubble led to Japan's 'Lost Decade' from the late 1990s and into the 2000s, with slow economic recovery and a shrinking ODA budget. This, quite significantly, coincided with the introduction of Beijing's 'Good Neighbor' and 'Going Out' campaigns, both of which pledged to expand development assistance to neighbouring countries. China then deployed the so-called 'Yuan diplomacy', channelling foreign aid and other state-backed funding to other countries to promote its political and economic interests (Kurlantzick, 2008; Liao, 2019). China's role as Asia's new infrastructure financier rose further as Xi Jinping came to power and expanded and further institutionalized the Going Out policy into BRI, with which Beijing has vowed to bring modern infrastructure to countries across Eurasia and beyond. While the intent of the BRI is hotly debated, an under-examined aspect of the initiative is how it has provoked an economic statecraft contest with capital rich countries other than the US. In much of East Asia, the BRI's primary competitor is Japan. Shortly after the announcement of BRI, Japan, under the then new Prime Minister Shinzo Abe, launched PQI, Japan's biggest post-recession foreign aid plan with a pledged value of ¥13.2 trillion (approximately US$110 billion) for 'quality infrastructure' development in Asia.

In the context of the intensification of China–Japan infrastructure financing competition, Vietnam deployed state spatial strategies geared towards its high-growth development plans while upholding its security objectives. Indeed, Vietnam's foreign policy has long been situated at the nexus of economic and security policy. In the early 1990s, *Doi Moi* began with the goal to pull the country back from the brink of collapse and towards economic modernization. Along with its economic reforms, Hanoi upheld a 'multi-directional' foreign policy by normalizing relations with neighbours and Western countries and cautiously opening its economy to the world to support its domestic development. While the reform has brought Vietnam remarkable economic accomplishments, throughout this time its communist leadership has consistently affirmed that maintaining sovereignty and territorial integrity remains their top priority (Vuving, 2006; Thayer, 2011). While economic pragmatism led Vietnam to normalize relations with Japan and China, security concerns informed Hanoi's distinct approach to these two tasks. On the one hand, on the eve of *Doi Moi*, Japan was among the earliest countries that Hanoi identified as a potential economic partner. Scholars have labelled Vietnam–Japan relations since the normalization as 'worry free' (Khong, 2012; Do and Dinh, 2018). On the other hand, the post-unified Vietnam has been characterized as, in Womack's (2004) words, 'paranoid' of Chinese influence, not only because of China's millennium-long domination of Vietnam but also due to their decade-long border conflict (1979–90). The latter resulted in significant deleterious economic consequences for Vietnam, so although the normalization of relations with China was a priority of the *Doi Moi* agenda, the opening of economic relations was a decade in the making as Hanoi took cautious steps in settling longstanding border disputes with its northern neighbour. While economic exchanges have expanded with China since the 2000s, Vietnam has been wary of the remaining SCS disputes and the security cost of its Chinese economic ties. As the following section will show, a similar dynamic also marked Vietnam's infrastructure collaboration with Japan and China from the 1990s to the present.

## Hedging in response to China–Japan infrastructural competition

Vietnam began opening to the world in the 1990s. During this period the country became heavily dependent on Japanese ODA for infrastructure development, while infrastructure collaboration with China was minimal and incremental. Soon after withdrawing troops from Cambodia, Hanoi settled debt disputes with Tokyo, which predated Vietnam's reunification. This paved the way for Japan's resumption of full-scale ODA correand for gaining other bilateral and multilateral donors to follow Tokyo's decisions. Japan soon became Vietnam's dominant donor throughout the 1990s, with the majority of its funding going to transportation and energy infrastructure. During this period, Japan's ODA made up about 40 per cent of Vietnam's overall income from foreign aid (Nguyen and Nguyen, 2010).

Vietnam–Japan infrastructure collaboration was not without challenges, however. Complaints ranged from slow fund disbursement to project delay to expensive Japanese contractors. Hanoi thus became concerned about over-reliance on Japan for its post-reform infrastructure development. Still, the biggest concern for Hanoi was Japan's post-bubble economic stagnation and looming budget crisis. As such, even though Japan boosted ODA to Vietnam, among other Southeast Asian countries, following the 1997 financial crisis, Hanoi began actively seeking new infrastructure financiers (Ohno, 2004).

Hanoi initially approached western donors in search of new infrastructure capital rather than Beijing, despite the launch of China's Good Neighbor and Going Out campaigns – in which Beijing first vowed to use economic diplomacy to improve relations with Southeast Asian states (Shambaugh, 2004). More importantly, Hanoi maintained a cautious stance in economic cooperation with Beijing while the primary task for both sides during this time was settling land border disputes. They signed several bilateral agreements in trade, taxation, and labour services that helped foster their infrastructure collaboration (Le, 2017). At this time, China's preferential loan deals for infrastructure development were politically symbolic but, in dollar terms, insignificant.

The beginning of the 2000s signalled Hanoi's shift towards pragmatism for it started to diversify foreign infrastructure partners. Meanwhile, China increasingly vied for Asia's contracting and infrastructure lending markets, and this coincided with the sharp fall of Japan's ODA budget in 2000, which contributed to the decline of Vietnam's infrastructure investment in the following years (Nguyen and Dapice, 2009). Meanwhile, there was a series of diplomatic breakthroughs with China, namely, the delimitation of maritime boundaries in the Gulf of Tonkin, the Vietnam–China Joint Statement of Comprehensive Cooperation in 2000, and the China–ASEAN Declaration of Conduct of Parties in the South China Sea in 2002, leading Hanoi to pragmatically enhance economic cooperation with China. In 2003, Hanoi and Beijing signed their first major loan agreement and another major one in 2005. Although the interest rates of Chinese loans – mainly commercial ones – were not as concessional as Japanese ODA loans, infrastructure collaboration with China appealed to Hanoi because of the cost-competitiveness of Chinese contractors and quick disbursement of funds (Le, 2017, 65–86). The 2005 Tendering Law was designed to take advantage of more Chinese loans and contractors for Vietnam's infrastructure development (Liao and Dang, 2020).

Vietnam–China relations continued to improve in the second half of the 2000s, following the completion of the demarcation of the Vietnam–China land border and Comprehensive Strategic Partnership in 2008. Alongside these accomplishments, both sides signed the largest loan deal yet – 1.2 billion Renminbi (US$173 million) – to support Vietnam's infrastructure development, including several large-scale thermal power projects that were long dominated by Japanese contractors. In 2009, Vietnam became the top overseas market for Chinese infrastructure contractors and exports of machinery.

However, Hanoi continued to hedge towards the Chinese infrastructure collaboration in two ways. First, the amount of the Chinese loans, approximately 0.67 per cent of Vietnam's government-backed external debt by the end of 2007, remained modest. Second, through the establishment of joint committees with multi-tier officials from top party leaders and vice-ministers to provincial leaders, Hanoi cautiously controlled the pace of cooperation and aimed to keep it low profile, particularly on major projects (Le, 2017, 73–5). This deliberation was necessary given continued anti-Chinese public sentiment in Vietnam, which manifested in a major protest in 2009 over Chinese investment in a mining project in the Central Highlands. The project precipitated widespread public discontent against Beijing's growing economic leverage over Vietnam (Morris-Jung, 2015). In response, Hanoi, while facing a crisis-hit economy at the time, vowed to investigate the project and paused several major Chinese infrastructure contracts.

Although Japan's ODA budget restricted the Vietnam–Japan infrastructure collaboration, Vietnam was diligent in expanding Japan's infrastructure support as part of its efforts in building comprehensive cooperation with Tokyo, shown by the Vietnam–Japan joint statement of the Higher Sphere of Enduring Partnership in 2004 and the 2009 finalization of the Strategic Partnership for Peace and Prosperity. Hanoi also lauded its support for the newly established Japan–Mekong Partnership and became the first government in the Mekong Delta region to host a summit under this framework. Throughout this time, Hanoi succeeded in getting Tokyo to support major economic infrastructure, including power plants, and national and provincial highways prioritized in its development plans. Japan's ODA to Vietnam rebounded in 2004; however, aid was suspended over a bribery scandal involving a JICA contractor. In response, the two countries agreed to a strategic partnership and established a joint anti-bribery committee. By 2009, Vietnam received more Japanese ODA than any other recipient in the world. Overall, the increasing China–Japan infrastructure financing competition and Vietnam's need for infrastructure capital drove Hanoi to expand collaboration with Asia's two giant economies.

## On a tightrope: the 2010s

Hanoi's pragmatism faced rising pressure in the early 2010s when China began a more assertive stance over territorial disputes. Two incidents caught Vietnam's attention: Beijing blocked discussions on the SCS issue at the 2010 ASEAN Regional Forum in Hanoi and a year later, China 'accidentally' cut a PetroVietnam survey vessel's cable in Vietnam's exclusive economic zone, well south of the disputed Paracel Islands. Facing continued economic uncertainties and an increased fiscal deficit, Hanoi settled the dispute after a series of high-level official exchanges with Beijing, leading to a new preferential loan deal with China. Yet, the value of this deal was far below those previously signed even though Hanoi's newly released master development plan outlined ambitious spatial objectives and required significant foreign investment or assistance. This indicated Hanoi's increased wariness of economic ties with China.

Anti-Chinese sentiment permeated a Vietnamese society increasingly concerned about its national security vis-à-vis China (Thayer, 2011). This concern manifested in widespread public discussions on two topics. The first was on Vietnam's soaring trade deficit with China and reliance on imported Chinese machinery and equipment. The discussion later shifted to focus on the possibility of 'back doors' installed in China-built public infrastructure in the wake of Chinese hackers' unruly attack on Vietnamese websites following the cable-cutting incident. In both discussions, Chinese contractors were portrayed as a security threat. Meanwhile, Vietnam–China relations took a nosedive after a Chinese state-owned oil firm moved an oil platform into the Vietnam-claimed exclusive economic zone (also known as the HD 981 incident) and Chinese lenders froze credits for several projects (Bowring, 2011). Many Vietnamese public figures, including officials, vocally opposed Chinese-built infrastructure in response, citing concerns surrounding national security, quality, and environmental issues. Many of these criticisms also faulted Hanoi for its hasty collaboration with China. In a rare occasion, the Ministry of Planning and Investment (2014) published a report titled *Vietnam's Dependence on China*, echoing this public sentiment, and calling for reduction of Chinese development funds.

Hanoi and Beijing eventually reconciled and resumed several stalled China-backed projects. Nonetheless, the HD 981 incident prompted Hanoi to deliberately pivot away from infrastructure collaboration with China. It unexpectedly selected Japan to fulfil a major thermal power project that was the subject of protracted negotiations with China. Hanoi later awarded other power projects in both northern and southern provinces to Japan and other countries. Vietnamese Transport Minister Dinh La Thang publicly threatened to terminate the Chinese contract for the Hanoi Metro Line 2A over the construction delay and a deadly incident. Thang also proclaimed his support for Japanese contractors for Hanoi's urban rail system, which Japan unsurprisingly won.

Hanoi's alignment with Japan coincided with the inauguration of the BRI. To promote the campaign and express Beijing's goodwill, President Xi visited Hanoi in early 2015, the first Chinese state visit to Vietnam in a decade. While Xi pledged funding support for Vietnam's infrastructure, he encountered strong pushback in Vietnam from the National Assembly and the public. Many projects in provinces bordering China for which Xi voiced support were ultimately unfulfilled (Voice of America News, 2015). Moreover, Hanoi cancelled the bidding on the Van Don–Mong Cai highway – a project with security implications for its linkage of the Vietnam–China border – to avoid awarding the project to Chinese contractors. Similarly, new laws on construction and environmental protection were released in 2014, which were meant to improve the 2005 Tendering Law through regulating the use of low-cost Chinese contractors. In addition, although Vietnam joined other Southeast Asian states in support of the BRI and became a founding member of the China-led Asian Infrastructure Investment Bank, the popular perception that the 'BRI [is] doing more harm than good to Vietnam' has stalled major BRI-designated projects in Vietnam (BBC News, 2017). These incidents indicated that Hanoi adjusted its geospatial objectives to better ensure its security interests.

It is important to note that Hanoi's economic hedging against China was in part because its infrastructure sector has attracted significant attention from foreign investors due to Vietnam's stunning growth rates during this period. Still, infrastructure cooperation with Japan remains most vital to Vietnam as such cooperation has become the linchpin of their deepened strategic relations. Hiroshi Fukuda, Japan's ambassador to Vietnam, stated candidly that 'Japan recognized the growing voice within Hanoi to be more independent of China and desired to deepen its ties to Hanoi through infrastructure cooperation' (Wright and Obe, 2015). Following Prime Minister Abe's call for a strategy-focused foreign policy and then announcement of PQI to rival BRI, Vietnam became the prime target of Japan's economic statecraft. Additionally, with Japanese investors moving out of China in response to the violent anti-Japanese protests following the 2012 Senkaku/Diaoyu Islands dispute, Tokyo also desired to cultivate Vietnam as a new investment partner. In return, Hanoi keenly identified Japan as a preferred foreign investor as well as a strategic partner. In 2017, Hanoi and Tokyo elevated their relations into an extensive strategic partnership and announced a series of major infrastructure projects. In addition to pivoting power contracts from China to Japan as mentioned earlier, Hanoi lobbied to maintain the preferential terms of ODA loans after Vietnam's graduation to 'middle-income' status. To convince Tokyo, the Vietnamese government took steps to improve its ODA governance mechanism, shown by its speedy handling of the 2014 bribery case involving the JICA-funded Vietnam Railway.

While Vietnam continued to look to Japan for major infrastructure development, this trend has not sparked concerns in the Vietnamese public as Chinese capital has. Since 2010, the annual average Japanese ODA disbursement to Vietnam has more than doubled and peaked in 2014 with a total of US$3.3 billion. With several mega-infrastructure projects breaking ground, including coal-fired power and deep-sea port construction, Japan became Vietnam's top foreign investor in 2017. Since 2015, the Japan-led ADB has also doubled down on lending to Vietnam for megaprojects under the public–private partnership arrangement (Fischler, 2017). In 2016, Japan's ODA to Vietnam accounted for approximately 40 per cent of its overall foreign aid income. The nexus of economic and security partnerships between the two countries was not a surprise in light of Japan's favourability among the Vietnamese, which is ranked the highest in East Asia (Pew Research Center, 2014).

Hedging has its limits. As Vietnam's new development plans show, in 2021–30 the country would need US$128.3 billion and US$65 billion respectively for its electricity and transportation development. In recent years, Vietnam has maintained its status as the second-largest infrastructure investor in Asia, only second to China (Yap and Nguyen, 2017). Moreover, Pham Minh Chinh, who became prime minister in May 2021, revealed ambitious goals to boost Vietnam's global competitiveness through implementing major infrastructure development. To satisfy its ever-expanding infrastructure demand, Hanoi, even with greater capital resources from Japan, needs to diversify and expand its capital pool and foreign partnerships.

Given its ambitious spatial objectives, Hanoi is unable to contemplate fully stopping infrastructure collaboration with China, the world's largest overseas infrastructure financier. As shown by Chinese state-owned firm

China Power's recent winning of construction contracts for solar power projects with 500MW power capacity in Vietnam, Hanoi continues to abide by economic pragmatism in its infrastructure policy decisions. Yet, it is worth noting that Hanoi is also actively expanding energy development partnerships with Japan as well as other countries in East Asia and beyond, including the US. This diversification is a tactic of soft economic hedging in preventing Chinese domination in Vietnam's energy sectors.

However, as the analysis shows, Hanoi must respond to a society that is increasingly questioning the normative and substantial value of economic pragmatism, which, in the public's eye, has eroded their country's sovereignty and security interests. The task of maintaining a balance between economic pragmatism and hedging is a difficult one – Hanoi's 2018 special economic zone (SEZ) plans illustrate this well. Under widespread rumours that Chinese investors with high legal autonomy would flood into Vietnam, massive protests took place across the country, forcing Hanoi to delay the plan. Prime Minister Pham, a key proponent of this plan, has suggested drafting a modified plan with greater administrative authority over foreign investors, a move indicating a policy shift driven by public pressure. Whether this plan to restructure the state and increase oversight over foreign investment can tame public demands remains to be seen. For now, Hanoi seems to be hedging its bets to dodge this public pressure, as recent Chinese-financed infrastructure projects are mostly under a joint venture/share acquisition to cover the identity of Chinese investors.

Two recent developments, though, point to increased difficulties facing Hanoi's balancing act regarding its infrastructure collaboration with China. On the one hand, Beijing projects an increasingly assertive foreign policy. In 2019, Vietnam and China engaged in a dispute over an offshore oil block in the SCS, which falls within Vietnam's exclusive economic zone. A year later, Beijing also held a similar line on a dispute involving Filipino fishing vessels. Moreover, Beijing's expanding economic sanctions against countries from South Korea to Australia on various issues have shown its willingness to flex its economic muscle to achieve political goals. With security concerns in mind, Hanoi, unlike Southeast Asian counterparts, was quick to omit China's telecom giant, Huawei, in a recent plan to expand foreign partnerships to build Vietnam's 5G networks. On the other hand, although Vietnam posted the highest economic growth rate in Southeast Asia in 2019, due largely to the windfall effect of the US–China trade tension and the resulting supply chain shift, the next year the country was struck by the COVID-19 pandemic and its growth slumped to its weakest since the *Doi Moi* reform. The stimulation of economic growth remains a national priority, and with infrastructure development at its centre, Hanoi cannot afford to exclude China as a partner.

## Conclusion

This chapter has made three primary contributions. First, for decades, studies of international relations have divided security and economy in their research agendas and have thus evolved into two distinct fields of literatures in international security and international political economy. Yet, as Mastanduno (1998) observes, although economic and security issues are highly intertwined in reality, their interrelations have been a 'neglected area of study'. Through the study of Vietnam's infrastructure policy vis-à-vis China and Japan, this chapter makes it clear that the nexus of security and economic policies is an important research agenda. Second, the impact of geopolitical competition between great powers needs to be understood alongside the agency of small states (Liu and Lim, 2019; Camba, 2020). As this chapter shows, how Vietnam assesses its respective relations vis-à-vis Japan and China determines how the China–Japan economic statecraft competition plays out. Third, the chapter also shows that this competition creates as many benefits as challenges for small states – or so say countries in the Global South in general – in their state spatial strategy crafting. It is increasingly delicate, if not impossible, for the governments of these countries – either in a democratic system or not – to strike a balance between their developmental and sovereignty priorities and their relations with great power counterparts and domestic constituencies.

### Disclaimer

The opinions in this chapter are those of the author and do not represent those of the US government.

### References

BBC News (2017) 'VN ở đâu trong "vành đai và con đường" của TQ [Where does Vietnam stands in the One Belt One Road initiative of China?]', BBC News, 11 May. Available from: https://www.bbc.com/vietnamese/business-39872817.

Bowring, G. (2011) 'Vietnam yields cautionary tale over Chinese investment', *Financial Times*, 24 November. Available from: www.ft.com/content/6ea71dd6-ccea-3779-87be-d4654fc9379b.

Brenner, N. (2004) *New State Spaces: Urban Governance and the Rescaling of Statehood*, Oxford: Oxford University Press.

Camba, A. (2020) 'The Sino-centric capital export regime: state-backed and flexible capital in the Philippines', *Development and Change*, 51(4): 970–97.

Do, T.T. and Dinh, J.L. (2018) 'Vietnam–Japan relations: moving beyond economic cooperation?', in H.H. Le and A. Tsvetov (eds) *Vietnam's Foreign Policy under Doi Moi*, Singapore: ISEAS–Yusof Ishak Institute, pp 96–116.

Fischler, N. (2017) 'China, Japan vie for Vietnam rail riches', *Asia Times*, 21 September. Available from: https://asiatimes.com/2017/09/china-japan-vie-vietnam-rail-riches/.

Goh, E. (2005) 'Great powers and Southeast Asian regional security strategies: omni-enmeshment, balancing and hierarchical order', Institute of Defence and Strategic Studies, Nanyang Technological University.

Khong, T.B. (2012) 'China–Vietnam–Japan: a strategic triangle', in L.P. Er and V. Teo (eds) *Southeast Asia between China and Japan*, Newcastle: Cambridge Scholars Publishings.

Kuik, C.-C. (2016) 'How do weaker states hedge? Unpacking ASEAN states' alignment behavior towards China', *Journal of Contemporary China*, 25(100): 500–14.

Kurlantzick, J. (2008) *Charm Offensive: How China's Soft Power Is Transforming the World*, New Haven, CT: Yale University Press.

Le, H.H. (2017) *Living Next to the Giant: The Political Economy of Vietnam's Relations with China under Doi Moi*, Singapore: ISEAS-Yusof Ishak Institute.

Liao, J.C. (2019) 'A good neighbor of bad governance? China's energy and mining development in Southeast Asia', *Journal of Contemporary China*, 28(118): 575–91.

Liao, J.C. and Dang, N.-T. (2020) 'The nexus of security and economic hedging: Vietnam's strategic response to Japan–China infrastructure financing competition', *The Pacific Review*, 33(3–4): 669–96.

Liu, H. and Lim, G. (2019) 'The political economy of a rising China in Southeast Asia: Malaysia's response to the Belt and Road Initiative', *Journal of Contemporary China*, 28(116): 216–31.

Mastanduno, M. (1998) 'Economics and security in statecraft and scholarship', *International Organization*, 52(4): 825–54.

Morris-Jung, J. (2015) 'The Vietnamese bauxite controversy towards a more oppositional politics', *Journal of Vietnamese Studies*, 10(1): 63–109.

Nguyen, X.T. and Dapice, D. (2009) *Vietnam's Infrastructure Constraints*. Cambridge, MA: Ash Institute for Democratic Governance and Innovation. Available from: https://ash.harvard.edu/files/vietnams_infrastructure_constraints.pdf.

Nguyen, X.T. and Nguyen, V.K. (2010) 'Impact of Japan's official development assistance on Vietnam's socio-economic development', *VNU Journal of Science, Economics, and Business*, 26: 1–10.

Ohno, I. (2004) *Fostering True Ownership in Vietnam: From Donor Management to Policy Autonomy and Content*, Tokyo: GRIPS Development Forum.

Pew Research Center (2014) 'How Asians view each other', Pew Research Center. Available from: www.pewglobal.org/2014/07/14/chapter-4-how-asians-view-each-other/.

Pitakdumrongkit, K. (2020) 'What causes changes in international governance details? An economic security perspective', *Review of International Political Economy*. doi:10.1080/09692290.2020.1819371.

Shambaugh, D. (2004) 'China engages Asia: reshaping the regional order', *International Security*, 29(3): 64–99.

Thayer, C., A. (2011) 'The tyranny of geography: Vietnamese strategies to constrain china in the South China Sea', *Contemporary Southeast Asia*, 33(3): 348–69.

Vietnam Ministry of Planning and Investment (2014) Thực trạng sự phụ thuộc của kinh tế Việt Nam vào Trung Quốc (Dependence of Vietnamese economy on China: Status Quo), CIEM Document Center. Accessed at http://www.vnep.org.vn/Upload/1-%20FULL%20Su%20phu%20th uoc%20cua%20KTVN%20vao%20TQ.pdf, December 27, 2017.

Voice of America News (2015) 'Vietnamese lawmakers grill PM over China aid', *Voice of America News*, 17 November. Available from: www.voanews. com/a/vietnamese-lawmakers-prime-minister-china/3062270.html.

Vuving, A. (2006) 'Strategy and evolution of Vietnam's China policy: a changing mixture of pathways', *Asian Survey*, 46(6): 805–24.

Womack, B. (2004) 'Asymmetry theory and China's concept of multipolarity', *Journal of Contemporary China*, 13(39): 351–66.

Wright, T. and Obe, M. (2015) 'Vietnam plays key role in China–Japan aid battle', *Wall Street Journal*, 27 March. Available from: www.wsj.com/ articles/vietnam-plays-key-role-in-china-japan-aid-battle-1427431451.

Yap, K.L.M. and Nguyen, D.T.U. (2017) 'In Asia's infrastructure race, Vietnam is among the leaders', Bloomberg News, 22 March. Available from: www.bloomberg.com/news/articles/2017-03-22/ in-asia-s-infrastructure-race-vietnam-is-among-the-leaders.

# 16

# Diversifying Dependencies? Hungary, the EU, and the Multifaceted Geopolitics of Chinese Infrastructure Investments

*Ferenc Gyuris*

The complex nature of the European Union's (EU) relationship with China was put on display when a trade deal reached in 2020 after years of painstaking negotiation collapsed in a flurry of recriminations. Shortly thereafter in 2021, the European Parliament voted to impose sanctions on officials in Xinjiang for the treatment of ethnic Uyghurs, and Beijing retaliated with sanctions against EU parliamentarians. Although this has made it unlikely that the trade deal will be ratified (Lau, 2021), it also tested the unity of EU member states, among which there is considerable debate surrounding relations with China. Notably for economic and political reasons, Hungary has cultivated closer ties with China as the government sought to position itself as a mediator between Beijing and Brussels. There was a genuine feeling in Hungary, particularly after the 2008 economic crisis, that enhanced relations with China would result in economic dividends. Cosying up to Beijing also provided Budapest with opportunities to snub Brussels. For example, Hungary was the only EU country whose ambassador to China abstained from signing a statement critical of the Belt and Road Initiative's (BRI) lack of transparency and subsidies for state-owned enterprises (Heide et al, 2018).

This chapter examines Hungary's attempt to leverage its strategic position as a member of the EU and its geography on the bloc's eastern periphery. While Hungary is firmly rooted in the EU and benefits from integration in European – mainly German – production networks, the right-wing Orbán government that came to power in 2010 turned to China in pursuit of domestic and regional political and spatial objectives. This orientation was precipitated by the hope that warm ties with Beijing would foster inward Chinese investment and increased international trade. This chapter contextualizes Hungary's 'eastward turn' as a reversal of the resolute faith in Western Europe, and 'the West' in general, that characterized Hungarian politics since the end of the Cold War. This faith was severely tested in the aftermath of the 2008 financial crisis and, in this context, the Hungarian government seized on integration with China's fast-growing economy as the answer to persistently sluggish growth and the EU's bureaucratic inertia. However, this chapter shows that this strategy's economic dividends have largely failed to materialize, while its infrastructural promise has also lagged far below expectations. As a result, domestic and foreign criticism of Hungary's relations with China has grown louder in recent years, and relations with China are currently becoming a central issue in Hungarian domestic politics.

## From post-Cold War to 2008 financial crisis: Hungary's eastward turn

Political elites in East Central Europe enthusiastically steered their countries towards 'the West' after the fall of the Iron Curtain. In Hungary, Poland, and Czechoslovakia (prior to the separation of Czechia and Slovakia), there was a consensus among consecutive governments and influential political parties that the path to a brighter future was through membership in NATO, the European Economic Community (later the EU), and as an ally of the US. The EU and US proved welcoming, at least from a political-institutional perspective. Brussels set about integrating countries in East Central Europe, while successive US administrations supported their accession in the bloc as well as NATO membership. For the EU and the US, support for liberal-democratic governments in the region was meant to prevent them from drifting into the sphere of influence of a resurgent Russia. Relations between East Central European countries and China were largely ignored by Brussels and Washington, and socialist-liberal Hungarian governments sought to cultivate improved economic ties with China beginning in 2003, and these efforts were intensified and politicized when the global financial crisis spread through the region in 2008 and peaked in 2009 (Salát, 2020).

The integration of East Central European countries with Western European production networks after the Cold War led to technology transfer and investment-fuelled economic growth. However, it also led to dependence on the EU (see Table 16.1) and the narrative that the financial crisis spilled over from Western Europe was popularized with more than

**Table 16.1:** The strong reliance of the Visegrád Group countries on EU member states in terms of trade and FDI, 2008 (%)

| Country | EU countries' share of merchandise trade | EU countries' share of inward FDI stock |
|---|---|---|
| Czechia | 74 | 89 |
| Hungary | 73 | 75 |
| Poland | 69 | 86 |
| Slovakia | 72 | 92 |

Source: Author's calculations based on data from UNCTADStat (https://unctadstat.unctad.org/) and OECD (https://stats.oecd.org/).

a little schadenfreude. Tensions that had festered for some time exploded during the financial crisis, with post-communist countries alleging that the EU had a 'dual-track' system that ensured they were second-class members. For example, the fact that old member states were allowed to postpone opening their labour markets to citizens of post-communist states until 2011 (Koikkalainen 2011) fostered bitter resentment.

In Hungary, the meteoric rise of the extreme right-wing and anti-EU Jobbik party illustrates how deeply some parts of Hungarian society resent Brussels. On the eve of the financial crisis in 2006 the Jobbik party gained a mere 2.2 per cent in parliamentary elections, but its share of votes rose to 14.8 per cent in 2009 EU parliamentary elections and 16.7 per cent in Hungary's 2010 national elections. More importantly, the right-wing Fidesz Party came to power in the wake of the financial crisis under the leadership of Viktor Orbán. This ushered in a period of radicalization of Hungary's political discourse, particularly surrounding the record of EU institutions. They were not only perceived as ineffective by many Hungarians, but also exploitative. The European Central Bank's President Mario Draghi (2012) promised to do 'whatever it takes' to save the Euro in 2012 when it was tested by the debt crisis that gripped the Greek economy. The European Commission's call on Eurozone members, including some post-communist countries, to contribute to the bailout of the Greek economy was met with widespread disbelief and even rage in Hungary, even though it was not yet a Eurozone member. Indeed, the average income in Hungary and other post-communist countries was far lower than in Greece and, furthermore, Hungarians and other East Central European nations had endured austerity and restructuring to align institutions with the EU and stabilize the economy after 1990. A general feeling was that the financial institutions in the EU's 'core' states had precipitated the financial crisis, and Brussels wanted the EU's 'eastern periphery' to subsidize its 'southern periphery' (Černiauskas and Raudseps, 2015).

The financial crisis precipitated a shift in regional politics, as East Central European countries enhanced intergovernmental cooperation. Most notably the Visegrád Group (V4) established in 1991 and comprised of Czechia, Hungary, Poland, and Slovakia, affirmed their collective identity and this manifested spatially with renewed regionalism. Prior to the financial crisis the primary state spatial strategy (see Brenner, 2004) of each of the V4 countries was to enhance integration with the European Community – namely Western Europe. For example, the EU Commission initiated the Trans-European Networks in 1993 to improve intra-continental connectivity followed by the Transport Infrastructure Needs Assessment (TINA) in 1997 to establish a state-of-the-art transport network. Although TINA targeted transport infrastructure development between post-communist East Central Europe and the European economic core *as well as* among V4 countries, it prioritized the former (Fleischer, 2016). For Hungary, these spatial plans underpinned its overall development strategy, which was geared towards integrating with pan-European production networks whose lead firms were based mainly in Germany. This led to significant investment from German automakers, yet after the 2008 financial crisis its spatial objectives included deepened spatial integration with the other V4 countries, particularly through a host of regional infrastructure projects. For example, Hungary and Slovakia connected their gas pipeline networks in 2014 by setting up a new 111-km long pipeline, filling the last section in a common network of gas pipelines through the region from Polish Baltic Sea ports to Slovenian and Croatian ports along the Adriatic Sea (Orient Press, 2014; Perényi and Remete, 2018).

In addition to strengthening cooperation with other V4 countries, the Hungarian government responded to the financial crisis by intensifying its economic and diplomatic links with countries outside of the EU. Budapest tried to re-activate relations with former 'friendly socialist states' that had been neglected since 1990, even outside of Europe, but most notably with Russia (Farkas et al, 2016). In 2014, the Hungarian government reached an agreement with Russia to revamp Hungary's only nuclear power plant, which was inaugurated in 1982 with Soviet technology and still generated more than half of the country's electricity. Moscow agreed to finance the upgrade primarily through a EUR10 billion loan (Index, 2014). Hungary subsequently joined the Russian-dominated International Investment Bank in 2015, and its headquarters was relocated from Moscow to Budapest in 2019. Hungary's friendly relations with Moscow – along with Orbán's EU-sceptic rhetoric – drew ire in Brussels and other capitals, and overshadowed Hungary's increasingly warm ties with China. Yet relations with Beijing were the key to economic and infrastructural aspirations emerging in the wake of the global financial crisis among policy makers in Budapest.

## Opening to China: economic and infrastructural ambitions

While Hungary–China cooperation began in the 1950s, it stagnated after the Sino-Soviet split. Relations were reset after Deng's reforms were introduced in the late 1970s, and Chinese expert delegations were often received in Hungary to scrutinize its economic reforms (Vámos, 2018). In recent years, the Orbán government was encouraged by the fact that Budapest was one of three stops on Premier Wen Jiabao's 2010 visit to Europe (along with Berlin and London). This was followed in 2012 by a Chinese initiative to partner with Central and Eastern European Countries, known as China-CEEC or 16+1 (later 17+1), designed to deepen trade relations.

The strengthening of diplomatic relations between Hungary and China – in the context of frustration with the EU – was reflected by profound changes in the way leading Hungarian politicians situated their society within Eurasia. Particularly after the 2010 elections, the government de-emphasized the European heritage of Hungarian society, and celebrated instead the Asian provenance of Hungarian tribes that settled the Carpathian or Pannonian Basin in the end of the 9th century CE. The intent of this discursive shift was to stress the 'natural' cultural affinities between Hungarians and 'the East' in broad sense, as a potential resource to foster cooperation with Asian countries including China (Balogh, 2020). It also supposedly distinguished Hungarians from their European counterparts and explained the latter's excessive bureaucratic proclivity in contrast to the pragmatism and superior work-ethic purportedly exhibited by Chinese. Indeed, Orbán asserted that there was a direct link between China's rapid economic growth and China's 'work-centric' social organization (Orbán, 2015a). This assertion was related to scepticism among Hungarian leadership surrounding the efficacy of liberal-democratic political systems, and it fuelled ongoing debates between Hungarian officials and several high-ranking politicians in EU institutions and member state governments. Thus, according to some right-wing Hungarian politicians, the erosion of liberal democracy represented the restoration of supposedly 'natural' characteristics whose roots could be traced to the Asian origins of Hungarian society. This narrative portrayed Brussels and the whole 'west' as arrogant and domineering, in contrast to the equal partnership seeking Chinese approach towards East Central Europe (Orbán, 2015b). These politicians hoped that partnership with Chinese counterparts would boost Hungary's economy and augment its infrastructural ambitions.

Excitement around the 'New Silk Road' bolstered notions of what China could do for Hungary, framing it as a historically rare opportunity for the country (Matolcsy, 2015). Proponents argued that it had the potential to dramatically increase trade with China, gain a measure of autonomy from Brussels, and strategically position Hungary as a central node within state-of-the-art Eurasian logistics networks. These goals were interrelated and once in power Orbán wasted little time in pursuing them in tandem. During bilateral negotiations with high-ranking Chinese officials in 2010 and 2011 Hungarian officials proposed that Hainan Airlines acquire a 49 per cent stake in Budapest's Liszt Ferenc International Airport and rescue the deficit-ridden Hungarian national airlines MALÉV. This was to be followed by the construction of a railway between the airport and downtown Budapest to be built by the China Railway Construction Corporation (CRCC) and financed with a EUR1 billion loan from the China Development Bank (Tevan, 2011). The implementation of this proposal stalled from the outset, highlighting the influence of European firms in Hungary. The Liszt Ferenc International Airport was partially owned by German-Spanish Hochtief Group, which rejected Hainan Airlines' offer and ultimately sold its stake to Canadian pension fund PSP Investments in 2013 (Varga, 2013). Perhaps due to the collapse of this segment of the proposal, Hainan Airlines did not invest in MALÉV, which ultimately went bankrupt (Associated Press, 2012). Hainan Airlines also ceased operating flights between Budapest and Beijing, and only returned to Liszt Ferenc International Airport in 2020 (MTI, 2020). Meanwhile, plans to connect the airport to the city by rail have been put on hold. Significant alterations to this project have been announced numerous times since 2011, and while construction is scheduled to begin in 2021 it is unclear if it currently involves any Chinese stakeholders.

While plans to upgrade the Liszt Ferenc International Airport have floundered, Budapest has had some success advancing its transnational spatial objectives. The Budapest–Belgrade railway project was the cornerstone of its vision of enhancing Hungary's centrality in Eurasian logistics networks. The plan called for upgrading the existing 350-km single-track electrified line between the two capital cities to a double-track connection with an increased speed limit (up to 160 km/h). The logic behind the project was that it would ultimately enhance Budapest's logistical connection with Piraeus Port in Athens, whose majority stake was acquired by China COSCO Shipping Company in 2009. Proponents of the project reasoned that it would unlock distant markets for Hungarian exports and allow Hungary to become a logistics hub for Chinese goods being imported into the EU (Farkas et al, 2016; Salát, 2020). The Hungarian government selected a Hungarian-Chinese consortium to build the 166-km long Hungarian section in 2019, with 85 per cent of the costs estimated to be approximately EUR2 billion to be covered by a long-term loan from the Export–Import Bank of China (Domokos, 2020). Details surrounding the terms of the loan are unknown because the contract and the project's impact assessments have been sealed until 2030 (Hopkins, 2020).

The lack of transparency surrounding the rail project provoked a backlash in Hungary. Further criticism was provoked in 2017 when the Hungarian representative of COSCO stated that the economic impacts of the railway upgrade would be marginal because a significant bottleneck would remain between Greece and Serbia. As a result, this route would be unable to compete with existing shipping routes that link Central Eastern Europe and China via ports in Western Europe such as Rotterdam and Hamburg (Szalai, 2017). Critics pointed out that the Hungarian section also bypasses major towns in the southeast, reducing the possibility of spillover effects and increased domestic passenger traffic. Therefore, allege critics, the project's primary beneficiaries are the Chinese stakeholders because they not only turn a profit, but they gain experience meeting EU standards and operating within its complex regulatory framework (Salát, 2020). Meanwhile, despite the secrecy surrounding the project, it has become clear that a small number of private construction firms closely related to the Hungarian government benefit from membership in the consortium (Domokos, 2020). The public backlash provoked by this project has led to increased scrutiny of other aspects of Sino-Hungary relations. Most notably, demonstrations were launched in 2021 despite COVID-19 to protest plans to establish an overseas campus of Shanghai-based Fudan University in Budapest. The project is to be financed by a loan from the China State Construction Engineering Corporation, and leaked documents appeared to show that the cost of construction was artificially inflated (Panyi, 2021). The uproar over Fudan University's proposed Budapest campus has become a symbol of Hungary's relationship with China, on which the upcoming 2022 parliamentary elections serves as a de facto referendum, inasmuch the opposition parties made clear, even in an official letter sent to President Xi in Beijing, that they will cancel the Fudan campus and the Budapest–Belgrade railway projects in the event they win the elections (HVG, 2021).

Regardless of the outcome Hungary's government will likely continue to embrace ambitious spatial objectives. Indeed, the national elites' obsession with such large-scale infrastructure projects is neither a novelty nor a direct outcome of Chinese economic influence in East Central Europe. Significant EU funding has been channelled into large-scale infrastructure projects in every country of the region regardless of the political persuasion of the party in power since the mid-2000s for several reasons. First, infrastructure is highly visible and can be hailed as an achievement during election campaigns. Second, influential interest groups such as Western European lead firms have argued that the improvement of transport infrastructure in the region was a prerequisite to further expansion of production networks. Third, infrastructure was notoriously neglected in communist regimes because it was classified as 'non-productive' (that is, not directly incorporated in production processes in a narrow sense), so there was no shortage of meaningful projects that would increase productivity. Finally, EU funding is complex and tied to fixed seven-year budgetary periods, and infrastructure projects allow for large amounts of money to be spent rather quickly in ways that are in accordance with regulatory frameworks.

Infrastructure projects with Chinese funding exhibit some fundamental differences from their EU-financed counterparts. Importantly, EU infrastructure finance functions as a subsidy in the region, because in addition to large-scale projects undertaken by large construction firms, they also cover small projects and tend to encourage the inclusion of small and medium-sized local firms. Thus, even if EU-funded infrastructure projects in the region tend to involve large German and French firms, the dominance of a single firm is uncommon and there is scope for local firms to play a role. This is uncommon in Chinese-funded projects, which are usually dominated by a single firm and lack mechanisms to include small- and medium-sized domestic enterprises. More importantly, a portion of EU-funded projects is often delivered transparently through relief grants, whereas Chinese funding is mainly provided by long-term loans whose terms are shrouded in secrecy.

In summary, Hungary can mobilize finance for infrastructure projects from Brussels or Beijing, although it has been far more successful attracting EU funding. Importantly, EU finance does not support projects that further Budapest's spatial orientation towards China, while Hungary risks running afoul of EU regulations if it draws heavily on Chinese sources of finance. Indeed, the EU Parliament has questioned the nature of the bidding process for the Budapest–Belgrade railway and the Commission is 'currently examining various issues linked to this tender procedure' (Breton, 2021). Quite simply, EU regulations inhibit Hungary from pursuing its Sino-centric spatial strategy, while EU-funded infrastructure projects are largely geared towards the deepening of Hungary's spatial integration with EU member states in general and 'core' member states in particular.

## Conclusion

This chapter traced the emergence of Hungary's Sino-centric foreign economic policy and showed how its position between Brussels and Beijing manifests spatially. Most significantly, it showed that Hungary's actual economic benefit from cooperation with Chinese stakeholders fell short of the aspirations that emerged around 2010 during the financial crisis, while Brussels will not finance infrastructure projects that further Budapest's Sino-centric spatial objectives. This has frustrated the Orbán administration because its enthusiasm for China comes at an increasing political cost while the anticipated economic benefits have not materialized.

Hungary has continued to maintain a negative balance of trade with China despite deepening bilateral relations after the 2010 parliamentary elections. In 2008 Hungary's balance of trade with China was −69 per cent, and this improved to −33 per cent in 2017 before plummeting to −62 per cent in

**Table 16.2:** Balance of merchandise trade with various partner countries in the Visegrád Group countries, 2017–19 (annual average, %)

| Country | All partners | China | US | Germany |
|---------|-------------|-------|-----|---------|
| Czechia | +5.1 | −78.6 | +1.5 | +14.8 |
| Hungary | +3.1 | −47.0 | +19.8 | +5.9 |
| Poland | +0.2 | −84.2 | −2.9 | +11.2 |
| Slovakia | +0.2 | −56.2 | +48.1 | +11.9 |

Source: Author's design based on UNCTAD data (https://unctadstat.unctad.org/EN/).

2019.[1] In the same period Hungary's trade with the US increased and it also enjoys a trade surplus with Germany. This is in line with V4 countries in general (see Table 16.2), and this has undermined China's diplomacy in the whole post-communist region. Most notably, the much celebrated 17+ 1 initiative was dealt a blow by the withdrawal of Lithuania. According to Vilnius this was a pragmatic decision related to the fact that access to the Chinese market remained restricted (Lo, 2021), but it was met with a warning from China's *Global Times* (2021): 'When such a small country is aggressive, proactively placing itself to become a tool of great power competition, it will invite trouble … Lithuania will reap what is hast sown.'

The diplomatic spat between China and Lithuania highlights how attempts by post-communist states to diversify their dependencies has drawn them into high-stakes risk-leaden geopolitics. The shift towards China that began around 2010 and was most notable in Hungary, has become embedded within an emerging global rivalry between the US and its partners – in this case the EU – and China. The threatening language against Lithuania highlights the risks inherent in attempts to balance relations with great powers, while a slowdown of China's outward investment after 2016 and EU initiatives to scrutinize FDI in order 'to prevent a foreign investor from acquiring or taking control over a company if such acquisition or control would result in a threat to their security or public order', including 'critical infrastructure' (EUR-Lex, 2020), have led countries in East Central Europe to politely distance themselves from China (and Chinese firms). Finally, this shift has been encouraged by US policy under the leadership of Joe Biden. While the Trump administration sought support from EU countries in its 'trade war' with China – particularly in high-tech sectors – it also sought to actively undermine EU unity. The Biden administration advanced a vision of a deeply integrated EU, as it has rallied traditional allies in its competition with China. In the case of Hungary, Washington's support for Brussels has complicated Orbán's brand of Euro-scepticism, and as memories of the 2008 financial crisis and Greek debt crisis fade, the calculations that animate foreign policy in Budapest may change once more. This would make sense from an economic standpoint – in seven out of the first fifteen years after EU accession in 2004, the surplus of national operative budgetary balance with the EU was higher than the annual growth rate of national GDI.[2] Nevertheless, at present Budapest continues to keep Brussels at arm's length. For example, the EU recently allocated EUR750 billion from the Next Generation EU recovery fund[3] launched by the European Council in July 2020 to support member states hit by the COVID-19 pandemic (Picek, 2020), and Hungary revised its application to access these funds from EUR16.5 billion to EUR7.1 billion after the EU Commission criticized its recovery plan for being non-transparent (Magyari, 2021).

Hungary currently has some of the highest levels of support for EU membership within the bloc, with 88.9 per cent regarding it as useful in 2019 in contrast to the 78.8 per cent EU average (Kolosi and Hudácskó, 2020). Even 77 per cent of government party voters also shared this view in 2020 (Ács 2020). In addition, the multi-party opposition to Orbán's government launched its campaign for the upcoming 2022 parliamentary elections with a pro-EU and pro-West message. Criticism of the government's collaboration with China is a key component of its campaign. When Orbán campaigned in the mid-2010s he popularized the slogan *Üzenjünk Brüsszelnek!* (Let's send a message to Brussels!) and opposition parties have repurposed it for this election: *Üzenjünk Pekingnek!* (Let's send a message to Beijing!) (Magyar Hang, 2021). At the time of writing the opposition enjoys a small lead in the polls, and regardless of the outcome, Hungarian politics has become inextricably bound up in the US–China rivalry.

## Notes

1. Author's calculations based on UNCTAD data. Available from: https://unctadstat.unctad. org/wds/ReportFolders/reportFolders.aspx?sCS_ChosenLang=en.
2. Author's calculations based on European Commission (2018) and World Bank data. Available from: https://data.worldbank.org/.
3. EUR390 billion were relief grants and EUR360 billion were loans to be repaid by 2058.

## References

Ács, G. (2020) 'Medián: rekordon a magyar EU-tagság támogatottsága', Euronews, 9 December. Available from: https://hu.euronews.com/amp/ 2020/12/09/median-rekordon-a-magyar-eu-tagsag-tamogatottsaga.

Associated Press (2012) 'Hungary's Malev Airline ceases operations', CNBC, 3 February. Available from: www.cnbc.com/2012/02/03/hungarys-malev-airline-ceases-operations.html.

Balogh, P. (2020) 'Clashing geopolitical self-images? The strange co-existence of Christian bulwark and Eurasianism (Turanism) in Hungary', *Eurasian Geography and Economics*, 1– 27. DOI: 10.1080/15387216.2020.1779772

Brenner, N. (2004) 'Urban governance and the production of new state spaces in western Europe, 1960–2000', *Review of International Political Economy*, 11(3): 447–88.

Breton, T. (2021) 'Answer given by Mr. Breton on behalf of the European Commission', European Parliament. Available from: www.europarl.europa. eu/doceo/document/E-9-2020-006114-ASW_EN.html.

Černiauskas, Š. and Raudseps, P. (2015) 'Poorer than Greece: the EU countries that reject a new Athens bailout', *The Guardian*, 9 July. Available from: www.theguardian.com/world/2015/jul/09/poorer-than-greece-the-eu-countries-that-reject-a-new-athens-bailout.

Domokos, E. (2020) 'Budapest-Belgrád-vasút: tíz évig nem tudható, hogyan költenek el 700 milliárdot', *Napi.hu*, 19 May. Available from: www.napi. hu/magyar-vallalatok/budapest-belgrad-vasut-titkositas-700-milliard-forint.706512.html.

Draghi, M. (2012) 'Speech by Mario Draghi, President of the European Central Bank at the Global Investment Conference in London 26 July 2012', European Central Bank. Available from: www.ecb.europa.eu/press/key/date/2012/html/sp120726.en.html.

EUR-Lex (2020) Communication from the Commission Guidance to the Member States concerning foreign direct investment and free movement of capital from third countries, and the protection of Europe's strategic assets, ahead of the application of Regulation (EU) 2019/452 (FDI Screening Regulation) 2020/C 99 I/01, 26 March. Available from: https://eur-lex. europa.eu/legal-content/EN/ALL/?uri=CELEX:52020XC0326(03).

European Commission (2018) *EU Budget 2018: Financial Report*, Luxembourg: Publications Office of the European Union.

Farkas, Z.A., Pap, N., and Reményi, P. (2016) 'Hungary's place on Eurasian rail land bridges and the eastern opening', *Hungarian Geographical Bulletin*, 65(1): 3–14.

Fleischer, T. (2016) 'The EU transport policy and the enlargement process', in A.O. Evin, E. Hatipoglu, and P. Balázs (eds) *Turkey and the EU: Energy, Transport and Competition Policies*, Deventer: Claeys and Casteels Law Publishers, pp 121–38.

*Global Times* (2021) 'Lithuania risks trouble with geopolitical move: Global Times editorial', 23 May. Available from: www.globaltimes.cn/page/202105/1224253.shtml.

Heide, D., Hoppe, T., Scheuer, S., and Stratmann, K. (2018) 'China first: EU ambassadors band together against Silk Road', *Handelsblatt*, 17 April. Available from: www.handelsblatt.com/english/politics/china-first-eu-ambassadors-band-together-against-silk-road/23581860.html.

Hopkins, V. (2020) 'Hungary to keep details of Beijing-funded rail link secret', *Financial Times*, 2 April. Available from: www.ft.com/content/251314b5-8d6a-4665-a14b-0110dd88754c.

HVG (2021) 'Hszi Csin-pingnek írtak levelet az ellenzéki miniszterelnök-jelöltek', 22 June. Available from: https://hvg.hu/itthon/20210622_karacsony_gergely_miniszterelnok_jelolt_kinai_nepkoztarsasag_level_fudan_egyetem_budapest_belgrad_vasutvonal.

Index (2014) '3,95-4,95 százalékos kamatú orosz hitelt veszünk fel Paksra', 5 February. Available from: https://index.hu/gazdasag/2014/02/05/megallapodtunk_paks_penzugyi_reszleteirol/.

Koikkalainen, S. (2011) 'Free movement in Europe: past and present', *Migration Information Source*, 21 April. Available from: www.migrationpolicy.org/article/free-movement-europe-past-and-present.

Kolosi, T. and Hudácskó, S. (2020) 'Az Európai Unióval és az euróval kapcsolatos vélemények nemzetközi összehasonlításban', in T. Kolosi, I. Szelényi, and I.G. Tóth (eds) *Társadalmi Riport 2020*, Budapest: TÁRKI, pp 453–61.

Lau, S. (2021) 'China throws EU trade deal to the wolf warriors', *Politico*, 22 March. Available from: www.politico.eu/article/china-throws-eu-trade-deal-to-the-wolf-warriors-sanctions-investment-pact/.

Lo, K. (2021) 'Lithuania quit 17+1 because access to Chinese market did not improve, its envoy says', *South China Morning Post*, 1 June. Available from: www.scmp.com/news/china/diplomacy/article/3135522/lithuania-quit-171-because-access-chinese-market-did-not.

Magyar Hang (2021) 'A főváros döntött, továbbra sem támogatja a Fudan megépítését', 17 May. Available from: https://hang.hu/belfold/2021/05/17/a-fovaros-dontott-tovabbra-sem-tamogatja-a-fudan-megepiteset/.

Magyari, P. (2021) 'Az egészségügyön vágott a legkisebbet, az egyetemek pénzén a legnagyobbat a kormány', 444.hu, 12 May. Available from: https://444.hu/2021/05/12/az-egeszsegugyon-vagott-a-legkisebbet-az-egyetemek-penzen-a-legnagyobbat-a-kormany.

Matolcsy, G. (2015) 'Magyarország a Selyemúton', Lecture at the 54th Roving Conference of Economists in Hungary, Kecskemét, 15 September. Available from: www.mkt.hu/wp-content/uploads/2016/09/Matolcsy_Gyorgy.pdf.

MTI (2020) 'Hivatalosan is köszöntötték a Hainan Airlines Budapest-Csungking járatát', Magyar Nemzet, 24 January. Available from: https://magyarnemzet.hu/gazdasag/hivatalosan-is-koszontottek-a-hainan-airlines-budapest-csungking-jaratata-7705334/.

Orbán, V. (2015a) 'Prime Minister Viktor Orbán's speech at the second world meeting of the Friends of Hungary Foundation', 9 May. Available from: https://akadalymentes.2015-2019.kormany.hu/en/the-prime-minister/the-prime-minister-s-speeches/prime-minister-viktor-orban-s-speech-at-the-second-world-meeting-of-the-friends-of-hungary-foundation.

Orbán, V. (2015b) 'A globalizáció új modellje', *Magyar Krónika*, 1 June. Available from: https://miniszterelnok.hu/magyarorszag-idealis-oszlopa-az-egy-ovezet-egy-ut-kezdemenyezesnek/.

Orient Press (2014) 'Átadták a magyar-szlovák földgázvezetéket', 28 March. www.orientpress.hu/cikk/atadtak-a-magyar-szlovak-foldgazvezeteket.

Panyi, S. (2021) 'The fight over Fudan: a Chinese university in Budapest sparks reckoning for Sino-Hungarian relations', Chinaobservers.eu, 7 June. Available from: https://chinaobservers.eu/the-fight-over-fudan-a-chinese-university-in-budapest-sparks-reckoning-for-sino-hungarian-relations/.

Perényi, Z. and Remete, B. (2018) 'V4 – Visegrád az EU-n belül', in A. Balaskó (ed) *A visegrádi négyek jelentősége, struktúrája és értékei*, Budapest: Külügyi és Külgazdasági Intézet, pp 34–51.

Picek, O. (2020) 'Spillover effects from Next Generation EU', *Intereconomics*, 55: 325–31.

Salát, G. (2020) 'An authoritarian advance or creating room for manoeuvre? The case of Hungary's China policy', *Stosunki Międzynarodowe – International Relations*, 56(2): 125–43.

Szalai, B. (2017) 'A kínaiaknak pont tökmindegy a Budapest–Belgrád-vasút', *Index*, 8 December. Available from: https://index.hu/gazdasag/2017/12/08/budapest_-_belgrad_cosco_kina_vasut/.

Tevan, I. (2011) 'Koncesszióban épülhet a ferihegyi gyorsvasút', *iho*, 11 December. Available from: https://iho.hu/hirek/koncesszioban-epulhet-a-ferihegyi-gyorsvasut-111217.

Vámos, P. (2018) 'A Hungarian model for China? Sino-Hungarian relations in the era of economic reforms, 1979–89', *Cold War History*, 18(3): 361–78.

Varga, G.G. (2013) 'Kanadai kézbe kerül Ferihegy', *Népszabadság*, 7 May. http://nol.hu/gazdasag/kanadai_kezbe_kerul_ferihegy-1384771.

# 17

# 'No One Stole Anyone Else's Cheese': The Politics of Infrastructural Competition in Kazakhstan

*Jessica Neafie*

'The cooperation of one or another country with Kazakhstan is not a zero-sum game. No one stole anyone else's cheese.'

Le Yucheng, Chinese Ambassador to
Kazakhstan (Farchy, 2014)

The Chinese Ambassador to Kazakhstan's quixotic statement about stolen cheese reflects a recognition of competition with the US for influence in Central Asia. It suggests that there is enough of Kazakhstan to go around, such that the government in Nur-Sultan (formerly Astana) can cultivate friendly relations with both the US and China. Nevertheless, Kazakhstan has taken centre stage in the US–China competition in multiple infrastructure sectors and this chapter interprets its delicate balancing act as a state spatial strategy designed to advance the infrastructural components of its national development vision. Inaugurated in 2014, *Kazakhstan 2050* emphasizes (trans-)national infrastructure development, and a host of state spatial projects have been launched under the banner of *Nurly Zhol*, which means 'bright path'. Its overarching spatial objectives are centred around the idea that Kazakhstan can become the *Eurasian Land Bridge*, a central hub for goods transportation and economic growth for the region. The manifestation of this territorial designation depends on the construction of a series of large-scale infrastructure projects focused on the transportation sector, which are linked to a series of economic diversification projects.

Since independence in 1991, Kazakhstan's role in Central Asia and the world has evolved as it became a regional pivot, and central to the interests of the US and China. The Kazakhstani government has worked closely with the US government since the 1990s on security issues, and it was the main staging point for the US to finance Afghanistan's security forces and transportation infrastructure development throughout the 2000s (US White House, 2018). While China also has security interests in Kazakhstan, its primary interest has been economic, as it views Kazakhstan as a source of natural resources and a transit route. American and Chinese foreign policies towards Kazakhstan were compatible, and sometimes even complementary, until recently when their relationship has become competitive, particularly in the field of infrastructure construction.

Xi Jinping announced the Belt and Road Initiative (BRI) during his 2013 visit to Kazakhstan, demonstrating the importance that China affords Kazakhstan and its place within the BRI. Meanwhile, the US increasingly sought to expand economic ties with Kazakhstan, thus centring Kazakhstan in geopolitical-economic competition, which offers Nur-Sultan the opportunity to leverage the country's strategic location as a pivot at the crossroads of Eurasia. By hedging between the US and China,[1] Kazakhstan has secured support for its spatial objectives,[2] which are centred around the idea that Kazakhstan becomes a central hub for goods transportation and an engine of economic growth for the region. These state spatial objectives and strategies animate *Kazakhstan 2050* and *Nurly Zhol*. Kazakhstan's centrality to the US–China rivalry affords it agency, but in some instances, this has undermined the realization of its spatial objectives.

## Kazakhstan as (geo)strategic pivot

In January 2020, during a five-nation tour that include Kazakhstan, US Secretary of State Mike Pompeo labelled China the 'central threat of our times'. He criticized China for the repression of ethnic minorities in Xinjiang Province and for undertaking development projects that put small, poor countries in debt distress (Associated Press, 2020; Tleuberdi, 2020). This attempt to drive a wedge between Nur-Sultan and Beijing represented a departure in US policy, which had focused overwhelmingly on security issues since Kazakhstan's independence in 1991. After the break-up of the Soviet Union the US focused on nuclear non-proliferation and securing the Soviet nuclear arsenal. These objectives gave way to anti-terrorism initiatives and the US sought support from Kazakhstan for its invasion of Afghanistan (US Department of State, 2021b). Thus, in the first two decades after the Cold War the US–Kazakhstan relationship was based on an informal defence partnership. The US provided assistance to train Kazakhstani troops as peacekeepers and to eliminate weapons of mass destruction under the

Cooperative Threat Reduction Program (Sanchez, 2020; US Department of State, 2021b).

In contrast to Washington's focus on security issues, China prioritized deepening economic ties with Kazakhstan. Trade between grew from US$368 million in 1992 to US$3.3 billion in 2003 (Yau, 2020). China tended to export finished consumer goods such as shoes, appliances, and toys, while 85 per cent of Kazakhstan's exports to China have been resources or raw materials (Chazan, 2020). This relationship deepened after the 2008 financial crisis when demand from other markets for Kazakhstan's exports declined significantly, and it was further strengthened after the announcement of the BRI in 2013. Since then, China has financed and built a series of infrastructure projects in Kazakhstan worth US$27.6 billion, many of which are joint initiatives with Kazakhstani firms (Simonov, 2019). BRI projects have also been fast-tracked, highlighting their geographic importance. The largest dry port in Central Asia – Khorgos – is situated on the Kazakh–China border, and it is here that Chinese trains transfer goods to wider gauged trains bound for Europe (Furlong and Kupka, 2018). From 2016 to 2017, the number of containers that passed through Khorgos doubled from approximately 25,000 to 50,000.

Despite close economic ties with Kazakhstan, China was careful to signal that it recognizes Russia's historical influence in Central Asia. Thus, Sino-Kazakh security initiatives have been limited to joint anti-terrorism exercises (16 since 2002) and military aid (Sukhankin, 2020). For much of the post-Cold War era China's emphasis on building strong economic ties with Kazakhstan complemented Washington's focus on security issues. This has changed recently, however, as the US has taken an interest in competing with China for influence in Central Asia through infrastructure. In September of 2020, the US Chamber of Commerce launched the US–Kazakhstan Business Council to advance economic and commercial cooperation between the US and Kazakhstan (Embassy of the Republic of Kazakhstan, 2020). The US State Department went further in 2021 when it established the Central Asia Investment Partnership to encourage investment in Kazakhstan and Uzbekistan (US Department of State, 2021a). Finally, the US directly signalled intentions to compete with China in Kazakhstan's infrastructure sector when the US International Development Finance Corporation committed US$500 million in infrastructure investment. The US policy shifts signal its intention to compete with Beijing in the field of geoeconomics, and its infrastructure funding comes at a time when China appears to be scaling back the BRI (Bloomberg, 2021).

As such, Kazakhstan represents a 'pivot' in the geopolitical-economic competition between the US and China. Halford Mackinder (1904) introduced the idea of the 'pivot area', which Brzezinski (1998, 22) defines as a state whose importance is derived from its 'sensitive location' and vulnerability to great powers. As such, great powers exert influence over pivots, yet small states may find themselves with multiple benefactors competing for attention and allegiance. In some instances this enhances the agency of smaller states, providing opportunities to pursue national interests and objectives (Brown, 1992; Cooley, 2012). Thus, a pivot state can pursue a multi-vector approach to foreign policy by creating 'overlapping spheres of influence ... that are competitive but positive sum' (Goh, 2008, 129). In the context of current US–China competition, Kazakhstan is a pivot, currently using careful diplomacy to maintain room to manoeuvre. In the case of Pompeo's visit, for example, Kazakhstani officials politely declined to agree with his assessment of Chinese foreign and domestic policy, yet the economic relationship between the US and Kazakhstan was deepened. The next section shows how Kazakhstan seeks to leverage its strategic position in pursuit of developmental and spatial objectives.

## Hedging and state restructuring between great powers

Nine months before Xi Jinping announced the BRI, Kazakhstani President Nursultan Nazarbayev proclaimed that the *Kazakhstan 2030* development objectives had been met. As such, the country needed to respond to new global challenges, and to that end, he announced *Kazakhstan 2050*. It is meant to serve as a comprehensive development strategy and re-work Kazakhstan's place in the world as it advances as an emerging economy. Nazarbayev explained that infrastructure construction will be a means of meeting these challenges through two 'new approaches':

> Firstly, we should integrate the national economy into the global environment, and secondly move towards regions within the country. It is important to focus attention on exit routes from the country and create transport and logistics facilities outside Kazakhstan. We must think outside the box and create joint ventures in the region and throughout the world – Europe, Asia, America – building ports in countries with direct access to the sea and developing transport and logistics hubs at nodal transit points. In that regard we need to develop a special program, 'Global Infrastructural Integration'. (Nazarbayev, 2012)

The preferred vehicle for achieving infrastructural ambitions in the medium term is *Nurly Zhol*, or 'bright path'. Launched in 2015, it was branded a 'new economic plan' that included infrastructure investments of US$14 billion to, as Nazarbayev (2015) asserts, connect all regions by railroads, highways, and air services. He describes 'the nine roads connecting with each other in Astana [as] the roots of life, spreading our capital's creative spirit. Improving interconnectivity between the regions will eventually lead to a greater domestic wellbeing. It will strengthen trade and economic ties between the regions. New markets will emerge from within the country'. While Kazakhstan embraced ambitious state spatial objectives that are consistent with the BRI, only 15 per cent of the total price tag will be publicly funded so significant external investment is required.

Kazakhstan leverages its status as a pivot to secure the capital required to achieve its spatial objectives. Bitaborova (2018, 151) explains that 'Central Asian states are not passive bystanders but *proactive agents* working to affect the course of the "game".' Indeed, Kazakhstan hedges by not relying on investment from any single country (Cinotto, 2020). This point was underscored by Kazakhstan's President Kassym-Jomart Tokayev at the 2020 Munich Security Conference, who stated that

> Kazakhstan does not share concerns about the excessive influence of the Chinese economy on the development of our country. On the contrary, we believe that close cooperation with China within the framework of the BRI provides many advantages and, ultimately, will increase the geopolitical significance of Central Asia as a whole. (Nurgaliyev, 2020)

One reason for Nur-Sultan's confidence in its ability to balance relations with China, the US, the EU, and Russia, is that it has been doing so for years. In the early 2000s, for example, Kazakhstan forced the Chinese National Petroleum Company (CNPC), a Chinese state-owned enterprise, to initially sell back 33 per cent of shares from its acquisition of PetroKazakhstan – a Canadian company – to the Kazakhstani national oil company KazMunaiGas (Cinotto, 2020). Kazakhstan was able to secure shared ownership of a refinery in Shymkent by implicitly threatening to sell shares to Russian entities. Ultimately this is just one occasion in which Kazakhstan achieved infrastructural objectives by hedging between more powerful neighbours.

The BRI presents an opportunity to obtain support from China for *Nurly Zhol*, yet given its size and scope, it is worth asking whether it poses a threat to Kazakhstan's multi-vector foreign policy. Indeed, when decreasing oil prices in 2014 jeopardized *Nurly Zhol* (Bitabarova, 2018), the BRI lent it new life. The two countries formally agreed to integrate it with the BRI in 2015 during Nazarbayev's visit to China. Transportation infrastructure construction has continued apace, and of particular importance is the Sino-Kazakh railway that links Xinjiang to Europe. From 2017–18 alone the amount of freight shipped on this route from China to Europe increased by 44 per cent (Standish, 2019b). The Khorgos Cooperation Center is a key node in the emergent Eurasian Land Bridge. Established as a strategic hub for cargo exports, it includes a Chinese subsidized special economic zone where Kazakhs and Central Asians can shop visa-free for duty-free cut-price products (Chazan, 2020). China's COSCO owns a 49 per cent stake in the dry port, while the Kazakh state railways corporation maintains the controlling stakes (Standish, 2019a). The port is operated by Dubai-based DP World (Ruehl, 2019). Thus, while the project is implemented in partnership with China, Kazakhstan has retained the majority share, and it is operated by an Emirati logistics firm.

China accepts Kazakhstan's multi-vector foreign policy, perhaps because of its strategic location. At times this can undermine the spatial objectives outlined in *Nurly Zhol*. For example, the Nur-Sultan light rail project, formerly Astana LRT, was a Chinese-financed scheme for the capital city, originally launched in 2011 and meant to be completed by the 2017 Expo Exhibition. A 22-km elevated railway line, it was meant to connect the airport to a new train station and pass through the city centre. Very little work was done from 2011 to 2015, but after the 2015 agreement to integrate *Nurly Zhol* and the BRI, China Development Bank (CDB) provided a loan of US$1.6 billion dollars or about 80 per cent of the cost (Koskina, 2019). The capital was transferred to a quasi-public Kazakh company, Astana LRT LLP, but by 2017 scant progress had been made (Koskina, 2019). Chinese contractors hired to work on the project were reportedly angered by mismanagement of the funds and the constantly changing terms in their contracts (Koskina, 2019). In 2018 the local bank that held the funds collapsed, and US$258 million of the US$313 million disbursed for the project had been re-allocated elsewhere (Gizitdinov, 2019). President Tokayev ordered an investigation by the anti-corruption agency and more than a dozen people have been fined or given prison sentences (Koskina, 2019; Chazan, 2020). The failure of the light rail project illustrates the limits of China's influence in Kazakhstan. In most cases CDB transfers funds directly to Chinese SOEs responsible for project construction and oversight. However, the Kazakhstani government was able to negotiate terms that required the transfer of the funds to a domestic bank, and CDB was unable oversee funds or prevent mismanagement. Not only did the funds disappear with minimal work completed, but the Chinese firm tasked with building the light rail went bankrupt in 2019.

Kazakhstan has also worked to incorporate the second part of the *Nurly Zhol* into its foreign relations: economic diversification. Historically Kazakhstan relied on extractive industries and raw materials exports (Kursiv, 2017; Office of the US Trade Representative, 2020), which rendered the country vulnerable to fluctuating energy and raw material prices. Policy makers fear the country could fall into a 'middle-income trap' if the economy is not diversified, so they seek to cultivate sectors with long-term growth potential (ADB, 2012). To this end, *Kazakhstan 2050* claims to embrace 'universal economic pragmatism', which amounts to a rather standard set of policies aimed at attracting foreign investment, upgrading key sectors, and identifying new markets. Importantly, rather than hoping market forces will spur these changes, these goals are meant to be achieved with state support (Nazarbayev, 2012). The state has sought to compel foreign investors interested in infrastructure and resource extraction to also invest in other sectors. For example, in BRI negotiations, Kazakhstan insisted that Chinese investment could not solely focus on the energy sector and should also engage in manufacturing and agriculture (Cinotto, 2020). This increased Kazakhstan's exports to China, of products such as wheat, meat, vegetable seeds, oil and honey (Kassenova, 2017). However, Chinese investors are prohibited from owning land, and they must partner with Kazakhstani businesses to invest in processing of agricultural products (Bizhanova, 2018; Xiaojing, 2018). To date these arrangements have increased the export of Kazakhstani agricultural products, though there has been less impact on the export of manufactured goods.

In contrast to China's emphasis on infrastructure investment and development, the US has focused on traditional economic development schemes and recently committed to strengthening economic and investment ties with Kazakhstan. The US Chamber of Commerce and the Kazakhstani government announced the new US–Kazakhstan Business Council in 2020. Its corporate chair is Chevron, the US multinational with the largest presence in Kazakhstan. However, there appears to be a consensus that the council will encourage US investment in new sectors of the Kazakhstani economy (so beyond oil and gas). Interestingly, the council notes that there is 'ample room for growth [of US investment] in sectors such as agriculture, energy, mining, and infrastructure'. Thus, the Kazakhstani state seeks to reach out directly to US firms in its bid to establish new relationships and diversify the economy, while the council serves as a forum for the US to compete with China in key sectors. Importantly, this new venture will also expand Kazakhstan's access to the US market by lowering tariffs (Rapoza, 2021).

In addition to strengthening bilateral ties, the US and Kazakhstan announced the Central Asia Investment Partnership along with Uzbekistan in 2021 (US Department of State, 2021a). This multilateral institution is designed to encourage American investment in Kazakhstan and Uzbekistan, and it has already committed over US$500 million in new investment projects in Kazakhstan alone. The Kazakhstani government has signed deals with American companies like Tyson Foods and a manufacturing subsidiary of Valmont Industries to set up new factories. While it has welcomed investment from the US, Kazakhstan has sought to maintain autonomy from Washington. The US encouraged a series of government reforms, and while the Kazakhstani government rhetorically supported the US anti-corruption measures, the government has avoided implementing US recommendations (Vanderhill et al, 2019). Indeed, Kazakhstan has even made it harder for US investors by increasing local content requirements and not complying with international arbitration rulings (Vanderhill et al, 2019). Despite refusing to reform and side-stepping commitments to democratize[3] the US continues to prioritize maintaining strong diplomatic and economic relations with Kazakhstan.

In sum, Kazakhstan's government acts in simultaneous cooperation and defiance of China and the US, maintaining a diplomatic and cooperative posture with both countries. In the case of China, Kazakhstan embraces the BRI but it demands control over certain aspects of projects in Kazakh territory that other countries are not afforded in their dealings with China. Similarly, its relationship with the US is aligned with its strategy to diversify the economy, while it has resisted embracing democratic reforms and refused to echo Pompeo's anti-China sentiments. While China and the US are surely not oblivious to these contradictions, they have both sought to maintain strong relations with Kazakhstan. This fiercely independent policy is possible due to Kazakhstan's status as a pivot, central to the regional foreign policy objectives of the US and China.

## Conclusion

Overall, Kazakhstan's multi-vector strategy has allowed the government to pursue domestic interests. Kazakh citizens are fearful of exploitation by stronger powers (Le Corre, 2019; Umarov, 2019), but evidence from the Khorgos dry port, Nur-Sultan light rail and economic trade negotiations suggest that Kazakhstani companies and government still maintain control of their capital and operations at the local/state level. Kazakhstan has balanced open investment policies with laws that favour domestic companies over foreign entities. Most importantly, it has leveraged its relationships with China and the US to establish Kazakhstan as the overland trade route from East Asia to Europe and, in this context, it is making increased efforts to diversify the domestic economy.

Kazakhstan's government is using its position as a pivot state in great power competition to emphasize its role as the Eurasian Land Bridge. This name is misleading, however, because it is not just a bridge connecting Europe and East Asia, but a central node in a network that connects the US, China, Europe, Turkey, Russia, South Asia, the Middle East, and Central Asia. Although balancing diplomatic imperatives among these diverse countries and regions is challenging, Kazakhstan has thus far done so in such a way that it has maintained its autonomy. Indeed, the way that Kazakhstan has exhibited agency in the pursuit of spatial objectives has important theoretical implications for the study of emerging middle powers. Its multi-vector strategy prevents its relationships with stronger powers from becoming clientelist, dependent, or exploitative as it hedges and balances the relationships against one another. This chapter has shown just some examples of how Kazakhstan's multi-vector approach to foreign policy allows it to secure support for infrastructure projects and realize state spatial objectives.

Kazakhstan's multi-vector approach draws competing states into local and regional organizations that create opportunities for promotion of, and investment in, local businesses and infrastructure development. Countries seeking a relationship with Kazakhstan are drawn into the *Nurly Zhol* initiative, and this has created opportunities for local businesses (Magistad, 2020; Rapoza, 2021). These growing economic relationships are also creating more opportunities for Kazakhstani consumers. Kazakhstan's multi-vector approach may foster economic growth and create jobs, while diversifying its economy. However, this requires infrastructure investment be spent more wisely than the past (such as in the case of the Nur-Sultan light rail).

While the government can pick and choose the policies and practices it wants to implement, the lack of oversight or pushback from either China or the US has meant that there is limited pressure to embrace best practices. Indeed, whether infrastructure projects are ultimately built depends largely on domestic politics. This chapter demonstrated that local control over spatial projects does not necessarily mean that they will be achieved. Both China and the US are fully aware that they are not the only players in this region, and as the key connection to East Asia, Europe, and Central Asia, Kazakhstan can hedge these relationships. At times individual politicians take advantage of this situation to pursue narrow self-interest. As a result, achievements under the *Nurly Zhol* initiative have been limited and if events surrounding the Nur-Sultan light rail project are repeated, the consequences may be far reaching. There are indeed other projects that may face a similar fate, and this could discourage foreign investment and undermine the goals surrounding economic diversification and spatial integration. Reforms to improve transparency and reduce corruption that the US has called for, as noted, have largely gone unheeded. Thus, Kazakhstan's status as a pivot is a double-edged sword. While it affords Kazakhstan the opportunity to maintain autonomy, it also allows it to resist reforms that could weaken the power of a small elite. This demonstrates how emerging middle powers may be able to hedge between great powers and pursue spatial objectives, but it also represents a warning to other countries. If Kazakhstan is to become a key component of the Eurasian Land Bridge, it must leverage its multi-vector approach to promote national interests that benefit economic and infrastructure development, but transparency and the rule of law are prerequisites for more broad-based and inclusive development.

## Notes

[1] On hedging between the US and China see Kuik (2020).

2   The notion of 'state spatial objectives' borrowed from Brenner (2004), who notes that they typically include 'state spatial projects' and their achievement requires 'state spatial strategies'.

3   For example, the failure to implement the 'Madrid Commitments' to further Kazakhstan's democratic reforms (OSCE, 2010).

## References

Asian Development Bank (2012) 'Kazakhstan: Country Partnership Strategy (2012–2016)'. Available from: www.adb.org/sites/default/files/linked-documents/cps-kaz-2012-2016-oth.pdf.

Associated Press (2020) 'Pompeo message in Europe, Central Asia trip: beware of China', *VOA News*, 1 February. Available from: www.voanews.com/usa/pompeo-message-europe-central-asia-trip-beware-china.

Bitabarova, A.G. (2018) 'Unpacking Sino-Central Asian engagement along the New Silk Road: a case study of Kazakhstan', *Journal of Contemporary East Asia Studies*, 7(2):149–73.

Bizhanova, M. (2018) 'Can the Silk Road revive agriculture? Kazakhstan's challenges in attaining economic diversifcation', in M. Laurelle (ed) *China's Belt and Road Initiative and its Impact in Central Asia*, Washington, DC: The George Washington University, Central Asia Program, 2018, pp 51–66.

Bloomberg (2021) 'How China's flagship Belt and Road project stalled out', 14 January. Available from: www.bloomberg.com/news/videos/2021-01-14/how-china-s-flagship-belt-and-road-project-stalled-out-video.

Brenner, N. (2004) *New State Spaces: Urban Governance and the Rescaling of Statehood*, Oxford: Oxford University Press.

Brown, C. (1992) *International Relations Theory: New Normative Approaches*, New York: Harvester-Wheatsheaf.

Brzezinski, Z. (1998) *The Grand Chessboard: American Primacy and Its Geostrategic Imperatives*, New York: Basic Books.

Chazan, Y. (2020) 'China BRI Ventures run into trouble in Kazakhstan', *Asia Sentinel*, 24 January. Available from: www.asiasentinel.com/p/china-bri-ventures-run-into-trouble.

Cinotto, F. (2020) 'State agency along the BRI: the case of Kazakhstan', *Cambridge Society for Social and Economic Development*, 9 December. Available from: https://camsed.uk/2020/12/09/state-agency-along-the-bri-the-case-of-kazakhstan/?fbclid=IwAR14ht7l3Q1wcjm4-WN2eD3-3TAo9G8ihMd7Jver7rKqJsIYRsL3kZchL2w/.

Cooley, A. (2012) *Great Games, Local Rules: The New Great Power Contest in Central Asia*, Oxford: Oxford University Press.

Embassy of the Republic of Kazakhstan (2020) 'US–Kazakhstan Business Council'. Available from: https://kazconsulny.org/us-relations/economic-cooperation/u-s-kazakhstan-business-council.

Farcy, J. (2014) 'Russia's neighbours: Primary colours', *Financial Times*, 9 June. Available from: ft.com/content/ad076a54-efbc-11e3-bee7-00144feabdc0.

Furlong, R. and Kupka, R. (2018) 'In a desert far away, China opens a gateway on its new Silk Road', *Current Time TV*, 18 January. Available from: www.rferl.org/a/kazakhstan-new-silk-road-china-exports-gamble/28970736.html.

Gizitdinov, N. (2019) 'China's $1.9 billion Belt-and-Road rail project goes off track', Bloomberg, 3 June. Available from: https://www.bloomberg.com/news/articles/2019-06-03/china-s-1-9-billion-silk-road-rail-project-goes-off-track.

Goh, E. (2008) 'Great powers and hierarchical order in Southeast Asia: analyzing regional security strategies', *International Security*, 32(3): 113–57.

Kassenova, N. (2017) 'China's Silk Road and Kazakhstan's Bright Path: linking dreams of prosperity', *Asia Policy*, 24: 110–16.

Koskina, A. (2019) 'Astana LRT: a project or a scam?', *CABAR*, 24 December. Available from: https://cabar.asia/en/astana-lrt-a-project-or-a-scam.

*Kursiv* (2017) 'Export of goods from Kazakhstan to China increased by 28% over the year', *Kursiv*. Available from: https://kursiv.kz/news/vlast-i-biznes/2017-10/eksport-tovarov-iz-kazakhstana-v-kitay-za-god-vyros-na-28.

Kuik, C-C. (2020) 'Hedging in post-pandemic Asia: what, how, and why?' *The Asan Forum*, Availabvle from: https://theasanforum.org/hedging-in-post-pandemic-asia-what-how-and-why/.

Le Corre, P. (2019) 'Kazakhs wary of Chinese embrace as BRI gathers steam', Carnegie Endowment for International Peace, 28 February. Available from: https://carnegieendowment.org/2019/02/28/kazakhs-wary-of-chinese-embrace-as-bri-gathers-steam-pub-78545.

Mackinder, H.J. (1904) 'The geographical pivot of history', *The Geographical Journal*, 23(4): 421–44.

Magistad, M.K. (2020) 'China's new Silk Road traverses Kazakhstan. But some Kazakhs are skeptical of Chinese influence', *The World*, 14 September. Available from: www.pri.org/stories/2020-09-14/chinas-new-silk-road-traverses-kazakhstan-some-kazakhs-are-skeptical-chinese.

Nazarbayev, N.A. (2012) 'Kazakhstan 2050 Strategy', Address by the Leader of the Nation. Available from: https://kazakhstan2050.com/2050-address#.

Nazarbayev, N.A. (2015) 'Nurly Zhol, bright path to the future'. Available from: https://afmrk.gov.kz/en/activity/strategy-and-program/state-program-on-infrastructure-development-nurly.html.

Nurgaliyev, B. (2020) 'China's Belt and Road Initiative: Kazakhstan and geopolitics', *The Astana Times*, 3 June. Available from: https://astanatimes.com/2020/06/chinas-belt-and-road-initiative-kazakhstan-and-geopolitics/.

Office of the US Trade Representative (USTR) (2020) 'US–Kazakhstan trade facts'. Available from: https://ustr.gov/countries-regions/south-cent ral-asia/kazakhstan#:~:text=Kazakhstan is currently our 81st,goods imports totaled %241.2 billion.

Organization for Security and Co-operation in Europe (OSCE) (2010) *Kazakhstan and so-called 'Madrid Commitments'*. Available from: www.osce. org/home/71600.

Rapoza, K. (2021) '"Strategic competition": US looks to Kazakhstan to expand ties', *Forbes*, 25 April. Available from: www.forbes.com/sites/ kenrapoza/2021/04/25/strategic-competition-us-looks-to-kazakhstan- to-expand-ties/?sh=39fa94e82c22.

Ruehl, H. (2019) 'The Khorgos hype on the Belt and Road', *The Diplomat*, 27 September. Available from: https://thediplomat.com/2019/09/the-khor gos-hype-on-the-belt-and-road/.

Sanchez, W.A. (2020) 'The future of US–Kazakhstan relations', *Georgetown Journal of International Affairs*, 28 May. Available from: https://gjia.georget own.edu/2020/05/28/the-future-of-us-kazakhstan-relations/.

Simonov, E. (2019) 'Half China's investment in Kazakhstan is in oil and gas', *China Dialogue*, 29 October. Available from: https://chinadialogue. net/en/energy/11613-half-china-s-investment-in-kazakhstan-is-in-oil- and-gas-2/.

Standish, R. (2019a) 'China's Central Asian plans are unnerving Moscow', *Foreign Policy*, 23 December. Available from: https://foreignpolicy.com/ 2019/12/23/china-russia-central-asia-competition/.

Standish, R. (2019b) 'China's path forward is getting bumpy', *The Atlantic*, October. Available from: www.theatlantic.com/international/archive/ 2019/10/china-belt-road-initiative-problems-kazakhstan/597853/.

Sukhankin, S. (2020) 'The security component of the BRI in Central Asia, Part 3: China's (para)military efforts to promote security in Kazakhstan, Uzbekistan and Turkmenistan', *China Brief*, 20(18). Available from: https:// jamestown.org/program/the-security-component-of-the-bri-in-central- asia-part-three-chinas-paramilitary-efforts-to-promote-security-in-kaz akhstan-uzbekistan-and-turkmenistan/.

Tleuberdi, M. (2020) 'Pompeo urges Kazakhstan to pressure China over Muslims in Xinjiang', Reuters, 1 February. Available from: www.reuters. com/article/uk-kazakhstan-usa-china-rights-idUKKBN1ZW066.

Umarov, T. (2019) 'What's behind protests against China in Kazakhstan?', Carnegie Moscow Center, 20 October. Available from: https://carnegie. ru/commentary/80229.

US Department of State (2021a) 'Joint statement on the announcement of the Central Asia Investment Partnership', 7 January. Available from: https:// kz.usembassy.gov/joint-statement-on-the-announcement-of-the-central- asia-investment-partnership/.

US Department of State (2021b) 'US relations with Kazakhstan'. Available from: https://www.state.gov/u-s-relations-with-kazakhstan/.

US White House (2018) 'United States and Kazakhstan: an enhanced strategic partnership for the 21st century', US White House Press Briefing, 16 January. Available from: https://kz.usembassy.gov/us-kz-21st-century-joint-statement/.

Vanderhill, R., Joireman, S.F., and Tulepbayeva, R. (2019) 'Do economic linkages through FDI lead to institutional change? Assessing outcomes in Kazakhstan, Azerbaijan and Kyrgyzstan', *Europe-Asia Studies*, 71(4): 1–23.

Xiaojing, X. (2018) 'Kazakhstan's capital Astana shakes economic funk due to infusion of Chinese investment', *Global Times*, 25 September.

Yau, N. (2020) 'Tracing the Chinese footprints in Kazakhstan's oil and gas industry', *The Diplomat*, 12 December. Available from: https://thediplomat.com/2020/12/tracing-the-chinese-footprints-in-kazakhstans-oil-and-gas-industry/.

# 18

# Outer Space Infrastructures

*Julie Michelle Klinger*

This chapter is about outer space infrastructures in Eurasia in the overlapping contexts of the Belt and Road Initiative (BRI), the final years of the US-led occupation of Afghanistan, and diverse approaches to development within and among Central Eurasian states. These contexts are situated within multiple histories: the footprint of Soviet-era space infrastructure, shifts in global extractive politics, and the more recent increase of public and private actors using outer space-based and space-linked technologies for a variety of purposes. The dynamics of the region defy reduction to a simplistic bipolar geopolitical framework suggested by discourses on a 'New Cold War' or 'Great Powers' clash that have proliferated in both Anglophone and Sinophone discourses (for example, Karagov, 2018; Westad, 2019; Zhao, 2019). As this chapter will show, there is, quite simply, too much going on.

Although critical social scientists have 'almost entirely demolished the myth that the Cold War was simply an East–West conflict dominated by the two superpowers' (Brazinsky, 2017, 2), addressing the heterogeneity of actors, interests, and institutions across the incredibly diverse region of Central Asia has proven challenging for policy makers and commentators speaking from circles in which 'Great Power' and 'New Cold War' discourses tend to emanate in Washington, DC, Beijing, and Moscow. The outer space element has even further eluded cogent commentary. If space activities are not analysed according to a new Cold War framework, they are absent from the discussion. The challenges seem to lie in placing the exceedingly delicate technologies orbiting the Earth at great speeds on the same analytical plane as concrete and steel of infrastructure projects unfolding on the ground. More fundamentally, it is important to resist the temptation to conflate the national origins of a project or investment stream with a singular set of national interests. Instead, an empirically grounded analysis reveals a plurality of actors, interests, and institutions involved. It also accounts for the linkages between Earth and outer space that characterize contemporary infrastructural processes.

The plurality of actors and interests involved in concrete and steel infrastructure projects in Eurasia has been substantially investigated elsewhere (Klinger and Muldavin, 2019; Grgić, 2019; Oliveira, 2019, among others). This work is instructive because satellites are also infrastructure. As major infrastructure investments, they are imbued with nationalism (Spiller, 2016). But also, in most cases, they are the outcome of multinational processes that blend public and private interests. Because they are essential to the communication, transportation, science, and surveillance activities that constitute many activities captured under the broad umbrellas of 'development' and 'security', developing or acquiring them is a central concern for the majority of the world's national governments. As of 2021, at least 106 national governments include space-related activities in their annual budgets (Oniosun, 2021). It is not too much to say that without satellites, and the remote sensing capabilities they provide, other sorts of contemporary infrastructures would not, or could not, be built.

Even if we acknowledge the fundamental roles of satellites to other more commonly studied issues that constitute the core of research on China's overseas activities, US activities in Central Eurasia, or national development, it is still possible to do so in a way that leaves intact the framings of a new Cold War or of a 'Great Power' competition. Indeed, these are precisely the sorts of political analyses that prevail outside of the domain of critical scholarship. Consider, for example, Anglophone commentaries on the growth of *BeiDou*, China's global navigation satellite system. Previously, the only countries with global navigation satellite systems had been the US (GPS), Russia (GLONASS), and the EU (Galileo). BeiDou entered into full service after the 55th and final satellite was launched in June 2020, providing millimetre-precision real-time navigation services to users throughout the world and exceeding the US GPS network by 20 satellites (Tabeta, 2020). The governments of over 40 Belt and Road partner states have signed agreements to embed the BeiDou system domestically. As of 2020, technology products exported from China to at least 120 countries, such as Huawei phones, vehicle-borne terminals, and wearable devices, are linked directly to the BeiDou system. In the Cold War framework, it is possible to read this as China 'taking over' or even 'out-competing' the US in the navigations market. But this too oversimplifies the complex dynamics playing out in and in relation to outer space. BeiDou had been under development since 1994 (Sun and Zhang, 2016), long before the US Congress froze space cooperation with China with the 2011 Wolf Amendment, which prohibits the use of federal funds to collaborate with, host, or coordinate bilaterally with China or Chinese-owned companies without certification from the Federal Bureau of Investigations (US Congress, 2012). BeiDou development was fast-tracked after the 2008 Sichuan earthquake, when rescue efforts were reportedly hindered by difficulties using GPS in central China (Tabeta, 2020). Although independence from the US GPS system for users in China and other countries was one important outcome, it is too simple to reduce this to bipolar competition.

However, this chapter wishes to do more than revisit headlines on the latest space 'race' that always seems to be 'heating up' with respect to China (Cao, 2021; Einhorn, 2021; Gan and Griffiths, 2021), and instead probes several contexts in Central Eurasia in order to provide a guide for complexifying the actors, interests, and institutions involved in this immense and varied region. To begin, the next section deconstructs an iconic image from Afghanistan to contextualize international politics in the region in general, and space politics in particular, in relation to extractive infrastructures. This is followed by three brief case studies of space activities in Pakistan, Turkmenistan, and Kazakhstan, chosen because they serve as illustrative examples of the multiple arrangements through which outer space politics play out. These examples decenter the roles of 'Great Power' or 'New Cold War' politics. Hence the two analytical moves made by this chapter – to put the grandiose discourses in their place, while also bringing the domain of outer space more centrally into discussions on infrastructure, investment flows, development, and security – enable us to discern with greater clarity the dynamics unfolding across the diverse regions of Central Eurasia, in subterranean, surface, and orbital spaces. The chapter closes with a brief discussion.

## Repositioning

To guide our thinking through the dual challenges of apprehending the complexity of actors and interests present in Central Eurasia, and to bring outer space politics out of the margins and closer to central concerns of security and development, this section considers an image taken by foreign correspondent Jonathan S. Landay (2009). At first glance, the image (Figure 18.1) has nothing to do, really, with outer space. But a closer look reveals more. It illustrates the interplay of diverse actors and interests in Central Eurasia and provides an entry point to thinking about the way in which infrastructures in orbital space are fundamental to the broader spatial transformations occurring on the ground.

The image shows a low cinder-block compound. The white paint is weathered and chipped, stained with dirt and rust. Faux minarets, the few architectural flourishes, stand not quite as high as the handful of satellite receiving dishes – for communication and likely also entertainment. Rusted fuel drums are lined up in front of the compound, and a fleet of all-wheel drive vehicles are visible in the background. A bright yellow construction vehicle with tank tread and an army jeep stand in the foreground of the

**Figure 18.1:** Compound to house road-building crews in 'central Afghanistan', 2009

Source: Photo by Jonathan S. Landay/MCT/Tribune News Service via Getty Images

muddy dirt road. These two vehicles show the twin purposes of the non-Afghani forces in the region: occupation and extraction. In some times and places, it is called development.

The location of this dirt access road in front of the compound, high in the mountains, was given as somewhere in central Afghanistan. Perhaps the photographer could not disclose the exact place, perhaps the editor determined that readers would not be interested in geographical precision. The purpose of the compound's construction was to house crews building a road for coal transport from wherever this is in 'central Afghanistan' to the Mes Aynak copper mine in Logar Province. The road construction was part of a 2008 deal won by Jiangxi Copper and the China Metallurgical Group Corporation over firms from the US, France, and Canada to develop a commercial mine on an immense deposit that has been extracted since the bronze age, and from which officials in the Afghani government hoped to extract the wages of development (Reuters, 2008; Zaheer, 2013). The path of the road was mapped out by survey teams from several countries, using remote sensing and ground-surveying techniques to determine the best route through the rugged terrain. As is often the case with road construction in remote regions, the most passable routes had long since been established by local people, many of whose homes and villages were destroyed or relocated to make way for the road (Dastgir et al, 2018; Rickard, 2020). In the story that Landay reported, US troops provided security to Chinese road crews, but the primary purpose of US deployment was not to protect China's investment interest, rather to stem 'Taliban infiltration' into Kabul. It is difficult to know how much of a threat this posed: the region was characterized as a 'Taliban stronghold' in the 1990s, but construction and archaeological teams had been working there in the meantime (Lawler, 2010). Afghan, French, and US military personnel used the compound while Chinese road construction crews were away for the winter.

Fast forward to mid-2021, just months before the abrupt US departure, and progress had stalled on the mining project. The road had yet to be finished, even though local people along its planned route were long since displaced (Jahanmal 2020). It seems that foreign interests had shifted away from infrastructure construction in favour of intrigue. Indian intelligence operatives uncovered an alleged ring of Chinese spies in Kabul who were cooperating with the Haqqani network – characterized in US media as a terrorist organization linked to the Taliban – to 'hunt down Uighur Muslims', which the Afghan government characterized as an act of betrayal (O'Donnell, 2021), the spectacle of which they attempted to leverage to pressure China to bring the copper mine online, lest they rescind the contract and reissue the tender (Ruttig, 2015). Afghan English-language media reported in March 2020 that the inactivity at the mine site had cost the Afghan government US$2 billion in lost revenues. Members of Afghan Security forces charged with guarding the deposit and providing protection to the few Afghan and Chinese workers performing maintenance on the machinery located on site have been killed in attacks. Archaeologists working nearby have also been killed. Some accounts blame the Taliban, others do not (Sediqi, 2018; Jahanmal, 2020). The stalled progress on the Mes Aynak copper mine, and the lost government revenues, stands in stark contrast to the production output and financial outlook of the Taliban, which generated hundreds of millions of dollars in illicit mining revenues in 2019 (Djani, 2020), which had placed the Afghan government in a losing battle with the Taliban to become financially and militarily independent (Bezhan, 2020). In anticipation of the US departure, Xi Jinping had directed high-level talks with Taliban leadership on peace, security, and economic development in the region (Keleman, 2019). This turned out to be a prescient move.

On the ground, people from the US, China, and many other places worked together to build extractive infrastructure to turn ancient sites and landscapes that host traditional livelihoods into the latest frontier for transnational mining and major infrastructure construction. Public and private sector interests overlap, at times, but not always in predictable ways. Development and militarism are at times inseparable, but it is important to see too how they are distinct. Under US-led occupation, the national government, mining officials and geologists entertained delegations from countries that show investment interest – the US, China, India, Saudi Arabia, Russia, France, to name a few – and displayed maps and satellite imagery to discuss various untapped geological endowments. Amidst this, activists and non-governmental organizations in Afghanistan lobbied their own government and advocated internationally to ensure that any resources extracted were done under terms that would benefit the people of Afghanistan (Noorani 2013; Alliance for Peacebuilding et al, 2014): a tall order when the occupying power has a history of brokering concessions to foreign companies, and the national government intended to use large-scale export-oriented mining as an engine to generate revenues not only for national development but also for security as it competed with the Taliban. Transnationally, financial flows from international aid programmes, multilateral banks, private investors, and a host of illicit actors passed through hands located throughout the world but with the purpose of transferring money and commodities in and out of the country (FATF, 2014).

On the ground, a diverse array of networks – private, public, illicit, humanitarian, terrorist, religious, financial, and technical – competed for power and legitimacy to write the region's future – as though one single future could be imposed on such an immense and varied space. Often, members of these competing networks hail from the same countries, though little can really be known about the interests embodied by a specific person without talking with them. Take the example of a Chinese, US, or Indian national: were they members of an alleged espionage ring? Or might they have been labourers drawn to the region for the higher wages to be garnered by working for a short time to support their families at home? Or both? Or an Afghan national: were they a government official intent on courting international investment? A member of an NGO intent on protecting people from further displacement? Or were they a member of the Taliban, working to establish financial and military autonomy to reassume control of the government? We might take this sensibility and apply it to satellite technology. It is difficult to know, just by looking at a satellite, what the cameras and antennas are for. It may have been launched by China, the US, or India, but is it for disaster management or surveillance? For transmitting television programmes or guiding missiles? Does it belong to the government, military, or a private firm, or some combination of the three? It is far easier to assume intrigue than to pursue precise answers: the simplification offered by the 'New Cold War' framework would reduce everyone to operatives and spies, but this obscures far more than it clarifies.

Although the photo and the specific details are unique to Afghanistan, the dynamics that characterize it are not – this is what enables the image to be instructive. It reveals more about the dynamics between the US and China in Central Eurasia than much of the 'Great Power' or 'New Cold

War' rhetoric that has become ubiquitous in recent years – a ubiquity that has grown in policy, media, and some scholarly circles despite the many and important contributions in Cold War historiography (for example, Romero, 2014). Complex entanglements among diverse interests and actors do not cleave along the neat bipolar lines represented in caricatures of competing hegemons. While simplifying frameworks can be revelatory for some particular world views, they must be considered in the context of actually occurring events unfolding on the ground.

We might also consider, then, how it is that the endowments of the Mes Aynak deposit have come to be known and characterized. To bridge the image with the broader question of space infrastructures within and above Central Eurasia, consider that scientists from Afghanistan, Turkey, South Africa, China, the US, and Japan have published scientific papers detailing remote sensing analyses of the geological characteristics of the Mes Aynak using multispectral images captured by cameras aboard satellites that are themselves enmeshed in complex multinational networks (Livo and Johnson, 2011; Yang and Han, 2012; Azizi et al, 2015; Mohammadi et al, 2020, among others). It is possible to produce detailed maps of this region without ever stepping foot there because of the mesh of observation satellites that orbit the Earth, many of which belong to entities that make their imagery freely available to the public. Nearly 1,000 active Earth observation satellites are in orbit, launched by private, public, military, and university entities from more than 50 countries (UCS, 2021). This is in addition to the approximately 2,000 active satellites used for communications, navigation and global positioning, space science, and reconnaissance and intelligence. These satellites, orbiting the Earth at tens of thousands of kilometres per hour, transmit their data to terrestrial and shipboard networks of ground-receiving stations, laboratories, universities, corporations, and militaries.

Several Central Eurasian countries have developed and launched their own satellites. Some independently, others in partnership with firms in the US, Russia, India, China, or European countries. Since the mid-20th century, launch sites in Central Eurasia have ferried satellites from dozens of countries to orbit. These space technologies mediate transnational information flows central to development, warfare, espionage, communication, and entertainment. They are better characterized by multipolar relations rather than bipolar. As such, they are useful for putting 'New Cold War' politics in their place.

## Case studies: space politics are multinational and multisectoral

To illustrate the diversity of space partnerships in Central Eurasia, we should move in three directions from Afghanistan: first, southward to Pakistan.

The case of Pakistan represents an instance of closer space collaboration between China and a Central Eurasian state, but a closer look contextualizes cooperation within a broader and more diversified set of relations. Second, to Turkmenistan which borders Afghanistan to the northwest, and which launched its first satellite in 2015. The multinational and multisectoral relations through which this satellite was produced better illustrate the prevailing norm in space politics, rather than the exception. It would be difficult to discern a kind of 'Great Powers' clash occurring herein, although US policies restricting the export of technologies to certain countries do influence which satellites get built by whom, as well as the location of their launch. The third case is far northward of Afghanistan, in Kazakhstan where the Baikonur Cosmodrome is located. Leased by the Kazakh government to Russia until 2050, it was the first operational spaceport in the world and illustrates the interplay between Cold War histories and contemporary processes. This is a case where the legacies of the Cold War are apparent in the infrastructure and landscape, and yet have also ceased to be primary determinants of space development at a national and international level.

There are, of course, many other examples. But taken together, these three show the inadequacy of a simplified new Cold War framework for characterizing the Central Eurasian space politics. This fact is noteworthy in itself: the 'Space Race' between the US and former USSR was, alongside atomic politics and proxy wars, a defining feature of the Cold War era. That it should fail to be so in this supposed era of a new Cold War, when space has become fundamental to the global economy, development, and security, indicates a serious shortcoming with the new Cold War framework for apprehending the present.

## Pakistan

In July 2018, China launched two remote sensing satellites (PRSS-1 and PakTES-1A) for Pakistan. The first, PRSS-1, is an optical remote sensing satellite developed by China's Academy of Space Technology and sold to Pakistan. PakTES-1A is an experimental satellite developed by Pakistan's Space and Upper Atmosphere Research Commission. Both satellites were built to meet remote sensing needs for government, research, and development institutions in Pakistan, with the joint purpose of monitoring the construction of the China–Pakistan Economic Corridor, which has proceeded despite local opposition and for which an unknown number of people have lost their lands (Chopra, 2018). In the early 2000s, PRSS-1 was initially conceived by SUPARCO engineers (the Space and Upper Atmosphere Research Commission, Pakistan's national space agency) as the first in a series of Pakistan's own global navigation satellite system, but in 2012 Pakistan instead adapted the BeiDou navigation system of China to complete this satellite (Ahsan and Khan, 2019). Such measures are common in space development; international collaboration is often more cost effective.

It is entirely possible to tell a version of this story that represents Pakistan's satellite development as dominated by China, but this erases at least two decades of science and technology history in Pakistan. Badr-1 was the country's first indigenously developed satellite (SUPARCO, n.d.), built with support from the Pakistan Amateur Radio Society in the 1980s. It was initially planned to be launched on the US Space Shuttle, but the 1986 Challenger explosion and subsequent delays from the US prompted a change in plans. Badr-1 launched from China's Xichang Launch Center in July 1990 (Lele, 2012). Badr-B, Pakistan's second satellite, was an Earth Observation Satellite designed by UK company Space Innovations and launched from Baikonur Cosmodrome in Kazakhstan in December 2001 (Krebs, n.d.). Pakistan's third satellite, and first geostationary satellite, was PakSat-1. It was manufactured and owned by Boeing, launched from Cape Canaveral in 1996, and was initially leased to Indonesia. After an electronics failure, Indonesia declared the satellite unusable, claimed insurance payments, and transferred the title to the California-based Hughes Space and Communications Company. In 2002, the government of Pakistan took it over in a full-time lease for USD$5 million, and it was relocated to the Pakistani-licensed orbital location at 38°E longitude (Siddiqui, 2012). Even though it is leased and controlled by Pakistan, its application is multinational. It provides communications services in Pakistan, North Africa, and parts of the Middle East.

This example shows that even in cases where, as in Pakistan, deepening investment and infrastructure relations with Beijing seem to be displacing those with Washington, when it comes to space politics, they are decidedly transnational and cosmopolitan, characterized by the involvement of public, private, and military actors. This case also illustrates the multiple uses and applications of satellite technologies. Satellites launched by a given country tend to be used by many parties, for a variety of purposes.

## Turkmenistan

The space engagements of Turkmenistan provide another illustrative example of the multinational and multisectoral character of space politics in Central Eurasia. Turkmenistan currently has one satellite jointly owned with Monaco, the TürkmenÄlem-52E/MonacoSAT. It is located in the geostationary orbital position licensed to Monaco, but it is operated by the Turkmenistan National Space Agency (Al-Ekabi and Lahcen, 2017). In 2009, the Monaco government issued a licence to private company Space Systems International Monaco (SSI-Monaco) to develop a satellite for the use of their 52°E orbital position. In November 2011, the Turkmenistan Ministry of Communications contracted with the French-Italian aerospace and defence company, Thales

Alenia Space, to build the first of their intended National System of Satellite Communications. Some of the transponders used by SSI-Monaco have been licensed to SES, a global satellite and telecommunications network provider headquartered in Luxembourg. This provides enhanced capacity for satellite television broadcasting over Kazakhstan, Uzbekistan, Southern Russia, the Middle East, North Africa, and Central and Northern Europe (SSI-Monaco, n.d.).

The Turkmenistan–Monaco satellite was built by a French-Italian company and was launched by a private US company, SpaceX, from Cape Canaveral in Florida in April 2015. It was originally intended to be launched from China, but the US International Traffic in Arms Regulations prohibited some of the US-made parts from being exported to China for any purpose, including the launch of a satellite made in another country (de Selding, 2015). US regulations cost China a launch contract, but this US–China dynamic is a small plot point in the larger story of Turkmenistan and the Principality of Monaco working with a range of private companies from different parts of the world to develop and launch a satellite that serves public and commercial interests over a large geographic area.

## Kazakhstan

Two of Kazakhstan's space activities are instructive here: satellite development and hosting the world's oldest spaceport. Although Kazakhstan's contemporary development is indelibly marked by the country's history as a former Soviet territory, it is important to understand these legacies alongside the other diverse currents that shape space politics within and beyond the country. Kazakhstan has been home to the Baikonur Cosmodrome since it was built in southern Kazakhstan in 1955 (Gruntman, 2019). It would be entirely possible to tell a story steeped in Cold War determinism to the effect that Kazakhstan did not launch its first satellite until 2006 because of its continued dependence on Russia. KazSat-1, the first geostationary satellite developed and launched to provide television and communications services, was built by the Moscow-based Khrunichev State Research and Production Space Center in cooperation with French-Italian firm Thales Alenia Space. It was launched from Baikonur. The second and third generation satellites, KazSat-2 and KazSat-3, were manufactured and launched with the same partners and for the purpose of expanding communications services in Kazakhstan (Sylkina et al, 2015).

The post-Cold War development history of Kazakhstan's space industry is transnational (Parks, 2012). The country's first Earth observation satellite was licensed in 2009 to be built by French and British firms and launched from the Russian air base, Domarovsky. More recently, Al-Farabi Kazakh National University in the city of Almaty developed a nano-satellite containing various experimental nano-scale systems, such as a remote sensing nano-camera. Several of the non-experimental components were developed by a SPUTNIX, one of Russia's first private space start-ups. The nano-satellite was part of a massive payload of 104 nano-satellites launched by India's Space Research Organization in February 2017 (Bharti, 2017). Along with satellites from India and Kazakhstan, other payloads hailed from Israel, the Netherlands, Switzerland, the United Arab Emirates, as well as 96 others from the US, 88 of which were from a single company. In April 2021, *The Astana Times* reported on a meeting between Prime Minister Askar Mamin and Sunil Mittal, the CEO of UK company One Web, to discuss contracting with the company to provide satellite internet to all of Kazakhstan, and to create 'a joint venture to locate a gateway station on the territory of Kazakhstan for the subsequent exclusive distribution of satellite Internet "One Web" to the countries of Central Asia' (Kuandyk, 2021). This would position the government of Kazakhstan as gatekeeper of internet access to the region. The engagements, contracts, and logistics of space for Central Eurasian states are multinational. While some occur within bipolar frameworks, others exceed them.

Baikonur Cosmodrome was originally built as a testing site for Soviet intercontinental ballistic missiles in 1955. The site was selected because it was surrounded by relatively flat terrain, which at the time was essential to maintain radio control of the missiles via a network of ground stations (Suvorov, 1997). It was also selected based on the calculation that the missile trajectories would not jeopardize heavily populated areas in the Soviet Union. This, of course, has rendered the surrounding areas, including cities and villages, vulnerable to impacts of several decades of pollution (Kopack, 2019). Later, when the site was used for space missions, rockets had to increase their minimum inclination to ensure that lower stage rocket boosters would not be dropped onto the territory of China. The boosters must fall somewhere, though, and this has meant onto the 'relatively unpopulated' lands in Kazakhstan. The long-term health and environmental impacts of rocket launches on people, animals, and plants in the region has been documented by Russian and Kazakh scientists (Abdrazak and Musa, 2015). Local economies, displaced by space-related activities, have resorted to recovering scrap metal to earn supplemental income (Kopack, 2019).

It was from Baikonur that the first satellite, Sputnik-1 and the first human in space, Yuri Gagarin, departed. In the decades since, astronauts from many countries have started their journeys from Baikonur, including many from the US, and hundreds of satellites have been launched for dozens of countries from this site. After the retirement of the US Space Shuttle in 2011, Russia became the sole transporter of astronauts to the International Space Station until 2019. When national governments, private firms, or space agencies issue tender for launch contracts, firms from Russia, China,

India, Japan, Brazil, the US, and France compete. For firms and agencies from Russia, most often it is Baikonur that is on offer; something that is possible because the Russian Federation Council, which is the upper house of the federal assembly of Russia, has leased Baikonur from the government of Kazakhstan for US$115 million per year until 2050 (Space Daily, 2010). But this arrangement is contentious, with the government of Kazakhstan refusing Russian proposals for 99-year leases or to reduce the cost of the rent. To reduce its dependence on Kazakhstan, the Russian government built the Vostochny Cosmodrome in Amur Oblast, located in Russia's Far East. It became operational in 2016. Thus, even in a space landscape built and shaped by the 20th-century space race and nuclear arms race, multinational and cosmopolitan practices prevail. Although a former Soviet territory, the government of Kazakhstan exercises agency in pursuing space partnerships and in shaping space politics.

## Closing discussion

During the Cold War, amid incredible tension and violence, space provided a basis for international cooperation. The competition between the US and USSR provided openings and leverage for newly independent states to extract resources and technology from the two hegemons, and to steer global policy making toward broader democratic governance. During the 20th-century space race, US and the former USSR expenditures on domestic space programmes included selected space partnerships and space infrastructure construction overseas as part of their empire building and diplomacy. The former USSR built launch sites, testing ranges, mines, and laboratories in Soviet satellite states. The US, to a more limited extent, entertained invitations from Eurasian governments to utilize their strategic geographical position to invest in launch infrastructure and some technology training (Siddiqi, 2010, 2015).

The former USSR trained astronauts from around Soviet-aligned countries, sending Afghanistan's only astronaut to date, Abdul Ahad Mohmand, to space in 1988 (Burgess and Vis, 2016), who later went to work as an accountant in Germany. While this has been extensively analysed in Anglophone literature as part of the Soviet diplomatic strategy, it did not constitute a direct form of competition with the US for the simple reason that the Cold War era US space programme was only interested in recruiting a very specific type of male US citizen for spaceflight (Weitekamp, 2004).

This marked difference of approach is instructive for the present. Whereas for China, satellite technologies, network agreements, and ground-receiving stations are incorporated within infrastructure spending priorities and international agreements, the US does not appear to take a similar approach: space-related activities remain outside the development, investment, and diplomatic priorities emphasized in Central Eurasia. The US government is not engaged in space partnerships with Central Asian states, aside from launch and transport agreements that take place via Kazakhstan's Baikonur Cosmodrome. Occasionally, a satellite launch contract is awarded to a private US company. Not only is the 'New Cold War' framework insufficient to grasp the complex dynamics and multinational processes comprising the politics of space infrastructure in Central Eurasia, but on a more fundamental level, it hardly counts as a bipolar competition if dozens of parties are working, together as well as independently, on a host of different initiatives.

## References

Abdrazak, P.K. and Musa, K.S. (2015) 'The impact of the cosmodrome Baikonur on the environment and human health', *International Journal of Biology and Chemistry*, 8(1): 26–9.

Ahsan, A. and Khan, A. (2019) 'Pakistan's journey into space', *Astropolitics*, 17(1): 38–50.

Al-Ekabi, C. and Lahcen, A. (2017) 'Chronology: 2015', in C. Al-Ekabi, B. Baranes, P. Hulsroj, and A. Lahcen (eds) *Yearbook on Space Policy 2015: Access to Space and the Evolution of Space Activities*, Vienna: Springer, pp 259–90.

Alliance for Peacebuilding, Afghanistan Watch, Afghan Development Association, British and Irish Agencies Afghanistan Group, Equal Access, Future Generations Afghanistan, Green Wave, Green Wish for Afghanistan, Global Rights, Global Witness, Heinrich Boll Stiftung, Natural Resource Governance Institute, Open Society Afghanistan, Publish what you pay, Salam Watandar, Sun Development and Environmental Protection Organization, The Liaison Office, Transparency International, Transparency International India, Transparency International UK, and more (2014) 'Open letter', 2 December.

Azizi, M., Saibi, H., and Cooper, G.R.J. (2015) 'Mineral and structural mapping of the Aynak-Logar Valley (eastern Afghanistan) from hyperspectral remote sensing data and aeromagnetic data', *Arabian Journal of Geosciences*, 8: 10911–18.

Bezhan, F. (2020) 'Exclusive: Taliban's expanding "financial power" could make it "impervious" to pressure, confidential report warns', RadioFree Europe. Available from: www.rferl.org/a/exclusive-taliban-s-expanding-financial-power-could-make-it-impervious-to-pressure-secret-nato-report-warns/30842570.html.

Bharti, P. (2017) 'India's ISRO successfully launched 104 satellites in a single mission', Indrastra Global, 15 February. Available from: https://www.indrastra.com/2017/02/NEWS-India-s-ISRO-Successfully-Launched-104-Satellites-Single-Rocket-PSLC-C37-003-02-2017-0045.html.

Brazinsky, G. (2017) *Winning the Third World: Sino-American Rivalry during the Cold War*, Chapel Hill, NC: University of North Carolina Press.

Burgess, C. and Vis, B. (2016) 'Afghanistan's cosmonaut-researcher', in C. Burgess and B. Vis (eds) *Interkosmos: The Eastern Bloc's Early Space Program*, London: Springer, pp 252–65.

Cao, S. (2021) 'A new space race heats up as three Mars probes from NASA, China, and the UAE arrive', *Observer*, 1 June. Available from: https://obser ver.com/2021/01/mars-rover-nasa-perseverance-china-tianwen-uae-arr ive-february/.

Chopra, R. (2018) 'Baloch activists protest against China–Pakistan economic corridor', *The Tribune*, 29 January. Available from: www.tribuneindia. com/news/archive/world/baloch-activists-protest-against-china-pakistan-economic-corridor-535451.

Dastgir, G., Kawata, K., and Yoshida, Y. (2018) 'Effect of forced relocation on household income and consumption patterns: evidence from the Aynak copper mine project in Afghanistan', *The Journal of Development Studies*, 54(11): 2061–77.

de Selding, P.B. (2015) 'Falcon 9 rocket launches satellite for Turkmenistan and Monaco', *Space News*, 27 April. Available from: https://spacenews. com/falcon-9-rocket-launches-satellite-for-turkmenistan-and-monaco/.

Djani, D.T. (2020) 'Eleventh report of the Analytical Support and Sanctions Monitoring Team submitted pursuant to resolution 2501 (2019) concerning the Taliban and other associated individuals and entities constituting a threat to the peace, stability, and security of Afghanistan', United Nations Security Council. Available from: https://www.securitycouncilreport. org/atf/cf/%7B65BFCF9B-6D27-4E9C-8CD3-CF6E4FF96FF9%7D/ s_2020_415_e.pdf.

Einhorn, B. (2021) 'The race for Mars takes China–US tensions into outer space', *Bloomberg Businessweek*, 21 January. Available from: https://www.bloomberg.com/news/articles/2021-01-20/ will-u-s-or-china-reach-mars-first-space-race-heats-up.

Financial Action Task Force (FATF) (2014) *Financial Flows Linked to the Production and Trafficking of Afghan Opiates*, Paris: Financial Action Task Force.

Gan, N. and Griffiths, J. (2021) 'China blames the US for hyping fears of uncontrolled rocket reentry as space race heats up', CNN, 1 June. Available from: www.cnn.com/2021/05/10/china/china-rocket-reaction-mic-intl-hnk/index.html.

Grgić, M. (2019) 'Chinese infrastructural investements in the Balkans: political implications of a highway project in Montenegro', *Territory, Politics, Governance*, 7(1): 42–60.

Gruntman, M. (2019) 'From Tyuratam missile range to Baikonur Cosmodrome', *Acta Astronautica*, 155: 350–66.

Jahanmal, Z. (2020) 'More than $2B lost in copper mining delays: company', *TOLO News*. Available from: https://tolonews.com/business/more-2b-lost-copper-mining-delays-company.

Karagov, S. (2018) 'The new Cold Way and the emerging Greater Eurasia', *Journal of Eurasian Studies*, 9(2): 85–93.

Keleman, B. (2019) 'China and the Taliban: pragmatic relationship', Central European Institute of Asian Studies, 26 June. Available from: https://ceias.eu/china-the-taliban-pragmatic-relationship/.

Klinger, J.M. and Muldavin, J.S.S. (2019) 'New geographies of development: grounding China's global integration', *Territory, Politics, Governance*, 7(1): 1–21.

Kopack, R.A. (2019) 'Rocket wastelands in Kazakhstan: scientific authoritarianism and the Baikonur Cosmodrome', *Annals of the American Association of Geographers*, 109(2): 556–67.

Krebs, G. (n.d.) 'Badr-B', Gunter's Space Page. Available from: https://space.skyrocket.de/doc_sdat/badr-b.htm.

Kuandyk, A. (2021) 'Kazakhstan develops satellite internet link up to One Web global satellite network', *The Astana Times*, 9 April. Available from: https://astanatimes.com/2021/04/kazakhstan-develops-satellite-internet-link-up-to-one-web-global-satellite-network/.

Landay, J.S. (2009) 'Afghan president steps up political confrontation with opponents', McClatchy DC Bureau, 7 March. Available from: www.mcclatchydc.com/news/nation-world/national/national-security/article24528526.html.

Lawler, A. (2010) 'Copper mine threatens ancient monastery in Afghanistan', *Science*, 329 (5991): 496–7.

Lele, A. (2012) *Asian Space Race: Rhetoric or Reality?* New Delhi: Springer India.

Livo, E.K. and Johnson, M.R. (2011) 'Analysis of imaging spectrometer data for the Aynak-Logar Valley are of interest', in S.G. Peters, T.V.V. King, T.J. Mack, and M.P. Chornack (eds) *Summaries of Important Areas for Mineral Investment and Production Opportunities of Nonfuel Minerals in Afghanistan: US Geological Survey Open-File Report 2011–1204*, Reston, VA: United States Geological Survey. Available from: https://pubs.usgs.gov/of/2011/1204/pdf/Front_matter_vol_I.pdf.

Mohammadi, F.A., Amin, Z.M., and Ahmad, A.B. (2020) 'Lineament assessment of Aynak copper mine using remote sensing approach', *IOP Conference Series: Earth and Environmental Science*, 540: 012034.

Noorani, J. (2013) 'Aynak: a concession for change', Integrity Watch Afghanistan.

O'Donnell, L. (2021) 'Afghanistan wanted Chinese mining investment. It got a Chinese spy ring instead', *Foreign Policy*, 27 January. Available from: https://foreignpolicy.com/2021/01/27/afghanistan-china-spy-ring-mcc-mining-negotiations-mineral-wealth/.

Oliveira, G. de L.T. (2019) 'Boosters, brokers, bureaucrats and businessment: assembling Chinese capital with Brazilian agribusiness', *Territory, Politics, Governance*, 7(1): 22–41.

Oniosun, T. (ed) (2021) *Global Space Budgets: A Country-Level Analysis*, Lagos: Space in Africa.

Parks, L. (2012) 'Satellites, oil, and footprints: Eutelsat, Kazsat, and post-communist territories in Central Asia', in L. Parks and J. Schwoch (eds) *Down to Earth: Satellite Technologies, Industries, and Cultures*, New Brunswick, NJ: Rutgers University Press, pp 122–40.

Reuters (2008) 'Jiangxi Copper says in $808 mln Afghanistan deal', Reuters, 10 May. Available from: www.reuters.com/article/jiangxicopper-afghanistan-idUKHKG15146720080528.

Rickard, S. (2020) 'Gender, agency and decision making in community engagement: reflections from Afghanistan's Mes Aynak mine', *The Extractive Industries and Society*, 7: 435–45.

Romero, F. (2014) 'Cold war historiography at the crossroads', *Cold War History*, 14(4): 685–703.

Ruttig, T. (2015) 'Copper and peace: Afghanistan's China dilemma', Afghanistan Analysts Network. Available from: www.afghanistan-analysts.org/wp-content/uploads/wp-post-to-pdf-cache/1/copper-and-peace-afghanistans-china-dilemma.pdf.

Sediqi, Q. (2018) 'Afghan archaeologist killed near Buddhist site, home to giant copper reserve', Reuters, 3 June. Available from: www.reuters.com/article/us-afghanistan-archaeologist/afghan-archaeologist-killed-near-buddhist-site-home-to-giant-copper-reserve-idUSKCN1IZ0ED.

Siddiqi, A. (2010) 'Competing technologies, national(ist) narratives, and universal claims: Toward a global history of space exploration', *Technology and Culture*, 51(2): 425–43.

Siddiqi, A. (2015) 'Science, geography, and nation: the global creation of Thumba', *History and Technology*, 31: 420–51.

Siddiqui, S. (2012) 'Lagging behind: 2040 – Pakistan's space od[d]yssey', *The Express Tribune*, 1 August. Available from: https://tribune.com.pk/story/415738/lagging-behind-2040-pakistans-space-oddyssey.

Space Daily (2010) 'Kazakhstan finally ratifies Baikonur rental deal with Russia', 12 April. Available from: www.spacedaily.com/reports/Kazakhstan_Finally_Ratifies_Baikonur_Rental_Deal_With_Russia_999.html.

Space Systems International Monaco (SSI-Monaco) (n.d.) 'Monacosat milestones'. Available from: www.ssi-monaco.com/overview/.

Spiller, J. (2016) *Frontiers for the American Century: Outer Space, Antarctica, and Cold War Nationalism*, New York: Palgrave Macmillan.

Sun, D. and Zhang, Y. (2016) 'Building an "Outer Space Silk Road": China's Beidou navigation system in the Arab World', *Journal of Middle Eastern and Islamic Studies*, 10(3): 24–49.

SUPARCO (2009) 'Paksat-1'. Available from: https://suparco.gov.pk/major-programmes/projects/paksat-1r/

SUPARCO (n.d.) 'Badr-1'. Available from: http://induction8402.blogspot.com/2011/03/badr-1-pakistans-first-experimental.html.

Suvorov, V. (1997) *The First Manned Spaceflight: Russia's Quest for Space*, Nova.

Sylkina, S.M., Dosymbekova, M.S., Baytukaeva, D.U., Begzhan, A.M., and Toktybekov, T.A. (2015) 'Development of space activity or the Republic of Kazakhstan: history and modern (political and legal aspect)', *Mediterrenean Journal of Social Sciences*, 6(5): 392–8.

Tabeta, S. (2020) 'China completes GPS rival Beidou with latest satellite launch', Nikkei Asia, 23 June. Available from: https://asia.nikkei.com/Business/China-tech/China-completes-GPS-rival-Beidou-with-latest-satellite-launch2.

Union of Concerned Scientists (UCS) (2021) UCS Satellite Database. Available from: https://www.ucsusa.org/resources/satellite-database.

US Congress. (2012) '112th Conrgess Public Law 55: Consolidated and Further Continuing Appropriations Act, Sec. 539. China', edited by Department of the Interior of the United States Geological Survey. Washington, DC: US Government Printing Office. Available from: https://www.govinfo.gov/content/pkg/PLAW-112publ55/html/PLAW-112publ55.htm.

Weitekamp, M.A. (2004) *Right Stuff, Wrong Sex: America's First Women in the Space Program*, Baltimore, MD: Johns Hopkins University Press.

Westad, O.A. (2019) 'The sources of Chinese conduct: are Washington and Beijing fighting a New Cold War?', *Foreign Affairs*, 98: 86–95.

Yang, S. and Han, X. (2012) 'Development of mineral resources in Afghanistan: history, current situation, and outlook', *Journal of Xinjiang Normal University (Social Sciences)*, 33(3): 13–20.

Zaheer, A. (2013) 'Aynak mine project offers opportunities but challenges persist', Pajhwok Afghan News, 26 November. Available from: https://mines.pajhwok.com/news/'aynak-mine-project-offers-opportunities-challenges-persist'

Zhao, M. (2019) 'Is a New Cold War inevitable? Chinese perspectives on US–China strategic competition', *The Chinese Journal of International Politics*, 12(3): 371–94.

# 19

# Conclusion: 21st-Century Third Worldism?

*Seth Schindler and Jessica DiCarlo*

The chapters in this volume demonstrate that competition between the US and China is truly global in scope and its expansiveness extends from the Earth's substratum, such as lithium mines in Argentina and uranium mines in Namibia, to the stratosphere. Contemporary great power rivalry has far-reaching consequences for people and places worldwide, and it increasingly serves as a reference point for issues that were unrelated until recently. For example, Joe Biden made the case for an extensive infrastructure bill on the grounds that it would enhance American competitiveness and steel the country for a long struggle against China and authoritarianism. Meanwhile, Beijing's crackdown on China's nascent tech sector took a geopolitical turn when ride-hailing app Didi was forced to de-list from the New York Stock Exchange and re-list in Hong Kong. In this sense, the intensification of great power competition is a pivotal and overarching political event of our time, against which the salience and meaning of other events and issues are shaped and judged. In both the US and China, politics in general are often refracted though the lens of geopolitical rivalry, and chapters in this volume demonstrate that this is also the case in many other countries. Indeed, this is what higher education in Hungary, Turkey's sovereign wealth fund, SEZs in Zambia, and the competition among various branches of the armed forces in Philippines have in common – politics surrounding all of these things are shaped by and can be understood in relation to (or in the context of) the US–China rivalry.

Scholarship on the expansion of China's global integration since the turn of the century has proliferated. This volume demonstrates that the rise of China can no longer be understood in isolation from great power rivalry. Indeed, geopolitical rivalries have enabled the explosion of large-scale infrastructure projects worldwide, as the US and China as well as other regional powers compete to orient people and places in ways that enhance the competitiveness of their respective national champions (and the value chains they anchor) and the resilience of their security systems and institutions. The centrality of infrastructure to this global competition is evident from flagship initiatives like the Belt and Road Initiative (BRI) and Build Back Better World (B3W). The former is far more articulate and advanced than the latter, but importantly, spatial planning and the financing and construction of large-scale infrastructure are central fields of competition.

Infrastructure also featured in the last prolonged contest between superpowers, the Cold War, but most states were forced to side with either the East or West Bloc. For example, the Aswan Dam was envisioned by technocrats in the West Bloc, but the dynamics of the Cold War in the Middle East shifted (see Lüthi, 2020) and the project was ultimately undertaken by the Soviet Union. At that time, most newly independent countries struggled to remain non-aligned and were associated with the East or West bloc. In contrast, today countries can simultaneously participate in the BRI, B3W, and regional initiatives, while also engaging multilateral institutions and private sector investors. Thus, the US and China are not competing to assemble blocs and 'contain' or 'encircle' one another, instead they seek to establish centrality within transnational networks through the financing and construction of large-scale infrastructure projects.

Network centrality affords the prospect of wielding power well into the 21st century, and, as chapters in this volume show, this high-stakes competition involves a host of middle and regional powers (such as Japan and India) as well as international institutions. The latter are fields of competition as well as agents – the US and China struggle to control and instrumentalize international institutions such as the WTO and the World Bank. Contemporary great power competition is a fast-moving target. New initiatives are announced with regularity – in 2019 the EU partnered with Japan to compete with the BRI, and this was followed by a similar agreement with India in 2021 (Peel, 2019; Peel et al, 2021). Shortly thereafter, the EU unveiled Global Gateway in an explicit response to the BRI. Even the UK seeks to compete with the BRI; just a week after the UK government cancelled a segment of a planned high-speed rail, it announced an initiative to offer developing countries loans for infrastructure. According to the Foreign Secretary, this will be an alternative to 'strings-attached debt from autocratic regimes', in order to build 'a network of liberty around the world with our friends and partners' (Hughes and Payne, 2021). Meanwhile, China hosted summits in Africa and Latin America. In the case of the former, the summit was followed by considerable debate as analysts parsed its declarations and China's pledges (see Olander, 2021). However, the scale and visibility of Chinese projects and finance has cemented its status as a viable development partner for many low- and middle-income countries. This reinforces findings in this volume – for the first time in decades low- and middle-income countries have scope to articulate ambitious developmental objectives and they can choose their partners. Many of the chapters demonstrate that countries actively resist pressure to choose between the US and China, and instead they hedge between them to secure support for spatial projects. While this much is apparent, it remains unclear if there is scope for more emancipatory politics to take shape among a bloc of non-aligned states capable of influencing the international order. It is to the possibility of the emergence of 21st-century Third Worldism that we now turn.

## Third World solidarity: 'Castro is a brother, Nasser is a teacher, but Tito is an example'

The quote is from Ahmed Ben Bella,[1] whose brief tenure as Prime Minister of post-revolutionary Algeria illustrated that the Cold War afforded newly independent states novel opportunities, but it also posed serious risks. The spirit of solidarity articulated by Ben Bella towards his contemporaries was born out of a shared struggle against European colonial powers and it was institutionalized at the Bandung Conference in 1955, where representatives of newly independent Asian and African states met to discuss the postcolonial future of the world. The world-historic significance of the Bandung Conference was captured by Alfred Sauvy, who declared that the 'ignored, exploited, scorned Third World, like the Third Estate, demands to become something as well'.[2] By comparing the two-thirds of the world's population that lived in Africa and Asia to the Third Estate – the 99 per cent of the French Revolution – Sauvy framed decolonized societies as an assertive and coherent social agent for the first time in history. Indonesian Prime Minister Sukarno (1955) was acutely aware that the tectonic plates underpinning politics on a planetary scale were in motion, and his opening speech at the Bandung Conference expressed the desire to forge a new international order:

> For many generations our peoples have been the voiceless ones in the world. We have been the unregarded, the peoples for whom decisions were made by others whose interests were paramount, the peoples who lived in poverty and humiliation. Then our nations demanded, nay fought for independence, and achieved independence, and with that independence came responsibility. We have heavy responsibilities to ourselves, and to the world, and to the yet unborn generations.

According to Sukarno the responsibility that befell African and Asian countries was twofold. First, they had to safeguard against new modes of colonialism whose 'modern dress' could take 'the form of economic control, intellectual control, actual physical control by a small but alien community within a nation. It is a skillful and determined enemy, and it appears in many guises'. The second responsibility was to 'inject the voice of reason into world affairs', and 'mobilise all the spiritual, all the moral, all the political strength of Asia and Africa on the side of peace'.

Sukarno's impassioned plea for Third World solidarity in pursuit of a peaceful international order free of (neo-)colonialism resonated across Africa and Asia. This political project was underpinned by the shared histories – namely exploitation at the hands of (mainly) European societies – of peoples across the majority world. In practice, however, leaders of many newly independent countries found it difficult to reconcile these two objectives and were forced to accept trade-offs. For example, Ahmed Ben Bella, Kwame Nkrumah, and Ahmed Sékou Touré fervently wanted to end white minority rule in southern Africa, and they were forced to choose between supporting armed liberation struggles, and regional peace and stability. This choice cut to the heart of the meaning of Third Worldism – was it first and foremost a political movement committed to combating imperialism, or a group of aligned nation states that sought to disrupt and remain aloof from great power competition?

The challenges facing African and Asian leaders were further complicated by the complexities of Cold War geopolitics. This was the meta-context that heightened risks but also occasionally allowed states to achieve objectives that would have otherwise been unattainable. The agency of smaller newly independent states in Africa and Asia, many of which championed Third Worldism, has been the subject of recent scholarship on the Cold War. Although orthodox interpretations framed the Cold War as a contest between great powers to which small states were beholden (Leffler, 2007), contemporary research has shown that the reverse was true in some instances. Lüthi (2020, 1) argues that instead of a single global Cold War, there existed 'regional, sub-systemic, Cold Wars', with contours defined by regional powers and small states. The implication is that 'great powers might have believed that they were pupped masters pulling the strings across the world, but the puppets had their own agendas and frequently pulled at their end of the strings to make the self-declared puppeteers dance' (Lüthi, 2020, 2). The ability of small states to articulate national developmental objectives while championing anti-imperialism explains why Castro, Nasser, and Tito, rather than Khruschev or Mao, inspired Ben Bella. He was initially able to operate very effectively in this complex environment – Algeria channelled support from China and Yugoslavia to African liberation movements, and the USSR committed to making it an exemplar of what Soviet economic and technical assistance could achieve, while France remained the largest single source of financial aid and the US contributed grain shipments (Byrne, 2016).

Algeria's post-revolutionary history illustrates the challenges that faced small states during the Cold War and ultimately served to undermine Third Worldism. As part of the deal through which France relinquished claims to Algeria, the latter's revolutionary leaders recognized the property rights of French citizens. Soon after achieving independence, however, the Algerian government abandoned this promise and launched a series of redistributive initiatives meant to compensate the population for years of colonial exploitation and the unspeakable horrors endured at the hands of the French military and paramilitaries (Byrne, 2016). Meanwhile, Ben Bella openly supported liberation movements across Africa, and Algiers's credentials as the 'Mecca of revolution'[3] was bolstered by the constant comings and goings of a who's who of 1960s revolutionary struggle including Che Guevara, Frantz Fanon, Yasser Arafat, as well as representatives of the South Vietnamese National Liberation Front and the Black Panthers (Chamberlin, 2012; Byrne, 2016; Malloy, 2017). These domestic and foreign policies would seemingly align Algiers with China and the Soviet Union, but Ben Bella aspired to maintain relations with the US as well. Furthermore, Byrne explains (2016, 95) that 'the Algerians knew that they were involved in a contest much more complicated than a bipolar zero-sum game between Moscow and Washington'. Indeed, not only did Algiers have to avoid being drawn too deeply into the East Bloc, it also had to contend with intra-bloc rivalries, which in practice meant balancing the interests of France against the US and the USSR against China.

The follow-up conference to Bandung was scheduled to be held in Algiers in 1965. If the Bandung Conference represented the beginning of Third Worldism, its follow-up in Algiers signalled the extent to which it had fragmented. Preparation for the conference was rife with conflict animated by the Sino-Soviet split (Friedman, 2015). It was cancelled after elite Algerian military units staged a coup and arrested Ben Bella without a struggle. The inability to hold a follow-up meeting to Bandung 'confirmed the end of decolonization's most idealistic and optimistic phase' (Byrne, 2016, 286). In the same year, a US-backed coup in Indonesia ousted Sukarno, the Vietnam War escalated significantly, and Mobutu seized power in the Republic of Congo. China was soon engulfed in the paroxysms of the Cultural Revolution. The notion of a Third Worldist political project – geared towards pursuit of world peace, non-alignment, and support for liberation struggles – seemed a distant memory as postcolonial states became more deeply integrated into the international political and economic order.

The integration of African and Asian states into international institutions and economic networks altered the structure of opportunities and constraints in which they operated. They continued to struggle as a bloc,[4] but these efforts were largely confined to the reform of international institutions rather than the transformation of the world order. Indeed, in the 1970s the political project of Third Worldism evolved and ultimately gave way to the economic project of international development (Lorenzini, 2019). For some, such as Algeria's new leadership, 'Cold War neutrality became economic; it was about alternative models of development, with the Global South supporting its own project to redress global economic inequality' (Lorenzini, 2019, 121). This involved a discursive shift from the 'Third World' whose members joined by choice due to their shared histories and political convictions, to the 'Global South' whose membership was determined by geography. This framing highlighted spatial inequality between North and South, and informed the proposal of the latter – whose members formed the G77 and often voted as a bloc in the UN – for the New International Economic Order. This included improved terms of trade for Southern countries, facilities to transfer technology and increase development assistance, and improve access to the markets of industrialized states (Lorenzini, 2019). Both the US and the USSR objected to this framework, albeit for very different reasons. The US opposed proposals that would curtail its hegemony over the international economic system, while the USSR was offended by the fact that it was grouped with the US and other capitalist countries as an industrialized state.

One result of increased economic integration in the 1970s was that countries in the Global South were able to obtain loans from private banks in the US and Europe tasked with managing the surplus generated by petroleum producers (Roos, 2019). Meanwhile, the post-war economic boom in the North Atlantic had atrophied, and in response to persistent stagflation the US Federal Reserve nearly doubled interest rates to 20 per cent. This unprecedented rate hike increased the cost of debt servicing and precipitated a debt crisis in Mexico that was soon repeated elsewhere in the Global South (Roos, 2019). Emergency financial assistance offered by the Bretton Woods institutions came with neoliberal strings attached (Killick, 1995), and Washington enjoyed considerable leverage because competition with the USSR and China for client states in the Global South was de-escalating throughout the 1980s. The USSR was bogged down in an unpopular and unwinnable war in Afghanistan, and inwardly focused on a series of reforms known as *Perestroika* launched by Mikhail Gorbachev (see Gorbachev, 1987; Kotkin, 2008). Similarly, China emerged beleaguered from the Cultural Revolution and Beijing prioritized economic growth over support for Third World revolutionary struggles (Friedman, 2015). On the eve of China's great transformation in 1979, Deng Xiaoping 'bluntly claimed that a strategy of economic openness, which included accepting aid from the US, was more likely to achieve economic growth' than socialism (Lorenzini, 2019, 116). Thus, although the Cold War raged through the 1980s – flashpoints included Afghanistan, Nicaragua, Ethiopia, Angola, and Grenada – the conflict increasingly ceased to be a context in which countries could enhance their agency by hedging among regional and great powers.

## Globalization, the unipolar order, and the end of Third Worldism

The dissolution of the USSR represented an end to the Cold War and the beginning of a unipolar era in which the US was the primary superpower. It also signalled the end of Third Worldism. According to Westad (2012, 132) the Cold War 'ended not only with the collapse of the Soviet Union – it also ended with the collapse of the Third World'. The period of intensified economic integration that followed 'split the Third World coalition apart' (Westad, 2012, 132), and in this brave new world all countries could be located in a universal process of modernization. To progress along this continuum less-developed countries had to implement a series of economic and institutional reforms, and in most cases, this 'structural adjustment' was imposed by the Washington Consensus institutions as conditions for emergency loans. Erstwhile socialist states in Africa that had championed Third Worldism were disciplined most sharply, but Latin American states were also restructured as were many Asian states in the wake of the 1997 financial crisis (Killick, 1995; Stiglitz, 2002; Soludo and Mkandawire, 2003). This was not only an economic order into which African, Asian, and Latin American states were subsumed, but a political one as well. The US emerged from the Cold War with swagger. By downplaying the role of internal struggles between Soviets committed to the union and nationalists who favoured autonomy in the dissolution of the USSR (see Plokhy, 2014), American policy makers convinced themselves that the end of the Cold War represented a monumental triumph of righteousness over evil that confirmed the superiority of their political and economic systems. This imbibed American policy makers with an unshakeable belief in the liberal international order and their role as its primary protagonist and defender. According to Ikenberry (2011, 234), '[b]y the end of the 1990s, a major consolidation and expansion of the U.S.-led international liberal order had been accomplished. The organizational logic of the Western order built during the Cold War was extended to the global level' and the US was this unipolar world's 'Liberal Leviathan'.

The unipolar world order proved unpopular. The US invasion of Iraq and revelations of a worldwide network of secret prisons run by US intelligence agencies seemed to substantiate Mao Zedong's well-known dictum that 'power grows out of the barrel of a gun'. Additionally, the institutional reforms advocated by the Washington Consensus simply did not propel countries through post-history's universal development trajectory. The Washington Consensus narrative was that the pain caused by austerity and structural adjustment in the short term was the price to pay for long-term prosperity. But societies that were most severely disciplined seemed to suffer the most and broad-based prosperity was perpetually postponed. Many governments grew impatient and began rolling back neoliberal reforms (Grugel and Riggirozzi, 2012), and even global policy makers began to question the wisdom of free market fundamentalism. One of the most ardent critics of the Washington Consensus was former Chief Economist of the World Bank Joseph Stiglitz (2002). By the mid-2000s there was a chorus of criticism emanating from within the international development establishment (see Rodrik, 2006). There was general agreement that neoliberal reforms had not worked as intended, and while Stiglitz and other closeted Keynesians came out against neoliberal free market orthodoxy, hardliners argued that the problem was that structural adjustment had not been implemented faithfully. This debate undermined the consensus among Washington's development policy-making community,[5] and in this context the highly financialized US housing market came crashing down and brought the global economy with it.

The 2008 financial crisis demonstrated the extent to which China's economy had become central to the global economy. It was no longer an 'emerging' economy or 'rising' power. According to Adam Tooze (2018, 251), '[i]n 2009, for the first time in the modern era, it was the movement of the Chinese economy that carried the entire world economy'. Indeed, as the collapse of the US housing market rippled across the global economy, both the US and China implemented expansive fiscal and monetary stimulus programmes. It was in this context that their relationship frayed. The accommodation that had animated their relationship throughout the unipolar era was based on the understanding that US firms would invest in China, which would result in a trade surplus in China's favour, and Beijing would park the proceeds in US Treasury bonds. Interdependence was not merely an arrangement established by officials in Beijing and Washington, but it was also the result of quotidian relations among producers and consumers in both countries. In the aftermath of the collapse of the US housing market, Beijing sought to limit its dependence on integration with the US economy and its financial system[6] by expanding in Sino-centric global production networks anchored by lead firms within China. The BRI represents China's attempt to territorialize and cement its centrality in the global economy through the establishment of maritime and terrestrial development corridors. These investments allowed China to employ its sizeable economic surplus in pursuit of a long-term spatial, economic, and political vision articulated by Xi Jinping, whose influence on China's economy and society cannot be understated.

While Sino-US relations were troubled after the financial crisis, they did not become openly hostile until Donald Trump became president. His 'America First' agenda was inconsistent and animated by 'China schizophrenia' (see Davis and Wei, 2020, 30), yet it set the tone. Despite bitter partisanship in Washington, lawmakers coalesced around the idea that the 'China challenge' necessitated a 'whole-of-government' response (see Schindler et al, 2021). Trump administration officials delivered a coordinated series of speeches that amounted to an informal declaration of (cold) war. For example, in an incendiary speech entitled 'Communist China and the Free World's Future', former Secretary of State Mike Pompeo stated that the US–China rivalry differs from the Cold War 'in kind'. One qualitative difference, according to Pompeo, is that China represents 'a complex new challenge that we've never faced before. The USSR was closed off from the free world. Communist China is already within our borders'.

In this context of rapidly deteriorating US–China relations a highly transmissible and dangerous novel coronavirus began to spread in 2019. Within months it had become a global pandemic that exposed the gulf that had widened between China and the rest of the erstwhile Third World. It is worth remembering that at the turn of the century there was anticipation surrounding the potential for 'large emerging market economies', namely Brazil, Russia, India, China – known as the BRICs[7] – to drive economic growth worldwide (O'Neill, 2001). The 2008 financial crisis revealed that China's economy was not only growing at a much faster rate than the others, but it was also transforming and Chinese firms were able to operate competently at the technological frontier. The response to the pandemic demonstrated that by 2021 it made more sense to talk about BRIS countries – China is now in a de facto G2 along with the US.[8] If US–China relations remain as tense as they have in recent years, future historians may interpret the post-Cold War era in which their economies were coupled as an interregnum between two periods marked by intense suspicion and hostility, the first lasting between 1949 until the 1970s, and the second beginning in the 2010s.

## 21st-century Third Worldism?

The chapters in this book demonstrated that countries have resisted attempts by the US and China to bind them into exclusive spheres of influence. In Argentina, for example, American firms are engaged in fracking, while lithium extraction is dominated by Chinese firms (Saguier and Vila Seoane, Chapter 10). Similarly, Kazakhstan has been able to wrangle favourable concessions from both China and the US (Neafie, Chapter 17), while Indonesia pursues terrestrial spatial objectives by partnering with China (such as on construction of high-speed rail) and also cautiously leans towards the US to secure its sovereign maritime space (Tritto et al, Chapter 14). Argentina, Kazakhstan, and Indonesia are not alone in actively resisting pressure to choose between the US and China (see Stromseth, 2019), but what remains unclear is whether these countries will coalesce into a bloc.

The world is clearly entering a period of great power competition, and it is worth remembering that the Third Worldism outlined in Sukarno's rousing speech at Bandung was animated by Cold War geopolitics. However, newly independent states were also drawn together by their shared history of colonial exploitation, and it is unclear what might constitute the shared basis for a 'non-aligned bloc' today. In further contrast to the context of Bandung in 1955, today there is an absence of totalizing ideological competition, nor are powerful states supporting armed revolutions. Indeed, it is difficult to envision anything that would approximate the Sandinista revolution or the *contra* counter-revolution. Thus, if the US–China rivalry does give rise to some sort of 21st-century Third Worldism, it would most certainly differ significantly from its mid-20th century vintage. Nevertheless, this raises the questions: what visions could animate this collective worldmaking? If infrastructure states can envisage a sort of emancipatory space, what would be the intellectual and ideological reference points? What role might non-state actors play? Finally, what would leaders of 21st-century Third Worldism demand? We end with these questions, with a view to the future.

## Notes

[1] Quoted in Byrne (2016), p 166.
[2] Quoted in Prashad (2007), p 11.
[3] This term was coined by Amilcar Cabral (Byrne, 2016, 202).
[4] This struggle largely unfolded in international institutions and the Third World states had few meaningful victories (Krasner, 1985).
[5] Dani Rodrik (2006) explored the breakdown of the 'Washington Consensus', and the emergence of what he called 'Washington Confusion'.
[6] Hale et al (2021) show that Wall Street and European investors are bullish on China despite geopolitical competition. American industrial capital is less sanguine. Meanwhile, policy makers fear that interdependence can be 'weaponized' (Farrell and Newman, 2019).
[7] South Africa was subsequently added by some commentators.
[8] The notion of a G2 comprised of China and the US has been used by Tooze (2018) and Yergin (2020).

## References

Byrne, J.J. (2016) *Mecca of Revolution: Algeria, Decolonization and the Third World Order*, Oxford: Oxford University Press.

Chamberlin, P. (2012) *The Global Offensive: The United States, the Palestine Liberation Organization, and the Making of the Post-Cold War Order*, Oxford: Oxford University Press.

Davis, B. and Wei, L. (2020) *Superpower Showdown: How the Battle Between Trump and Xi Threatens a New Cold War*, New York: Harper Business.

Farrell, H. and Newman, A.L. (2019) 'Weaponized interdependence: how global economic networks shape state coercion', *International Security*, 44(1): 42–79.

Friedman, J. (2015) *Shadow Cold War: The Sino-Soviet Competition for the Third World*, Chapel Hill, NC: University of North Carolina Press.

Gorbachev, M. (1987) *Perestroika: New Thinking for Our Country and the World*, London: Fontana/Collins.

Grugel, J. and Riggirozzi, P. (2012) 'Post-neoliberalism in Latin America: rebuilding and reclaiming the state after crisis', *Development and Change*, 43(1): 1–21.

Hale, T., Agnew, H., Mackenzie, M., and Sevastopulo, D. (2021) 'Wall Street's new love affair with China', *Financial Times*, 28 May. Available from: www.ft.com/content/d5e09db3-549e-4a0b-8dbf-e499d0606df4.

Hughes, L. and Payne, S. (2021) 'UK seeks to counter China's influence with new development investment arm', *Financial Times*, 28 November. Available from: www.ft.com/content/93de6cc1-451a-465d-8233-8c9b903cedd4.

Ikenberry, G.J. (2011) *Liberal Leviathan: The Origins, Crisis, and Transformation of the American World Order*, Princeton, NJ: Princeton University Press.

Killick, T. (1995) *IMF Programmes in Developing Countries: Design and Impact*, London: Routledge.

Kotkin, S. (2008) *Armageddon Averted: The Soviet Collapse 1970–2000 (Updated Edition)*, Oxford: Oxford University Press.

Krasner, S.D. (1985) *Structural Conflict: The Third World against Global Liberalism*, Berkeley, CA: University of California Press.

Leffler, M. (2007) *For the Soul of Mankind: The United States, the Soviet Union, and the Cold War*, New York: Hill and Wang.

Lorenzini, S. (2019) *Global Development: A Cold War History*, Princeton, NJ: Princeton University Press.

Lüthi, L. (2020) *Cold Wars: Asia, the Middle East, Europe*, Cambridge: Cambridge University Press.

Malloy, S.L. (2017) *Out of Oakland: Black Panther Party Internationalism during the Cold War*. Ithaca, NY: Cornell University Press.

O'Neill, J. (2001) *Building Better Global Economic BRICs*, Global Economics Paper No. 66, London: Goldman Sachs.

Olander, E. (2021) 'No one's really sure what to make of China's FOCAC financial pledge diplomacy finance', China Africa Project. Available from: https://chinaafricaproject.com/analysis/no-ones-really-sure-what-to-make-of-chinas-focac-financial-pledge/.

Peel, M. (2019) 'Japan and EU sign deal in riposte to China's Belt and Road', *Financial Times*, 27 September. Available from: www.ft.com/content/dd14ce1e-e11d-11e9-9743-db5a370481bc.

Peel, M., Flemming, S., and Findlay, S. (2021) 'EU and India plan global infrastructure deal', *Financial Times*, 21 April. Available from: www.ft.com/content/2e612c38-aba9-426a-9697-78e11ab1c697.

Plokhy, S. (2014) *The Last Empire: The Final Days of the Soviet Union*, London: Oneworld Publications.

Prashad, V. (2007) *The Darker Nations: A People's History of the Third World*, New York: The New Press.

Rodrik, D. (2006) 'Goodbye Washington Consensus, hello Washington Confusion? A review of the World Bank's economic growth in the 1990s: learning from a decade of reform', *Journal of Economic Literature*, 44 (4): 973–87.

Roos, J. (2019) *Why Not Default? The Political Economy of Sovereign Debt*, Princeton, NJ: Princeton University Press.

Schindler, S., DiCarlo, J., and Paudel, D. (2021) 'The new cold war and the rise of the 21st century infrastructure state', *Transactions of the Institute of British Geographers*. doi.org/10.1111/tran.12480.

Soludo, C.C. and Mkandawire, T.P. (2003) *African Voices on Structural Adjustment: A Companion to Our Continent, Our Future*, Ottawa: International Development Research Centre.

Stiglitz, J. (2002) *Globalization and Its Discontents*, New York: W.W. Norton and Company.

Stromseth, J. (2019) *Don't Make Us Choose: Southeast Asia in the Throes of US-China Rivalry*, Washington, DC: Brookings.

Sukarno (1955) 'Bandung Conference opening address', Bandung, Indonesia, 18 April.

Tooze, A. (2018) *Crashed: How a Decade of Financial Crises Changed the World*, London: Allen Lane.

Westad, O.A. (2012) 'Two finals: how the end of the Third World and the end of the Cold War are linked', in G. Lundstad (ed) *International Relations since the End of the Cold War: New and Old Dimensions*, Oxford: Oxford University Press.

Yergin, D. (2020) *The New Map: Energy, Climate, and the Clash of Nations*, London: Allen Lane.

# Index

References to figures appear in *italic* type.
References to endnotes show both the
page number and the note number (54n1).